Learn
SOLIDWORKS

Second Edition

Get up to speed with key concepts and tools to become an accomplished SOLIDWORKS Associate and Professional

Tayseer Sadiq J Almattar

BIRMINGHAM—MUMBAI

Learn SOLIDWORKS
Second Edition

Group Product Manager: Rohit Rajkumar

Business Development Executive: Chayan Majumdar

Senior Editor: Hayden Edwards

Content Development Editor: Aamir Ahmed

Technical Editor: Simran Udasi

Copy Editor: Safis Editing

Project Coordinator: Rashika BA

Proofreader: Safis Editing

Indexer: Subalakshi Govindhan

Production Designer: Alishon Mendonca

Marketing Coordinator: Elizabeth Varghese

First published: November 2019

Second edition: January 2022

Production reference: 2140222

Published by Packt Publishing Ltd.

Livery Place

35 Livery Street

Birmingham

B3 2PB, UK.

ISBN 978-1-80107-309-7

www.packt.com

To my parents and siblings for their unbounded love, support, and care…

– Tayseer Almattar

Contributors

About the author

Tayseer Almattar holds a bachelor's degree in mechanical engineering and a **Master of Design** (**MDes**) degree in international design and business management. He is a believer in the power of design in enabling sustainable business innovation. He is also the founder of TforDesign.

SOLIDWORKS is Tayseer's software choice in generating 3D designs. He has been a SOLIDWORKS user for over a decade and has published multiple related online learning programs attracting thousands of learners from across the globe. With this book, Tayseer has brought together his design and training experience to produce a unique and practical SOLIDWORKS learning experience in writing.

I want to thank the people who made this book possible, the amazing reviewers, and the Packt team.

About the reviewers

Deepak Gupta graduated in 2000 from Indo Swiss Training Centre (Chandigarh, India). With over 20 years of rich experience working in various industries, he has worked in different roles with different product lines. His main areas of experience and interest are design and manufacturing processes along with team and project management.

He is working as technical director at Vishnu Design Services, an engineering outsourcing service provider based in Chandigarh, India, offering a wide range of CAD design, drafting, and design automation engineering services. A certified SOLIDWORKS Expert, SOLIDWORKS Champion, and user for the last 15 years, he is passionate about working with SOLIDWORKS. Besides working for Vishnu Design Services, he enjoys writing his own blog, *Boxer's SOLIDWORKS Blog*, where he writes tips, tricks, and tutorials and provides news about SOLIDWORKS. In addition to that, he has participated in various SOLIDWORKS world conferences as a press member.

An important aspect of his life is family – his parents, his wife, his son, and other family members. He loves to travel and make friends.

I would like to thank my wife, Swati Gupta, for the love, kindness, and support she has shown during the past few weeks it has taken me to review this book. Furthermore, I would also like to thank my parents for their endless love and support. Last but not least, I would like to thank Packt Publishing and its team for choosing me as one of the reviewers of this book and helping me to complete the review of this book.

Mohsina Zafar holds a bachelor's degree in mechatronics and control engineering (UET Lahore, Pakistan) and is currently doing a master's with a focus on deep learning. Her journey with SOLIDWORKS started 6 years ago when she utilized the software as her main design tool for different projects. Today, she provides CAD design services and writes blogs related to the software. She also provides technical support to thousands of SOLIDWORKS students around the world.

Mohsina believes in learning at all ages and is keen to spread the knowledge she has attained. When not working, she loves to go for a walk and spend time with family.

I would like to thank my mother and my husband for their unwavering support and belief in me.

Table of Contents

Preface

Section 1 – Getting Started

1

Introduction to SOLIDWORKS

Introducing SOLIDWORKS	4	Associate certifications	12
SOLIDWORKS applications	4	Professional certifications	13
Core mechanical design	5	Professional advanced certifications	14
Sample SOLIDWORKS 3D Models	6	Expert certifications	14
Understanding parametric modelling	8	Summary	15
		Questions	16
Exploring SOLIDWORKS certifications	11	Further Reading	16

2

Interface and Navigation

Technical requirements	18	Main components of the SOLIDWORKS interface	23
Starting a new part, assembly, or drawing file	18	The Command Bar	24
What are parts, assemblies, and drawings?	18	The Feature Manager Design Tree	26
		The Canvas/Graphics Area	27
Opening a part, assembly, or drawing file	21	The Task Pane	28

The document's measurement system   29

Different measurement systems   29

Adjusting the document's measurement system   30

Summary   32
Questions   32

Section 2 – 2D Sketching

3

SOLIDWORKS 2D Sketching Basics

Technical requirements   36
Introducing SOLIDWORKS sketching   36

The position of SOLIDWORKS sketches   36
Simple sketches versus complex sketches   38
Sketch planes   40

Getting started with SOLIDWORKS sketching   41

Getting into the sketching mode   41
Defining sketches   43
Geometrical relations   44

Sketching lines, rectangles, circles, arcs, and ellipses   45

The origin   46
Sketching lines   46
Sketching rectangles and squares   52
Sketching circles and arcs   56
Sketching ellipses and using construction lines   61
Fillets and chamfers   66

Under defined, fully defined, and over defined sketches   70

Under defined sketches   70
Fully defined sketches   72
Over defined sketches   73

Summary   74
Questions   75

4

Special Sketching Commands

Technical requirements   80
Mirroring and offsetting sketches   80

Mirroring a sketch   80
Offsetting a sketch   84

Creating sketch patterns   90

Defining patterns   90
Linear sketch patterns   92

Circular sketch patterns   100

Trimming in SOLIDWORKS sketching   107

Understanding trimming   107
Using power trimming   108

Summary   112
Questions   112

Section 3 – Basic Mechanical Core Features – Associate Level

5

Basic Primary One-Sketch Features

Technical requirements	120	Understanding and applying fillets and chamfers	144
Understanding features in SOLIDWORKS	120	Understanding fillets and chamfers	144
Understanding SOLIDWORKS features and their role in 3D modeling	120	Applying fillets	145
		Applying chamfers	149
Simple models versus complex models	121	Modifying fillets and chamfers	157
Sketching planes for features	122	Applying partial fillets and chamfers	158
Understanding and applying extruded boss and cut	123	Understanding and applying revolved boss and revolved cut	163
What are extruded boss and extruded cut?	123	What are revolved boss and revolved cut?	163
Applying extruded boss	124	Applying revolved boss	165
Applying extruded cut and building on existing features	132	Applying revolved cut	170
Modifying and deleting extruded boss and extruded cut	139	Modifying revolved boss and revolved cut	174
		Summary	**174**
		Questions	**175**

6

Basic Secondary Multi-Sketch Features

Technical requirements	180	Understanding and applying swept boss and swept cut	195
Reference geometries – additional planes	180	What are swept boss and swept cut?	195
Understanding planes, reference geometries, and why we need them	180	Applying swept boss	197
		Applying swept cut	204
Defining planes in geometry	182	Modifying swept boss and swept cut	208
Defining a new plane in SOLIDWORKS	186	Understanding and applying lofted boss and lofted cut	210

What are lofted boss and lofted cut?	211	Guide curves	226
Applying lofted boss	213		
Applying lofted cut	222	**Summary**	**232**
Modifying lofted boss and cut	226	**Questions**	**232**

Section 4 – Basic Evaluations and Assemblies – Associate Level

7

Materials and Mass Properties

Technical requirements	**238**	**Assigning materials and evaluating and overriding mass properties**	**243**
Reference geometries – defining a new coordinate system	**238**	Assigning materials to parts	244
		Viewing the mass properties of parts	247
What is a reference coordinate system and why are new ones needed?	238	Overriding mass properties	254
How to create a new coordinate system	240	**Summary**	**255**
		Questions	**256**

8

Standard Assembly Mates

Technical requirements	**260**	Applying the coincident and perpendicular mates	267
Opening assemblies and adding parts	**260**	Applying the parallel, tangent, concentric, and lock mates	274
Defining SOLIDWORKS assemblies	260	Under defining, fully defining, and over defining an assembly	279
Starting a SOLIDWORKS assembly file and adding parts to it	261	Viewing and adjusting active mates	281
Understanding mates	265		
Understanding and applying non-value-oriented standard mates	**266**	**Understanding and applying value-driven standard mates**	**282**
Defining the non-value-oriented standard mates	266	Defining value-driven standard mates	283
		Applying the distance and angle mates	283

Utilizing materials and mass
properties for assemblies 287

Setting a new coordinate system for
an assembly 287

Material edits in assemblies 288

Evaluating mass properties
for assemblies 290

Summary 292

Questions 293

Section 5 – 2D Engineering Drawings Foundation

9

Introduction to Engineering Drawings

Understanding engineering
drawings 300

Interpreting engineering
drawings 302

Interpreting lines 302

Interpreting views 304
Axonometric projections 312

Summary 313

Questions 313

Further Reading 314

Project 1:

3D-Modeling a Pair of Glasses

Technical requirements 316

Understanding the project 316

3D-modeling the individual
parts 318

Creating the individual parts 318
Creating a mirrored part 331

Creating the assembly 335

Summary 339

10

Basic SOLIDWORKS Drawing Layout and Annotations

Technical requirements 342

Opening a SOLIDWORKS
drawing file 342

Exploring and Generating
orthographic and isometric
views 345

Selecting a model to plot 346

Generating orthographic and isometric
views 348

Adjusting the drawing scale and the
display 359

Communicating dimensions
and design 367

Using the Smart Dimension tool   368
Centerlines, center marks, notes, and
hole callout annotations   370

Utilizing the drawing sheet's information block   378
Editing the information block   379
Adding new information to the

information block   382

Exporting the drawing as a PDF or image   383
Exporting a drawing as a PDF file   384
Exporting the drawing as an image   385

Summary   387
Questions   387

11
Bill of Materials

Technical requirements   392
Understanding BOMs   392
Understanding a BOM   392

Generating a standard BOM   394
Inserting an assembly into a
drawing sheet   395
Creating a standard BOM   397

Adjusting information in the BOMs   400
Adjusting listed information in the BOM 400

Sorting information in our BOMs   405
Adding new columns   407

Utilizing equations with BOMs   409
What are equations in SOLIDWORKS
drawings?   409
Inputting equations in a table   410

Utilizing parts callouts   416
Manual Balloon command   420

Summary   423
Questions   423

Section 6 – Advanced Mechanical Core Features – Professional Level

12
Advanced SOLIDWORKS Mechanical Core Features

Technical requirements   430
Understanding and applying the draft feature   430
What are drafts?   430
Applying drafts   431

Understanding and applying the shell feature   436
What is a shell?   436
Applying a shell   436

Understanding and utilizing the Hole Wizard 442

What is the Hole Wizard and why use it? 442
Utilizing the Hole Wizard 443

Understanding and applying features mirroring 450

What is mirroring for features? 450
Utilizing the Mirror command to mirror features 451

Understanding and applying the rib feature 455

Understanding ribs 456
Applying the Rib command 456

Understanding and utilizing multi-body parts 461

Defining multi-body parts and their advantages 462
Generating and dealing with a multi-body part 463
Separating different bodies into different parts 468

Understanding and applying linear, circular, and fill feature patterns 471

Understanding feature patterns 471
Applying a linear pattern 472
Applying a circular pattern 479
Applying a fill pattern 483

Summary 491
Questions 491

13

Equations, Configurations, and Design Tables

Technical requirements 498
Understanding and applying equations in parts 498

Understanding equations 498
Applying equations in parts 500
Modifying dimensions with equations 504

Understanding and utilizing configurations 507

What are configurations? 508
Applying configurations 508

Understanding and utilizing design tables 515

What are design tables? 515
Setting up a design table 517
Editing a design table 522

Summary 525
Questions 525

Section 7 – Advanced Assemblies – Professional Level

14

SOLIDWORKS Assemblies and Advanced Mates

Technical requirements 534

Understanding and using the profile center mate 534

Defining the profile center advanced mate 534
Applying the profile center mate 535

Understanding and using the width and symmetric mates 539

Defining the width advanced mate 539
Applying the width advanced mate 540
Defining the symmetric advanced mate 543
Applying the symmetric advanced mate 545

Understanding and using the distance range and angle range mates 549

Defining the distance range and angle range 549
Applying the distance range mate 550
Applying the angle range mate 553

Understanding and using the path mate and linear/linear coupler mates 556

Defining the path mate 556
Applying the path mate 557
Defining the linear/linear coupler 562
Applying the linear/linear coupler 563

Summary 568
Questions 569

15

Advanced SOLIDWORKS Assemblies Competencies

Technical requirements 572

Understanding and utilizing the Interference Detection and Collision Detection tools 572

Interference Detection 573
Collision detection 577

Understanding and applying assembly features 580

Understanding assembly features 580

Applying assembly features 581

Understanding and utilizing configurations and design tables for assemblies 584

Using manual configurations 585
Design tables 588

Summary 591
Questions 591

Project 2:
3D-Modeling an RC Helicopter Model

Technical requirements	596	Exploring the individual parts	599
Understanding the project	596	Creating the assembly	628
3D-modeling the individual parts	598	Summary	637

Index

Other Books You May Enjoy

Preface

SOLIDWORKS is one of the most used pieces of software for 3D engineering and product design applications. These applications cover areas such as aviation, automobiles, consumer product design, and more. This book takes a practical approach to mastering the software at a professional level. The book starts with the very basics, such as exploring the software interface and opening new files. However, step by step, it progresses through different topics, from sketching and building complex 3D models to generating dynamic and static assemblies.

This book takes a hands-on approach when it comes to covering different tools in SOLIDWORKS. Whenever a new tool is introduced, we will go through a practical exercise of using it to create sketches, 3D part models, assemblies, or drawings. When required, we will provide you with supporting files that you can download to follow up on the concepts and exercises in your own time. In addition, it includes two comprehensive projects linking the different parts of the book together through practical applications. If you are a complete beginner in SOLIDWORKS, it will be best to follow the book from start to finish, like a story. However, you can also jump between chapters.

Who this book is for

This book targets individuals who would like to get started with SOLIDWORKS and feel comfortable using the software. They could be aspiring engineers, designers, makers, drafting technicians, or hobbyists. This book is also designed for individuals interested in becoming **Certified SOLIDWORKS Associates (CSWAs)** or **Certified SOLIDWORKS Professionals (CSWPs)**.

The book does not require a specific background to follow it, as it starts from the basics of what SOLIDWORKS is and how to use it. However, basic theoretical background knowledge of what 3D modeling is would be helpful.

What this book covers

Chapter 1, Introduction to SOLIDWORKS, covers what SOLIDWORKS is and the applications that utilize the software. It also explores the professional certifications that SOLIDWORKS offers.

Chapter 2, Interface and Navigation, teaches you how to navigate around the SOLIDWORKS interface.

Chapter 3, SOLIDWORKS 2D Sketching Basics, covers what sketching is in SOLIDWORKS. It also covers how you can sketch basic entities such as lines, circles, rectangles, arcs, and ellipses.

Chapter 4, Special Sketching Commands, covers commands that enable us to sketch more efficiently. These include the mirror, offset, trip, and pattern commands.

Chapter 5, Basic Primary One-Sketch Features, explores the most basic features used for generating 3D models from sketches. Each of these features requires you to have one sketch to apply it. The features include extruded boss and cut, revolved boss and cut, fillets, and chamfers.

Chapter 6, Basic Secondary Multi-Sketch Features, explores another set of basic features that require more than one sketch to apply. They include swept boss and swept cut and lofted boss and lofted cut. It also explores reference geometries and how to generate new planes.

Chapter 7, Materials and Mass Properties, explores structural materials for your 3D parts. It also teaches you how to calculate mass properties such as mass, volume, and the center of gravity.

Chapter 8, Standard Assembly Mates, explores what assemblies are in SOLIDWORKS. You will learn how to generate simple assemblies using the standard mates: coincident, parallel, perpendicular, tangent, concentric, lock and set distance, and angle.

Chapter 9, Introduction to Engineering Drawing, explores what engineering drawings are and how to interpret them according to commonly recognized international standards.

Project 1, 3D - Modeling a Pair of Glasses, presents a comprehensive practical exercise linking the topics in chapters 2 to 9 to 3D model a pair of glasses.

Chapter 10, Basic SOLIDWORKS Drawing Layout and Annotations, teaches you how to generate basic engineering drawings using SOLIDWORKS drawing tools.

Chapter 11, Bills of Materials, explores what bills of materials are and how to generate and adjust bills of materials with SOLIDWORKS drawing tools.

Chapter 12, Advanced SOLIDWORKS Mechanical Core Features, covers the advanced features used to generate more complex 3D models. These include the draft feature, shell feature, Hole Wizard, feature mirroring, rib features, multi-body parts, and feature patterns.

Chapter 13, Equations, Configurations, and Design Tables, explains applying equations to link different dimensions within the model. You will also learn how to utilize configurations and design tables to generate multiple variations of a single part within one SOLIDWORKS file.

Chapter 14, SOLIDWORKS Assemblies and Advanced Mates, covers using advanced mates to generate more dynamic assemblies. These include the profile center, symmetric, width, distance and angle range, path, and linear/linear coupler mates.

Chapter 15, Advanced SOLIDWORKS Assemblies Competencies, explores additional assembly features to better evaluate and generate more sound and flexible assemblies. These include the interference and collision detection tools, assembly features, configurations, and design tables for assemblies.

Project 2, 3D - Modeling an RC Helicopter, presents a comprehensive practical exercise covering topics from across the book to 3D model a remote-control helicopter.

To get the most out of this book

You will need to have access to the SOLIDWORKS software for most of the chapters. Some chapters will also require you to have access to Microsoft Excel on the same machine.

You should practically follow all the steps and examples in this book in SOLIDWORKS as you are reading the book. This is because the book was designed to give you hands-on practical experience.

You require no previous knowledge or skills to follow this book. However, having a basic, theoretical understanding of what 3D modeling is would be helpful.

Download the example code files

You can download the example code files for this book from GitHub at `https://github.com/PacktPublishing/Learn-SOLIDWORKS-Second-Edition`. If there's an update to the code, it will be updated in the GitHub repository.

We also have other code bundles from our rich catalog of books and videos available at `https://github.com/PacktPublishing/`. Check them out!

Code in Action

The Code in Action videos for this book can be viewed at `https://bit.ly/3IUs7eO`.

Download the color images

We also provide a PDF file that has color images of the screenshots and diagrams used in this book. You can download it here: `https://static.packt-cdn.com/downloads/9781801073097_ColorImages.pdf`.

Conventions used

There are a number of text conventions used throughout this book.

`Code in text`: Indicates code words in text, database table names, folder names, filenames, file extensions, pathnames, dummy URLs, user input, and Twitter handles. Here is an example: "To recall the part configurations, we can use the title format `$configuration@partName<instance>`."

Bold: Indicates a new term, an important word, or words that you see onscreen. For instance, words in menus or dialog boxes appear in **bold**. Here is an example: "Once we have the parts at the colliding position, we can use the **Smart Dimension** command to get the exact collision angle or distance measurements."

> **Tips or important notes**
> Appear like this.

Get in touch

Feedback from our readers is always welcome.

General feedback: If you have questions about any aspect of this book, email us at `customercare@packtpub.com` and mention the book title in the subject of your message.

Errata: Although we have taken every care to ensure the accuracy of our content, mistakes do happen. If you have found a mistake in this book, we would be grateful if you would report this to us. Please visit `www.packtpub.com/support/errata` and fill in the form.

Piracy: If you come across any illegal copies of our works in any form on the internet, we would be grateful if you would provide us with the location address or website name. Please contact us at copyright@packt.com with a link to the material.

If you are interested in becoming an author: If there is a topic that you have expertise in and you are interested in either writing or contributing to a book, please visit authors.packtpub.com.

Share Your Thoughts

Once you've read *Learn SOLIDWORKS Second Edition*, we'd love to hear your thoughts! Scan the QR code below to go straight to the Amazon review page for this book and share your feedback.

https://packt.link/r/1-801-07309-0

Your review is important to us and the tech community and will help us make sure we're delivering excellent quality content.

Section 1 – Getting Started

This section introduces you to all the foundations and background you will need to start your SOLIDWORKS journey. This includes the foundation of parametric modeling, what SOLIDWORKS and its certifications are, and how to navigate around the SOLIDWORKS interface.

This section comprises the following chapters:

- *Chapter 1, Introduction to SOLIDWORKS*
- *Chapter 2, Interface and Navigation*

1
Introduction to SOLIDWORKS

SOLIDWORKS is a **Three-Dimensional (3D)** design application. This is a **Computer-Aided Design (CAD)** software that runs on Windows computer systems. It was launched in 1995 and has grown to be one of the most common pieces of software used globally regarding engineering design.

This book covers the fundamental skills for using SOLIDWORKS. It will take you from knowing nothing about the software to acquiring all the basic skills expected of a **Certified SOLIDWORKS Professional (CSWP)**. En route, we will also cover all the skills needed for the more basic **Certified SOLIDWORKS Associate (CSWA)** level. In addition to knowing what the tools are, you will also need to develop software fluency, which you will gain gradually as you practice using the software for different applications. Both the tools and the fluency are essential to acquiring any official SOLIDWORKS certifications. If you are new to SOLIDWORKS, we recommend that you follow the book like a story, from *Chapter 1, Introduction to SOLIDWORKS*, onward. If you are already familiar with SOLIDWORKS, feel free to jump between chapters.

This chapter will provide you with a brief introduction to what SOLIDWORKS is and the fields it can support. Equipped with this knowledge, we will learn about all the features and capabilities of SOLIDWORKS and will have a clearer idea of what types of certifications or fields you can strive for. Learning about applicable certifications will enable you to plan your personal SOLIDWORKS development.

The chapter will also explain the governing principle with which SOLIDWORKS functions: parametric modeling. Equipped with a knowledge of SOLIDWORKS' operating principles, we will be able to deepen our understanding of how the software works and what to expect from it. Understanding the software's operating principles will help us manage the different software commands that are used when building 3D models.

The following topics will be covered in this chapter:

- Introducing SOLIDWORKS
- Understanding parametric modeling
- Exploring SOLIDWORKS certifications

Introducing SOLIDWORKS

SOLIDWORKS is a 3D design software that's officially capitalized to SOLIDWORKS. It is one of the leading pieces of engineering 3D design software globally. Today, more than 2 million organizations use SOLIDWORKS to bring in products and innovations, which represent a large proportion of over 6 million SOLIDWORKS users in total. In this section, we will explore the different applications that SOLIDWORKS supports.

SOLIDWORKS applications

SOLIDWORKS mainly targets engineers and product designers. It is used in a variety of applications and industries. Some of these industries are as follows:

- Consumer products
- Aerospace construction
- High-tech electronics
- Medicine
- Oil and gas
- Packaging
- Machinery

- Engineering services
- Furniture design
- Energy
- Automobiles

Each of these industries utilizes SOLIDWORKS for its design applications to some extent. Within SOLIDWORKS, several disciplines correspond to different design and analysis approaches. They are as follows:

- Core mechanical design
- Two-dimensional (2D) drawings
- Surface design
- Sheet metal
- Sustainability
- Motion analysis
- Weldments
- Simulations
- Mold making
- Electrical

Even though the preceding list highlights some possible domains where SOLIDWORKS can be applied, it is not necessary for a single individual to master them all. However, they do demonstrate the capabilities enabled by the software and the fields it can serve. This book will focus on addressing applications within the core mechanical design disciplines. These disciplines will cover the most common usage scenarios for SOLIDWORKS.

Core mechanical design

Core mechanical design skills are the most commonly used foundational design application for SOLIDWORKS users. This includes the fundamental 3D modeling features that are essential for modeling mechanical components; this book will focus on this type of design application. Mastering this will enable you, as a learner, to draft complex parts and assemblies. These can include engines, furniture, and everyday consumer products such as phones and laptops.

We will cover all the knowledge and skills needed to achieve the two major SOLIDWORKS certifications under the core mechanical design discipline. These are the **Certified SOLIDWORKS Associate (CSWA)** and **Certified SOLIDWORKS Professional (CSWP)** levels. Also, mastering core mechanical design concepts can be considered as a prerequisite to learning most other specialized modeling disciplines, such as sheet metal and mold making. Because of that, we will only cover a common foundation for mechanical core design in this book. Later in this chapter, we will discuss all the certifications and levels in more detail in the *Exploring SOLIDWORKS Certifications* section.

Now that we know what SOLIDWORKS is and the different applications and disciplines it covers, we will cover the principle under which the software operates: **parametric modeling**.

Sample SOLIDWORKS 3D Models

As SOLIDWORKS caters to a variety of fields, it is possible to create 3D models with varying complexity using the software. Here, you can find samples of 3D models from different fields that have been made using SOLIDWORKS:

Figure 1.1 – A 3D Model of "Gallon." Image courtesy of TforDesign

Figure 1.2 – Gears assembly for a pump. Image courtesy of TforDesign

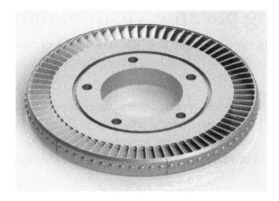

Figure 1.3 – A turbine rotor. Image courtesy of TforDesign

Figure 1.4 – Geometric bookshelf design. Image courtesy of TforDesign

Figure 1.5 – A mechanical seal. Image courtesy of TforDesign

These models are selections from different fields that can show the flexibility and the range of possible applications. In reality, SOLIDWORKS is a tool, and it will remain up to you as to what you will use it for.

Understanding parametric modelling

Parametric modeling is the core principle that SOLIDWORKS operates on. It governs how SOLIDWORKS constructs 3D models and how a user should think when dealing with SOLIDWORKS.

In parametric modeling, the model is created based on relationships and a set of logical arrangements that are set by the designer or draftsman. In the SOLIDWORKS software environment, they are represented by dimensions, geometric relationships, and features that link different parts of a model to each other. Each of these logical features is called a **parameter**.

For example, a simple cube with a side length of 1 mm would contain the following parameters:

1. **Four lines in one plane** with the following relationships listed and noted in the following diagram in writing:

- All two-line endpoints are merged at the same point. This is presented with the **merged** parameter in the following diagram.

- Two opposite angles are **right angles** (90 degrees).

- Two adjacent lines are equal to each other in length.

- The length of one line is **1 mm**, as follows:

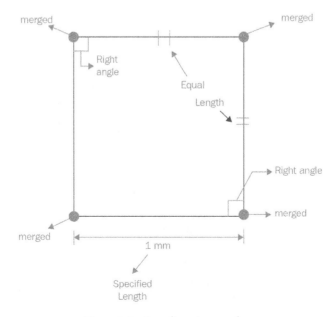

Figure 1.6 – Four lines in one plane

2. A **Vertical Extrusion** that is perpendicular to the square defined in the first set of parameters. This extrusion is by an amount equal to the length of the square's side (1 mm). This vertical extrusion will result in the shape shown in the following diagram:

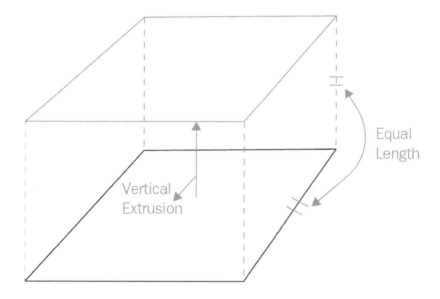

Figure 1.7 – Extruding four base lines upward to make a cube

The parameters listed here show how software such as SOLIDWORKS interprets and constructs 3D models. Another term that is commonly used to refer to those parameters is **design intent**. The user of the software should specify all those parameters to create a cube or any other 3D model. Creating 3D models based on parameters/design settings has many notable advantages. One major advantage is the ease of applying design updates. Let's go back to our cube to see how this works.

Notice that in the preceding cube, we have specified the length of only one side in the base square; the other specifications are all relationships that fix and highlight the fact that the model is a cube (equal, parallel, and perpendicular sides). Those parameters make all the parts of our cube inter-connected based on what we decide is important. Thus, updating the length of the side of the cube will not sabotage the cube's structure. Rather, the whole cube will be updated while keeping the parameters intact.

To clarify this, we can revisit the cube we just made to update it. In the same model, let's change the dimension we identified earlier from **1 mm** to **5 mm**:

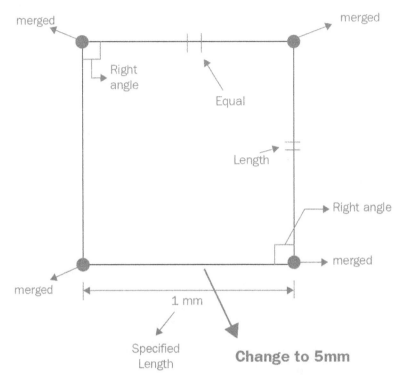

Figure 1.8 – Adjusting the elements in a parametric design propagated to the different parts

With that single step, the cube is fully modified, with all the sides changing to 5 mm in length. Again, this is because our cube parameters must have equal perpendicular and parallel sides. Given that we have defined our intended parameters/design settings for the software, all of those will be retained, resulting in the whole cube model being updated with one single adjustment.

This can be contrasted with pure direct modeling methods. In pure direct modeling, the user creates the cube more abstractly by drawing each line separately and constructing a cube of a certain size. Even though creating the initial cube might be faster, updating it would require updating all of the elements separately as they don't relate to each other with any intent or logical features. This would result in considerably more time and effort being invested in creating variations, which is an essential requirement for industrial applications.

Other advantages of parametric modeling are as follows:

- The ease of modifying and adjusting models throughout the design and production cycles.

- The ease of creating families of parts that have similar parameters.

- The ease of communicating the design to manufacturing establishments for manufacturing.

All the advantages of parametric modeling make it a popular modeling method for technical applications relating to engineering or product design. On the other hand, direct modeling can perform better in more abstract applications, such as modeling more artistic objects used in gaming or architecture. Understanding parametric modeling will enable us to use the software more easily as we are aware of its limitations, as well as how the software interprets the commands we apply. As we go through this book, we will expand our understanding of parametric modeling as we tackle more advanced functions, such as design tables and other features.

Now that we know more about SOLIDWORKS and parametric modeling, we will discuss the certifications offered by SOLIDWORKS.

Exploring SOLIDWORKS certifications

SOLIDWORKS provides certifications that cover different aspects of its functionality. As a user, you don't need to gain any of those certifications to use the software; however, they can prove your SOLIDWORKS skills. SOLIDWORKS certifications are a good way of showing employers or clients that you have mastery over a certain aspect of the software that would be required for a specific project.

Certifications can be classified under four levels: associate, professional, professional advanced, and expert. Associate certifications represent the entry level, expert certifications represent the highest level, and professional and professional advanced represent the middle levels, respectively. The following subsections list the certification levels provided by SOLIDWORKS. Note that SOLIDWORKS adds or removes certifications over time.

You can check the SOLIDWORKS certification program for more information. You can find the link to the program in the *Further Reading* section.

Associate certifications

Associate certifications are the most basic ones offered by SOLIDWORKS. Some of those certifications require hands-on testing, while others require the student to have theoretical knowledge related to the certification topic. Brief details pertaining to each certification are as follows:

- **CSWA**: This is the most popular SOLIDWORKS certification. It covers the basic modeling principles involved in using the software. This certification allows the user to prove their familiarity with the basic 3D modeling environment in SOLIDWORKS. It touches on creating parts, assemblies, and drawings. The test for this certification is hands-on, so the student will need to have SOLIDWORKS installed before attempting the test.

- **Certified SOLIDWORKS Associate – Electrical (CSWA-E)**: This covers the general basics of electrical theory, as well as aspects of the electrical functionality of SOLIDWORKS. This certification test does not involve practical work, so the student will not need to have SOLIDWORKS installed.

- **Certified SOLIDWORKS Associate – Sustainability (CSWA-Sustainability)**: This covers theoretical principles of product-sustainable design, such as cradle to cradle. To take this certification, SOLIDWORKS software is not required.

- **Certified SOLIDWORKS Associate – Simulation (CSWA-Simulation)**: This covers basic simulation principles based on the **Finite Elements Method (FEM)**. This mainly includes stress analysis and the effect of different materials and forces on solid bodies. This is a hands-on test, so the student is required to have SOLIDWORKS installed.

- **Certified SOLIDWORKS Associate – Additive Manufacturing (CSWA-AM)**: This is one of the newer certifications offered by SOLIDWORKS, due to the emergence of the common use of additive manufacturing methods such as 3D printing. This certification covers basic knowledge regarding the 3D printing market. This is not a hands-on test, so the student does not need to have the SOLIDWORKS software installed.

Professional certifications

Professional certifications demonstrate a higher mastery of SOLIDWORKS functions beyond the basic knowledge of the certified associate. All the certifications in this category involve hands-on demonstrations. Thus, the student is required to have access to SOLIDWORKS before attempting any of the tests. Brief details pertaining to each certification are as follows:

- **CSWP**: This level is a direct sequence of the CSWA level. It demonstrates the user's mastery over advanced SOLIDWORKS 3D modeling functions. This level upgrade focuses more on modeling more complex parts and assemblies.

- **Certified SOLIDWORKS Professional – Model-Based Definition (CSWP-MBD)**: MBD is one of the newer SOLIDWORKS functionalities. This certification demonstrates the user's mastery of MBD functions, which enable the communication of models in a 3D environment rather than in a 2D drawing.

- **Certified PDM Professional Administrator (CPPA)**: PDM stands for Product Data Management. This certification focuses on managing projects with a wide variety of files and configurations. Also, it facilitates collaboration in teams working on the same design project.

- **Certified SOLIDWORKS Professional – Simulation (CSWP-Simulation)**: This is an advanced sequence of the CSWA-Simulation certificate. It demonstrates a more advanced mastery of the simulation tools provided by SOLIDWORKS, as well as the ability to evaluate and interpret more diverse simulation scenarios.

- **Certified SOLIDWORKS Professional – Flow Simulation (CSWP-Flow)**: This is another advanced sequence of the CSWA-Simulation certificate. However, it focuses on the ability to set up and run different fluid flow simulation scenarios.

- **Certified SOLIDWORKS Professional API (CSWP-API)**: API stands for application programming interface. This certificate addresses the user's skill in programming and automating functions within the SOLIDWORKS software.

- **Certified SOLIDWORKS Professional CAM (CSWP-CAM)**: CAM stands for computer-aided manufacturing. SOLIDWORKS provides a suite of CAM tools that can facilitate the manufacturing of parts by enabling the user to simulate and plan different manufacturing processes. The CSWP-CAM certificate assesses your ability to use those tools in SOLIDWORKS.

Professional advanced certifications

Professional advanced certifications address very specific functions within SOLIDWORKS. Often, these certifications apply to more specific industries compared to the CSWP certificate. All these certificates are advanced specializations of the CSWP certificate.

The advanced certificates offered by SOLIDWORKS are as follows:

- **Certified SOLIDWORKS Professional Advanced – Sheet Metal (CSWPA-SM):** This focuses on applications related to sheet metal. This includes bending sheet metal into different shapes, as well as conducting different related analyses.

- **Certified SOLIDWORKS Professional Advanced – Weldments (CSWPA-WD):** This focuses on applications related to welding. This includes welding both sheet metals and different formations such as frames.

- **Certified SOLIDWORKS Professional Advanced – Surfacing (CSWPA-SU):** This focuses on modeling surfaces of irregular shapes, such as car bodies and computer mice.

- **Certified SOLIDWORKS Professional Advanced – Mold Making (CSWPA-MM):** This focuses on making molds for productions. This includes molds for both metal and plastic parts.

- **Certified SOLIDWORKS Professional Advanced – Advanced Drawing Tools (CSWPA-ADT):** This focuses more on generating 2D engineering drawings to help communicate models to different parties. These can include internal quality teams or external manufacturers.

Expert certifications

Expert certifications are the highest level of certification offered by SOLIDWORKS. Obtaining an expert certificate indicates your mastery of a large array of functions in the software. Also, expert certificates are the only ones with required prerequisites. Two expert certificates are offered, as follows:

- **Certified SOLIDWORKS Expert (CSWE):** This demonstrates mastery over all SOLIDWORKS modeling and design functions. To qualify for this exam, the user must have the CSWP certificate, in addition to four CSWPA certificates.

- **Certified SOLIDWORKS Expert in Simulation (CSWE-S):** This demonstrates mastery over all the areas of the SOLIDWORKS Simulation software. To qualify for this exam, the user must have the CSWP, CSWA – Simulations, and CSWP – Simulations certificates.

A SOLIDWORKS user doesn't need to obtain all these certifications. It is rare to find one person with all these certificates. This is because each certification level can address very different needs and serve different industries and/or positions. Also, some certification levels are more in demand than others as they are more essential and, hence, used in more industries. Sequentially, the certifications can be viewed as follows:

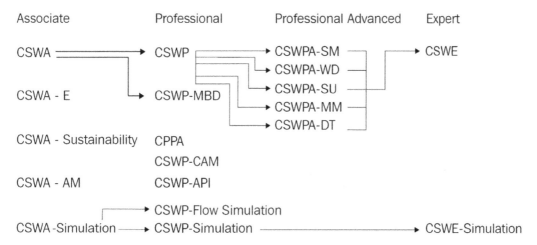

Figure 1.9 – A map of the different SOLIDWORKS certifications

This book covers the two most essential, sequential certification levels: **Certified SOLIDWORKS Associate (CSWA)** and **Certified SOLIDWORKS Professional (CSWP)**. These two certifications cover the common usage scenarios within SOLIDWORKS.

Summary

In this chapter, we learned about what SOLIDWORKS is, how parametric modeling works, and the different certifications offered by SOLIDWORKS. This will help us set our expectations and create our future development roadmap concerning SOLIDWORKS. It will also help us to understand the capabilities of the software and its vast scope.

In the next chapter, we will cover the SOLIDWORKS interface and its navigation. This will enable us to navigate the software and identify the different components that exist in its interface.

Questions

Answer the following questions to test your knowledge of this chapter:

1. What is SOLIDWORKS?

2. Name some industries that utilize SOLIDWORKS.

3. How is parametric modeling defined?

4. What are the major advantages of parametric modeling?

5. What is the difference between parametric modeling and direct modeling?

6. What are the SOLIDWORKS certifications and why are they important?

7. What are the main categories of certification levels offered by SOLIDWORKS?

> **Important Note**
> The answers to the preceding questions can be found at the end of this book.

Further Reading

More information about the certifications offered by SOLIDWORKS can be found here: `https://www.solidworks.com/solidworks-certification-program`.

2
Interface and Navigation

In this chapter, we will look at SOLIDWORKS and its software interface, as well as its main components. In addition, we will cover how to navigate through the software interface so that you will be able to easily find your way around the software in the upcoming chapters. We will also talk about the document's measurement system in terms of the different standard units it uses globally, such as feet, inches, centimeters, and millimeters for measurements of length. Interacting and setting up an interface with the software and setting up our measurement system will be the first two actions we will perform in any new project.

The following topics will be covered in this chapter:

- Starting a new part, assembly, or drawing file
- Main components of the SOLIDWORKS interface
- The document's measurement system

Technical requirements

In this chapter, you will need to have access to SOLIDWORKS.

The project files for this chapter can be found in this book's GitHub repository: `https://github.com/PacktPublishing/Learn-SOLIDWORKS-Second-Edition/tree/main/Chapter02`.

Check out the following video to see the code in action: `https://bit.ly/3EVROJv`

Starting a new part, assembly, or drawing file

This section addresses the three types of SOLIDWORKS files: parts, assemblies, and drawings. Here, we'll briefly cover what each file is for and how we can use each of them; however, more about each type of file will be covered throughout this book.

What are parts, assemblies, and drawings?

As we just mentioned, SOLIDWORKS files fall into three distinctive categories: parts, assemblies, and drawings. Each file type corresponds to a certain deliverable when we're making a product. By deliverable, we mean whether we need to deliver a **three-dimensional** (**3D**) part file, a 3D assembly file, or a **two-dimensional** (**2D**) engineering drawing. To illustrate these three file types, let's break down the simple cylindrical box shown in the following diagram:

Figure 2.1 – A cylindrical box assembly consisting of two parts

We can induce three distinctive categories from the preceding cylindrical box diagram: **parts**, **assemblies**, and **drawings**. Let's take a look at each of these here:

- **Parts**: Parts are the smallest elements that make up an artifact. They are the first step in building any product in SOLIDWORKS. Since SOLIDWORKS is used to create 3D software, all of its parts are 3D. Also, each part can be assigned to one type of material. Our cylindrical box contains two parts: a main **cylindrical** container and a **cap**, as shown in the following diagram:

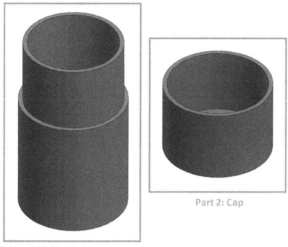

Figure 2.2 – Separating the parts comprising the cylindrical box assembly

After creating the two parts separately in two different part files, they can be put together into an assembly file.

- **Assemblies**: SOLIDWORKS assemblies are where you will be able to join more than one part together to make an assembly. Most of the artifacts we use in our everyday life contain more than one part, linked together. Some examples include cars, phones, water bottles, tables, and more. In our cylindrical box example, the assembly will look like this:

Figure 2.3 – A closed cylindrical box assembly

The main purpose of SOLIDWORKS assemblies is to check how different parts—which are often created separately—interact with each other. This will help us evaluate whether or not the parts fit together correctly. It also helps the design and engineering teams evaluate the look of the product as a whole. In addition, through SOLIDWORKS assemblies, we can simulate the movements of mechanical products.

- **Drawings**: SOLIDWORKS drawings allow you to create 2D engineering drawings out of your parts or assemblies. Engineering drawings are the most common way to communicate designs on paper. They often show dimensions, tolerances, materials, costs, parts **identifiers (IDs)**, and so on. Engineering drawings are often required when designs need to be reviewed by certain parties. Also, they are often required if you wish to talk about your designs with clients or manufacturing/prototyping establishments. For our cylindrical box, an engineering drawing might look like this:

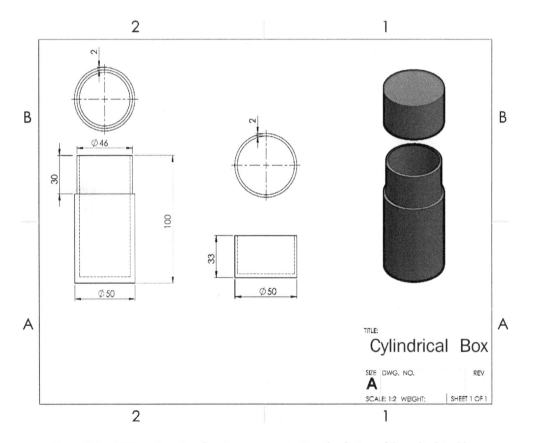

Figure 2.4 – A 2D engineering drawing communicating the design of the cylindrical box

All three types of files—parts, assemblies, and drawings—are essential to SOLIDWORKS users. This is because they are all necessary for the creation of products.

Now that we understand what parts, assemblies, and drawings are, let's look at how we can open them in SOLIDWORKS.

Opening a part, assembly, or drawing file

Now that we know the difference between parts, assemblies, and drawings, we will explore how to start each type of file. Once you open SOLIDWORKS 2022, a **Welcome** window will appear, along with some shortcuts. One of those options is starting a new **Part**, **Assembly**, or **Drawing** file. These options are highlighted in the following screenshot. Once you click on any of these options, that type of file will be opened:

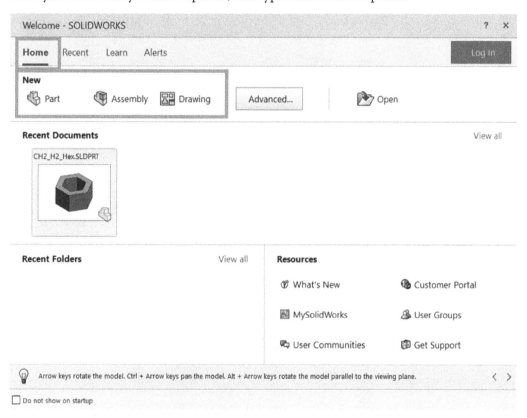

Figure 2.5 – Default Welcome window once SOLIDWORKS is launched

If the **Welcome** message does not appear, there is another way to open a new file, as follows:

1. Click on **File** in the top-left corner of SOLIDWORKS.

2. Select **New...**, as shown in the following screenshot:

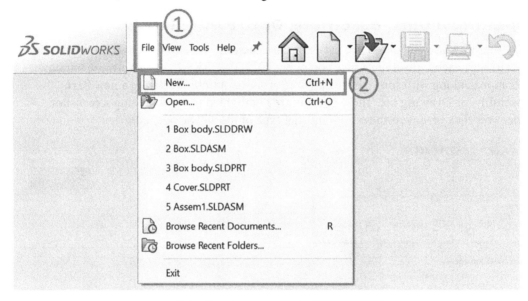

Figure 2.6 – Opening a new file in SOLIDWORKS

3. After selecting **New...**, you will be able to pick one of the three options—that is, to either start a new **Part**, **Assembly**, or **Drawing** file, as shown in the following screenshot. You can select the type of file you want to work with and click **OK**. Alternatively, you can double-click on the file type you would like to start with.

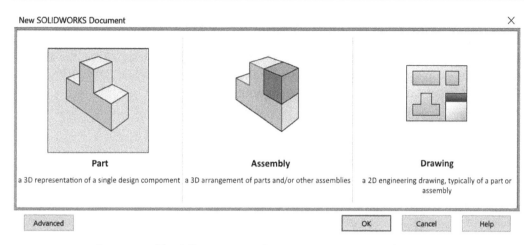

Figure 2.7 – The different options for a new SOLIDWORKS document

In this book, first, we will focus on creating parts, then assemblies, and—finally—drawings. Being able to distinguish between the different types of files is very important as everything we do afterward will be built on top of the file type we choose. Now that we understand how to open parts, assemblies, and drawings in SOLIDWORKS, let's look at how to use the software's interface further.

Main components of the SOLIDWORKS interface

In this section, we will discuss the main components of the SOLIDWORKS interface. These main components are the **Command Bar**, the **Task Pane**, the **Canvas/Graphics Area**, and the **FeatureManager Design Tree**.

Being familiar with these components is essential if we wish to use the software to a good extent. For a practical follow-up, you can download the SOLIDWORKS part linked with this chapter, which will be used to explain the main components of the SOLIDWORKS interface.

In this chapter, we will be focusing on the interface that's used when we need to deal with parts, instead of assemblies and drawings. However, the main components of the interface are the same when we deal with each file type.

When opening a part in SOLIDWORKS, regardless of whether it is new or existing, you will be faced with the view shown in the following screenshot. We will cover the four main categories of this screen: the **Command Bar**, the **FeatureManager Design Tree**, the **Task Pane**, and the **Canvas/Graphics Area**. These are the main sections of SOLIDWORKS that we'll be interacting with and referring to throughout this book.

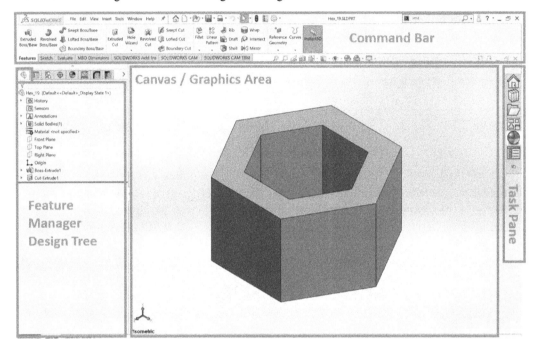

Figure 2.8 – A breakdown of the SOLIDWORKS interface

We will look at the **Command Bar**, the **FeatureManager Design Tree**, the **Canvas/ Graphics Area**, and the **Task Pane** in more detail in the following sections.

The Command Bar

The **Command Bar** is located at the top of the screen. It contains all the SOLIDWORKS commands that are used for building models. It contains different categories of commands, and each category contains a set of different commands. A close-up of the **Command Bar** is shown in the following screenshot:

Figure 2.9 – A breakdown of the Command Bar

Different categories (tabs) of commands correspond to different functions. For example, in the **Sketch** category/tab, you will find all the commands that we will need in the sketching phase. In the **Features** category/tab, you will find all the commands that we will need in order to go from the sketching phase and start creating a 3D model. The categories that are shown in the preceding screenshot are not the only ones SOLIDWORKS provides, but they are the most common ones we will use. To show the hidden **Commands** categories, we can do the following:

1. Right-click on any of the **Commands** categories, and then expand the **Tabs** menu. You will get the view shown in *Figure 2.9*, which contains more **Commands** categories, such as **Surfaces**, **Weldments**, and **Mold Tools**.

2. Select the categories you want to be shown. By doing this, these categories will be added to the **Command Bar**, as illustrated in the following screenshot:

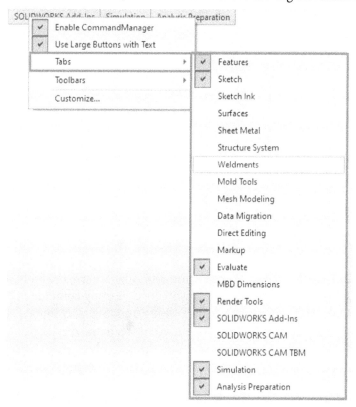

Figure 2.10 – List of command categories that can be added to the Command Bar

This concludes our overview of the **Command Bar**, which contains the different commands we will use as we build 3D models. Now, we will look at the **FeatureManager Design Tree**.

The Feature Manager Design Tree

The **FeatureManager Design Tree** details everything that goes into creating your parts. The following screenshot shows the **FeatureManager Design Tree** for the part we explored in this chapter. We can simplify the **FeatureManager Design Tree** by splitting it into four parts, as illustrated in the following screenshot:

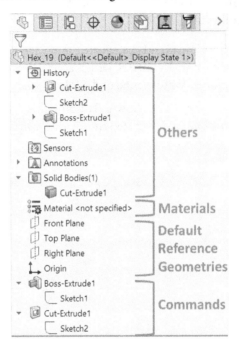

Figure 2.11 – A breakdown of the FeatureManager Design Tree

The four parts of the **FeatureManager Design Tree** are listed here:

- **Commands/Features**: These are the commands that are used to build the model. This includes sketches, features, and any other supporting commands that were added during the modeling phase (since we are building a 3D model). In the preceding screenshot, two features were used to create the model, as indicated by **Commands**. The first is **Boss-Extrude1** and the second is **Cut-Extrude1**. Note that these commands are listed in the order of when they were applied.

- **Default Reference Geometries**: The SOLIDWORKS canvas can be understood as endless space. These **Default Reference Geometries** are what can fix our model to a specific point or plane. Without these, our model will be floating in an endless space without any fixtures. Throughout this book, we will start our models from these default references. There are three planes (**Front Plane**, **Right Plane**, and **Top Plane**), in addition to the origin.

- **Materials**: Realistically, all of the artifacts we have around us are made of a certain material. Some examples of materials include plastic, iron, steel, and rubber. SOLIDWORKS allows us to assign which structural material the part will be made of. In the preceding screenshot, the **Material** feature is classed as **<not specified>**.

- **Others**: This section includes other aspects of our model's creation, such as **History**, **Sensors**, **Annotations**, and **Solid Bodies**. We will explore them later in this book.

> **Note**
>
> Through the book, we will use the term *design tree* as a shortcut to **FeatureManager Design Tree**.

The design tree helps us to easily identify how the model was built and in which sequence. This makes it easier for us to modify existing models. Now, let's look at the canvas.

The Canvas/Graphics Area

The canvas provides a visual representation of the model we have at hand. It contains three main components, as illustrated in the following screenshot:

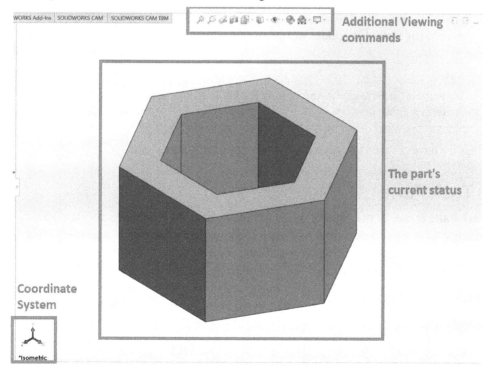

Figure 2.12 – A breakdown of the Canvas/Graphics Area

Let's break down the components, as follows:

- **Coordinate System**: This shows the orientation of the model in relation to the default coordinate system in terms of the x, y, and z axes. They are interactive and can be used to position the viewing angle of the model. By clicking on the various axes, you can arrive at that viewing orientation.

- **The part's current status**: This shows the current status of the part at work. This is updated with every construction command that's used to build the model.

- **Additional viewing commands**: These provide alternative views of the model, such as the wireframe view and section view. It also provides shortcuts that we can use to modify the scene, the appearance of the model, and hide/show various properties.

When controlling the model in the canvas, using a mouse with a scroll wheel is recommended due to the functionalities the scroll wheel has. Here are two ways the scroll helps model control:

- When the cursor is within the canvas, rolling the scroll wheel will allow us to zoom in and out of the cursor's location.

- When the cursor is within the canvas, pressing on the scroll wheel and moving the mouse will rotate the model in a certain direction. For example, if we move the mouse to the right, the model will pivot to the right.

> **Note**
> We will use the term *canvas* throughout the book. However, many would use the term *graphics area*.

Now that we have covered the canvas, let's talk about the **Task Pane**.

The Task Pane

The **Task Pane** shows to the right of our interface by default. It contains shortcuts for the different tools we will be using in order to enhance the efficiency of our work. This includes access to common online resources and forums, as well as different tools, such as appearance adjustments and the **View Palette** (mainly for drawing files). In this book, however, we won't be using linked resources while making parts or assemblies. We will use the **View Palette** in *Chapter 10, Basic SOLIDWORKS Drawing Layout and Annotations*.

Now that we know about the major components of the SOLIDWORKS interface, we will learn how to adjust the measurement system of our open document.

The document's measurement system

Since SOLIDWORKS is an engineering software, all of the models are constructed in relation to user-provided (user-input) measurements. To facilitate communication, SOLIDWORKS uses standard systems that are currently used in the industry, including the **International System of Units (SI)**, the **imperial system**, and variations of each.

Different measurement systems

When modeling with SOLIDWORKS, the user must take note of the measurement system that is set in the document. A measurement system is a set of common agreed-upon units that facilitate how we communicate quantities in terms of length, mass, volume, and so on. Some examples of such units are meters and inches, which are measurements of length.

These often correspond with internationally recognized systems such as the SI and the imperial system. The SI system is also commonly known as the **metric system**. Currently, it is used in most countries around the world. Another common system is the imperial system, which is mostly used in the **United States (US)**.

The following table compares the major units that are used in the SI and imperial systems:

	Imperial unit	**SI unit**
Length	Inches (inch)	Meters (m)
Mass	Pounds (lb.)	Kilograms (kg)
Time	Seconds (s)	Seconds (s)

Figure 2.13 – A comparison between the imperial and SI unit systems

Before we start modeling anything in SOLIDWORKS, we must decide on which system to use. The unit system we use often depends on the standards that have been adopted by the organization we work for or by the requirements of our clients.

Adjusting the document's measurement system

Now that you have decided which system to use, you must set it up on the software. You can adjust the unit of measurement by following these steps:

1. Open a new part file.

2. In the bottom-right corner, you will find the current/default measurement system in an abbreviated form. Click on the displayed measurement system, as shown in the following screenshot:

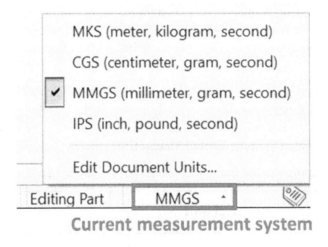

Figure 2.14 – The set and default options for unit systems

3. Choose from the default settings.

You can create your own measurement system by selecting **Edit Document Units...**. This will open the following window, where you can select custom options and customize and implement your own custom units:

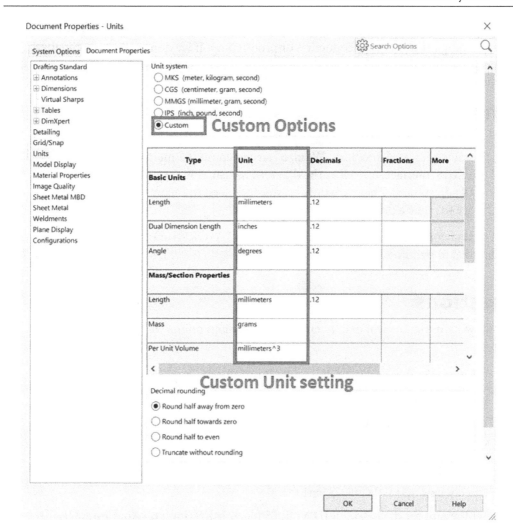

Figure 2.15 – Customizing the unit system

Note that you can change the set units at any time during the modeling process. This will convert all the units that were already set in the file. For example, let's assume that the document's length measurement was set to **IPS (inch, pound, second)** and a line was drawn to measure 2 inches. If we change the measurement system later to **MMGS (millimeter, grams, second)**, the length of the line will be automatically converted into 50.8 millimeters. This is because 1 inch is equal to 25.4 millimeters.

Note

The same procedure on adjusting the measurement system applies to the part, assembly, and drawing file types.

Knowing how to deal with measurement systems is essential, as our design aim will be to produce a tangible object for production or prototyping. If we don't follow the required settings from the start, our final 3D model may not have a tangible value.

Summary

In this chapter, we learned how to start the different types of SOLIDWORKS files—that is, parts, assemblies, and drawings. We also learned about the main components of the SOLIDWORKS interface, as well as the different measurement systems that are available and how to adjust them. These are the first steps we need to follow when we plan to use the software to make a project and its foundations.

In the next chapter, we will start working with SOLIDWORKS sketching. Sketching is foundational to building any 3D model.

Questions

The following questions will help to emphasize the main points we have learned in this chapter:

1. What are the three types of files a SOLIDWORKS user can create?
2. What is the difference between SOLIDWORKS parts, assemblies, and drawings?
3. What is contained in the canvas?
4. What does the design tree show?
5. What is the difference between SI units and imperial units?
6. Open a new SOLIDWORKS part file.
7. Set the unit for the new SOLIDWORKS file to **MKS (meter, kilogram, second)**.

> Important Note
> The answers to the preceding questions can be found at the end of this book.

Section 2 – 2D Sketching

Making 2D sketches in SOLIDWORKS is essential to applying any feature or creating any 3D model. Without a strong foundation in sketching tools, you will not be able to construct 3D models. This section covers all the sketching foundations you will need as a SOLIDWORKS associate and professional.

This section comprises the following chapters:

- *Chapter 3, SOLIDWORKS 2D Sketching Basics*
- *Chapter 4, Special Sketching Commands*

3
SOLIDWORKS 2D Sketching Basics

The foundation of any 3D SOLIDWORKS model is a 2D sketch. This is because SOLIDWORKS builds 3D features based on the guidance of 2D sketches. This chapter will get you started with SOLIDWORKS 2D sketching. We will cover multiple sketching commands that will allow you to sketch shapes such as rectangles, triangles, circles, and ellipses. You will also learn how to combine those different sketches and create more complex shapes. Then, we will explore the different levels at which SOLIDWORKS defines a sketch. Mastering SOLIDWORKS 2D sketching is essential if we wish to build a 3D model.

The following topics will be covered in this chapter:

- Introducing SOLIDWORKS sketching
- Getting started with SOLIDWORKS sketching
- Sketching lines, rectangles, circles, arcs, and ellipses
- The state of sketches: under defined, fully defined, and over defined

Technical requirements

In this chapter, you will need to have access to SOLIDWORKS.

Check out the following video to see the code in action: `https://bit.ly/3GKC1xu`

Introducing SOLIDWORKS sketching

In this section, we will discuss what **SOLIDWORKS sketches** are. We will inform you of the importance of SOLIDWORKS sketching functions and how to view them when modeling with SOLIDWORKS. SOLIDWORKS sketches are the base of each SOLIDWORKS model. Thus, it is important to master SOLIDWORKS sketching first.

The position of SOLIDWORKS sketches

Sketches are typically viewed as fast drafts of a certain shape. For example, the following diagrams show a hand-drawn sketch of a square and a sketch of a cube, respectively. The main point of these is to provide a rough idea of an object.

In this hand-drawn sketch, we are communicating the idea of a square, without specifying how big that object is:

Figure 3.1 – A hand-drawn sketch of a square

Similarly, the following hand-drawn sketch communicates the idea of a cube, without specifying how big the cube is:

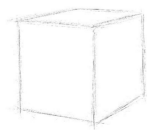

Figure 3.2 – A hand-drawn sketch of a cube

SOLIDWORKS sketches are a bit different. In SOLIDWORKS, a sketch is a fully dimensional and exact shape that's mostly given in two dimensions. SOLIDWORKS also has a 3D sketches function that is more commonly used with surface modeling. In this book, we will only use 2D sketching.

The following diagram shows a SOLIDWORKS sketch of a square with a side dimension of **50** mm. Note that the sketch is different than the one we looked at previously; it is an exact square and not an approximation of a square:

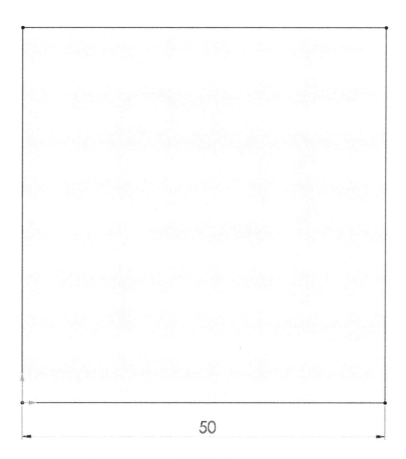

Figure 3.3 – A SOLIDWORKS sketch of a square

SOLIDWORKS sketches are the starting points of any 3D model. They are the basic guiding elements for 3D SOLIDWORKS features. For example, if we want to make a cube, we have to start by drawing the preceding square sketch. After that, we can extrude it to generate a cube, as shown in the following diagram.

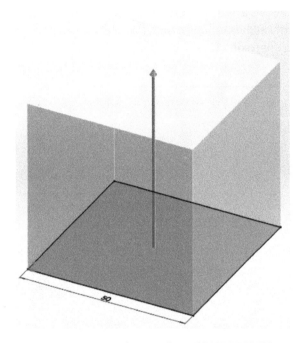

Figure 3.4 – Extruding a cube in SOLIDWORKS

Note that you can see our initial sketch at the bottom of the cube.

When we create a 3D shape in SOLIDWORKS, we often start by creating a 2D sketch and then apply a feature to it. Then, we keep iterating those two steps as the 3D shape becomes more complicated. This is why it is very important that we master SOLIDWORKS sketching before anything else.

The preceding diagram also shows the common sequence of a SOLIDWORKS model, starting with a sketch rather than a feature. The sequence then repeats as the model becomes more complex.

Simple sketches versus complex sketches

SOLIDWORKS has many ready-made commands that we can use to create simple sketch shapes such as lines, squares, circles, ellipses, arcs, and so on. We will learn about these sketching commands in this chapter. However, it is important to understand that all complex sketches are a combination of different simpler sketches.

Let's illustrate this with an example. The following figure shows a sketch of a relatively complex shape:

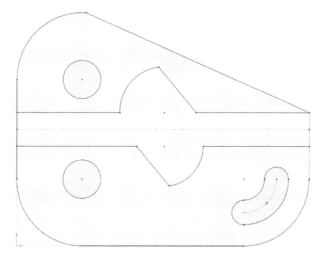

Figure 3.5 – A complex sketch in SOLIDWORKS

Note that the complex shape is made up of a combination of different simple elements. This image can be easily made with four different sketching commands: lines, arcs, circles, and slots. The following figure shows how we can break down the complex sketch into these four sketching commands. All the unmarked elements of the sketch are repetitions of the marked ones. We can refer to each of these elements (a line or an arc, for example) as a sketch entity:

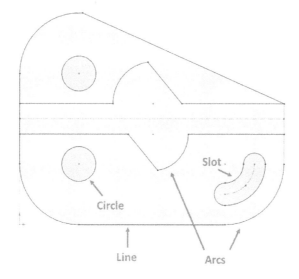

Figure 3.6 – Complex sketches are combinations of simple sketch entities

Simplifying complex sketches so that they're simple elements is a skill that takes time to build. Since we've just started sketching, we may face difficulties simplifying complex sketches. This book will help you develop this skill, but you will become better at it with experience.

Now that we know what sketches are, we can start defining the elements that are involved in our sketches. We will start with sketch planes.

Sketch planes

Sketch planes are flat surfaces that we can use as bases for our sketches. They are important because they give our sketches a solid location. If we didn't have them, our sketches would float undefined in 3D space.

SOLIDWORKS provides us with default sketch planes when we start a new part. We can find them listed in the design tree. The default sketch planes are the **Front Plane**, the **Top Plane**, and the **Right Plane**. They are listed in the following screenshot:

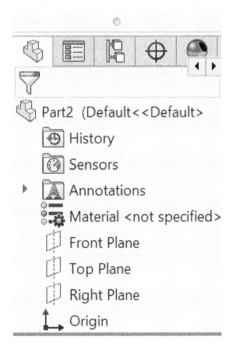

Figure 3.7 – The sketch planes as shown in the software interface

To help you visualize these default planes, imagine a box and the planes being the top, the front, and the right-hand sides of it. The following figure shows the three default planes in the shape of a box:

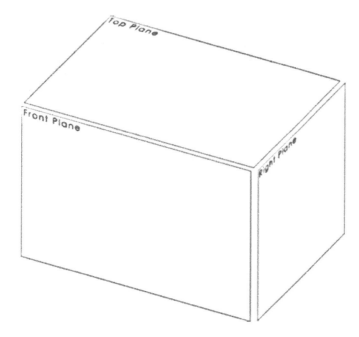

Figure 3.8 – The three base planes reshaped as a box

These three planes are not the only ones that are available to us. In addition to the default planes, we can use any other 2D straight surface as a sketch plane as well. We also have the option of creating our own planes based on different geometrical references. We will explore these possibilities later in the book. Now that we've identified our sketch planes, we can start using those planes to create our sketches.

Getting started with SOLIDWORKS sketching

In this section, we will discuss how to start sketching, what it means to define a sketch, and what the major geometrical relations that exist in SOLIDWORKS sketching are. These topics will be our practical introduction to getting into SOLIDWORKS sketching.

Getting into the sketching mode

To start a sketch, we need to have a part file open. Then, we can follow these steps to get into the sketching mode:

1. Select one of the default sketch planes: **Front**, **Top**, or **Right**.
2. In the **CommandManager**, select the **Sketch** option (which is marked as **2** in *Figure 3.9*). This will open up the **Sketch commands category**, which will show all the commands related to sketching.

3. Select the **Sketch** command (which is marked as **3** in *Figure 3.9*). This will allow us to enter the sketching mode. When we're in the sketching mode, we can apply different sketching commands, such as the marked **Simple sketching commands**:

Figure 3.9 – The Sketch tab and the Sketch command

Now that we are in the sketching mode, we can see multiple sketching commands on the command bar. This includes commands for sketching simple shapes such as lines, rectangles, circles, polygons, arcs, and others. Let's go over these commands in more detail. However, since we are already in the sketching mode, take some time to randomly click on those commands and try them out on the canvas.

There are two alternative ways to start the sketching mode, as follows:

*   Select the **Sketch** command first and then select the sketch plane.
*   Click on the **Sketch** shortcut that appears when you click on one of the planes. The following screenshot highlights this shortcut:

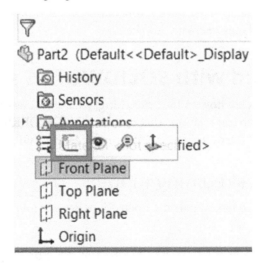

Figure 3.10 – Another way to get into the Sketch mode is to use the popup after selecting a plane

Before we start applying these sketching commands, we will discuss what it means to define a sketch.

Defining sketches

Now that we know how to start sketching, we will learn about what defines a sketch. Constructing a sketch will require two elements: the first is the sketch entities, such as lines, arcs, and circles, while the second is the dimensions and relations that define the sketch entities. Remember that SOLIDWORKS is a form of engineering software that aims to support the design and manufacture of products. Defining sketches ensures the design intent integration in the design. Thus, defining shapes is very important.

To define sketches, we can use dimensions and relations. Let's start by defining them:

1. **Dimensions**: These represent distances and angles that can be defined with a numerical value. Some examples are as follows:

   - Lengths of lines
   - The diameters and radii of circles and arcs
   - An angle between two lines

2. **Relations**: These represent geometric relations between the different parts of a sketch. Some examples are as follows:

   - A line that is has a **horizontal** relation to a sketch plane
   - Two lines are **perpendicular** to each other
   - Two circles are **concentric** to each other
   - A line is a **tangent** to a curve

To illustrate this, let's look at a visual comparison between two lines. One is **Fully defined** and one is **Under defined**:

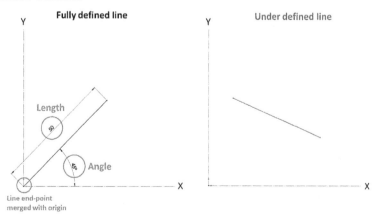

Figure 3.11 – A visual comparison between a fully defined and an under-defined line

Of the two preceding lines, the one on the left is fully defined while the one on the right is under-defined. The line on the left is defined as such because of the following:

- One end coincides with the origin of the coordinate system.

- The length of the line was defined as 50 millimeters.

- The angle between the line and the x axis is 45 degrees.

These dimensions and relations are what make a sketch defined. It may not look like there's much of a difference between them when we look at the two lines, assuming that we drew them on paper. However, when we work with a computer program, we need to let the computer know everything about the line (or any other sketch entity); otherwise, it won't know what we want.

To make it easier to distinguish between fully defined and under defined sketches, SOLIDWORKS color-codes them. **Black** parts are fully defined, while **blue** parts are under defined. In addition to color-coding, SOLIDWORKS also indicates the status of our sketch below the canvas. The following screenshot shows an indication of a **Fully Defined** sketch. Other classifications include **Under Defined** and **Over Defined**:

| 0.43in | -2.36in | 0in | Fully Defined | Editing Sketch2 | IPS | ▲ |

Figure 3.12 – The sketch status classification as found at the bottom of the interface

> **Note**
> You can change the default color code through the settings if needed.

Sketches are defined with measurements and relations, where measurements refer to numbers and relations refer to geometrical relations. We will explore geometrical relations in the next section. Also, in the section titled *Under defined, fully defined, and over defined sketches*, we will be discussing those terms in more depth.

Geometrical relations

The following table summarizes the majority of the geometrical relations we will come across while working with SOLIDWORKS sketching. You don't need to memorize all of these relations for now. Simply read through the following table and use it as a reference as we continue to develop our SOLIDWORKS skills. The relations in the following table have been organized in alphabetical order. In the SOLIDWORKS interface, we will deal with these icons more and more since they will appear in the sketches themselves, as well as when we choose which relation to apply them to:

Relation's Name	Relation Icon	Relation Function
Coincident		Coincident relations occur between points and lines, arcs, circles, and so on. This would make a point lie in other sketch entities, such as lines.
Colinear		This makes two or more lines lie in one direction.
Concentric		This can make two or more circles or arcs share the same center.
Coradial		This applies to two or more arcs if the different arcs share the same center and radius.
Equal	=	This makes two or more lines or arcs equal to each other in terms of length.
Equal Curve Length		This relation occurs between an arc or circle and a line. It makes the perimeter equal to the line in terms of length.
Fix		This fixes the selected sketch entity to where it exists at the time of setting the relation.
Horizontal		This makes lines horizontal to the sketch plane. In addition, it can make more than one point lie in a horizontal line.
Intersection		This will position a point at the intersection point of two lines. This includes the extension of the lines, as well as the lines themselves.
Merge		This can merge more than one point together into one point location.
Midpoint		This can position a point so that it's in the middle of a line.
Parallel		This can make more than one line parallel to another.
Perpendicular	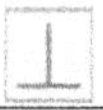	This can make two lines perpendicular to each other.
Tangent		This relates to circles, arcs, and other curved entities. It can turn a line tangent into a curved entity. It can also make more than one curved entity tangent to each other.
Vertical		This makes lines vertical to the sketch plane. In addition, it can make more than one point lie in a vertical line or in relation to another point.

Figure 3.13 – The geometrical relations for sketching

These icons and their relations will repeatedly show up as we are creating sketches. Thus, it is important that we know what they mean. We will start using these relations in the next section, that is, when we start sketching different shapes.

Sketching lines, rectangles, circles, arcs, and ellipses

In this section, we will discuss the major sketching functions and how to use them. These include sketching lines, rectangles, circles, arcs, and ellipses. We will address each of these sketching commands separately and find out how to define each one.

The origin

On the canvas, you should be able to see a small red dot with arrows, as shown in the following figure. This dot is located exactly where the two red arrows meet and represent the origin point of the canvas. It is also the only defined and fixed point in our SOLIDWORKS infinite canvas:

Figure 3.14 – The SOLIDWORKS graphical representation of the origin

Because it is the only fixed point, it is very important to always link our sketches to that origin point. Otherwise, our sketch will always be under-defined in the infinite canvas.

Sketching lines

To illustrate how to sketch lines, we will sketch the following shape. Note that the sketch is fully defined, and so SOLIDWORKS will show it filled in with black after it's been sketched. Also, take note of the relations and dimensions (in millimeters) shown in the following figure:

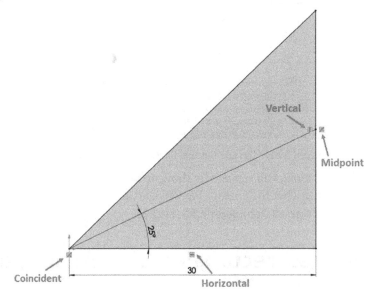

Figure 3.15 – The final output of our sketching exercise

Let's go ahead and start constructing the shown sketch. To do that, we will go through two stages: outlining and defining.

In the outlining stage, our aim will be to draw a rough outline of the final shape. Here, we will not pay much attention to dimensions or relations. We will follow these steps in the outlining stage:

1. Start a new part file.

2. Set the document's measurement system to **MMGS (millimeter, gram, second)** if it is not already.

3. Navigate to the **Sketch** mode using the **Top Plane**.

4. Select the **Line** sketch command as indicated in the following figure.

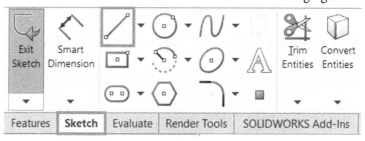

Figure 3.16 – The location of the line command

5. Move the mouse cursor onto the canvas and click on the origin. To start drawing the triangle, move your cursor to the right and, once the line is drawn, left-click on your mouse. Note that the shape of the cursor will change into a pen.

6. Next, move the mouse more upward to create the second line, and then left-click on the mouse again.

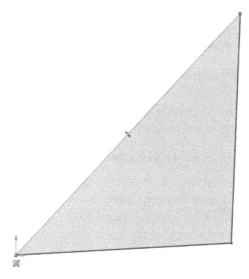

Figure 3.17 – Three under-defined lines forming a triangle

7. Move the cursor again to draw a line linking back to the origin. This concludes the creation of a shape that looks like a triangle, as shown in *Figure 3.17*.

8. Now, we can draw the last line, which links the origin point to any point on the vertical line so that we end up with the following sketch. Note that the sketch on the screen is blue, indicating that it is under-defined.

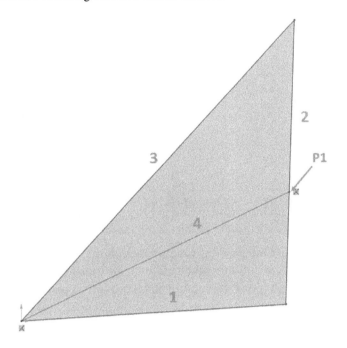

Figure 3.18 – The sketch with all the sketch entitles

9. Exit the **Line** sketching command by pressing *Esc* on the keyboard.

In the defining stage, we will work on defining our outline with the necessary dimensions and relations to fully define our sketch. Note that some of the relations are set automatically by SOLIDWORKS, according to how we place our lines. Follow these steps for the defining stage referring to the numbers indicated in *Figure 3.18*:

1. Click on line **1**, as indicated by the preceding sketch.

2. A new panel will appear on the left, in place of the design tree. It will be titled **Line Properties**, as shown in the following screenshot. Under **Add Relations**, click on **Horizontal**. This will add a horizontal relation to the line. You will see a small icon appear next to the line, showing that the line is horizontal:

Figure 3.19 – The Line Properties manager showing the applicable geometric relations

---

Note

SOLIDWORKS can apply relations automatically if they can be inferred by how you sketch an entity. For example, if you try to sketch a horizontal line, SOLIDWORKS will apply the relation horizontal to the line automatically. In this case, you will see the relation listed under **Existing Relations** in the **Line Properties**.

---

3. Click on line **2** and add a **Vertical** relation.

Note

To deselect a line, click anywhere else on the canvas.

---

4. Press and hold down *Ctrl* on your keyboard, click on the endpoint **P1**, then click on line **2**. From the properties that appear on the left, select the **Midpoint** relation.

> **Tip**
>
> We can select multiple sketch entries at the same time by clicking and holding down *Ctrl* on the keyboard and selecting multiple entities.

> **Note**
>
> By following the preceding steps, we will have set up all the relations for our sketch. Note that many parts of the sketch are still blue. Try to click and hold different parts of the sketch, points, or lines and move the mouse around. All the sketch elements will move in a way that preserves all the relations that have been set.

5. On the command bar, select the **Smart Dimension** command, as shown in the following screenshot:

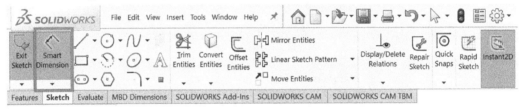

Figure 3.20 – The Smart Dimension command

6. Go back to the canvas and click on line **1**. A dimension will appear, displaying the current length of the line. Left-click on an empty space in the canvas once more. You will be prompted to enter a length value for the line. Type in the value 30 and then click on the green checkmark, as shown in the following screenshot. After that, you will notice that the line's length changes to match the new length. Also, note that the line, including its endpoints, turns black:

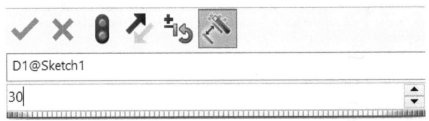

Figure 3.21 – The Smart Dimension interface with space to input the dimension value

7. Click on line **1** again, and then click on line **4**. Note that the specified dimension changed to the angle between the two lines, which is the dimension we want to specify. Left-click anywhere on the canvas once more to confirm the dimension's location. In the box, type 25, which indicates the degree, and then click the green checkmark.

> **Note**
>
> To delete a dimension, select it and press *Delete* on the keyboard. Alternatively, we can right-click on the dimension and select **Delete** from the options that are available.

8. Exit the smart dimension mode by pressing *Esc* on the keyboard or clicking on the **Smart Dimension** command on the command bar.

Note the shown sketch in the exercise is shaded indicating an enclosure. You can turn this feature on and off by toggling the **Shaded Sketch Contours** command on the command bar, as shown in the following screenshot.

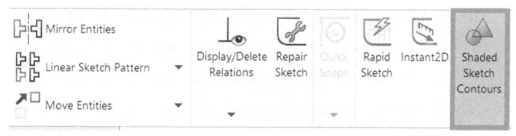

Figure 3.22 – The Shaded Sketch Contours option shades enclosed sketch entities

At this point, we will see that the sketch is fully black, indicating that it is fully defined. This concludes our first sketching exercise. In this simple exercise, we have covered many essential sketching features that we'll keep using when we model with SOLIDWORKS throughout this book, including the following:

- How to start sketching
- How to sketch lines
- How to set up lengths and angles using smart dimensions
- How to set up geometric relations such as vertical, horizontal, and midpoint

These sketching commands are essential to sketching using SOLIDWORKS sketching tools. In the next section, we will use these skills to sketch rectangles and squares.

Sketching rectangles and squares

In this section, we will learn how to sketch rectangles and squares. To illustrate this, we will sketch the following diagram. We have already covered most of the concepts we'll need in order to complete the sketch, including how to get into the sketching mode and how to define a sketch. In this example, the dimensions are in **IPS (inch, pound, second)**, where the length is in inches. Note that we coded the sketch with **R1** and **R2**, indicating rectangles 1 and 2, and **S1**, indicating square 1:

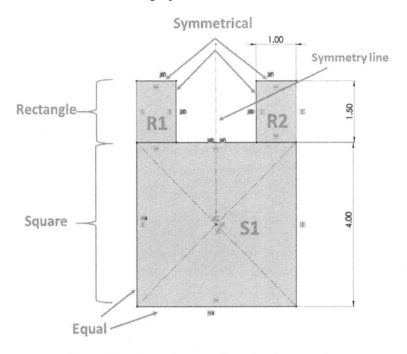

Figure 3.23 – The final output of our sketching exercise

To create this sketch, we will go through the same two stages that we went through previously: outlining and defining.

For outlining, we will draw an arbitrary outline of the final shape we want to draw. Let's get started:

1. Start a new part and make sure that the document measurement system is set to **IPS (inch, pound, second)**. Note that we are using a different measurement system for practice purposes here.

2. Navigate to the **Sketch** mode using any of the sketch planes (for example, **Front Plane**).

3. On the command bar, select the drop-down menu next to the rectangle shape and click on the **Center Rectangle** command. In this exercise, we will use the two rectangle commands, both of which are shown in the following screenshot: **Center Rectangle** and **Corner Rectangle**.

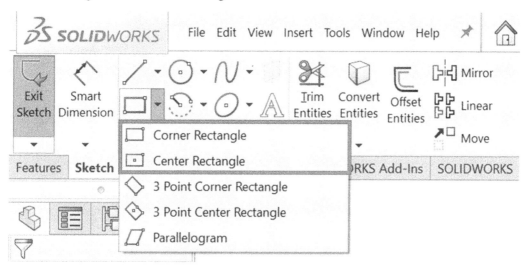

Figure 3.24 – The sketching commands Corner Rectangle and Center Rectangle used in the exercise

Both commands create rectangles; the difference is how those rectangles are created. A **Center Rectangle** is created with two clicks: one indicating the center and the other indicating a corner. The **Corner Rectangle** is created with two clicks, indicating the opposing corners. Note that the small figures also show us how to draw that particular type of rectangle by showing us the sequence of clicks that are needed.

4. After selecting **Center Rectangle**, click on the origin, move the mouse to the side, and click away from the origin to form an approximate shape of a square. We are doing this to create square **S1**.

> **Tip**
>
> You can delete any part of the sketch by highlighting or selecting that part and pressing *Delete* on the keyboard. Alternatively, you can right-click and select **Delete**.

5. Select the **Corner Rectangle** command and draw the two rectangles, that is, **R1** and **R2**. For **R1**, click on the top-left corner of square **S1**. Then, move the mouse up and away from the first click, then click again to form the rectangle. Do the same for **R2**:

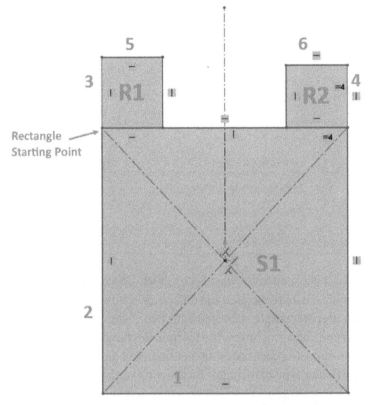

Figure 3.25 – The final result after the outlining stage

6. The last step is to add the symmetry line. The symmetry line is a **centerline** that we will be able to use to create symmetrical relations between parts on opposite sides of that line. To create a centerline, click on the drop-down menu next to the line command and select **Centerline**. The centerline goes through the middle of the sketch, starting at the origin and going upward.

> **Tip**
>
> Since we are creating the centerline, note that we can move the line endpoint slowly at an angle until the vertical relation appears, at which point, you can lock it on that. This is one way in which SOLIDWORKS interprets the relations we want to apply and applies them for us. We can use this approach to make sketching faster.

For our defining stage, we will define the outlined shape by applying relations and dimensions. Follow these steps to do so:

1. Select lines **1** and **2**, and apply the **Equal** relation. This condition is what will make a normal rectangle into a square.

2. Select lines **3** and **4**, as well as the symmetry line, and apply the **Symmetric** relation. In this case, SOLIDWORKS will automatically interpret the centerline as the symmetry line since it is located in the middle. You can also do the same with lines **5** and **6**.

3. Using the smart dimension, set the given lengths so that they match what's shown in the following figure. We can use the smart dimension in the same way we did when we sketched a line:

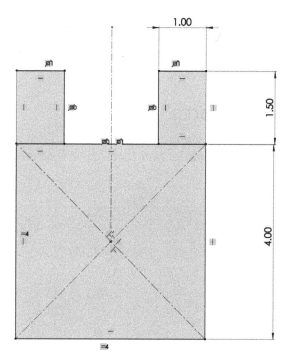

Figure 3.26 – The final sketch being fully defined

At this point, we should have a fully black shape that is fully defined. Also, take note of the status of the sketch, which is shown at the bottom of the SOLIDWORKS screen. It is indicated as **Fully Defined**, as shown in the following screenshot:

| 0.43in | -2.36in | 0in | Fully Defined | Editing Sketch2 | | IPS | ˄ |

Figure 3.27 – The sketch referenced as fully defined

This concludes our sketching exercise. In this simple exercise, we have covered many essential sketching features that we will capitalize on throughout our SOLIDWORKS interaction, including the following:

- How to sketch squares and rectangles

- How to set up equal-length lines

- How to draw a centerline and set up symmetrical sketch entities

At this point, we already have many of the sketching basics under our belt. Now that we're advancing, we won't need as much guidance as all the commands will start becoming second nature to us. Before moving on, take the time to experiment with creating other types of rectangles, such as the **3 Point Corner Rectangle** and the **3 Point Center Rectangle**. In addition, take some time to experiment with creating a **Parallelogram**. We can find all of these shapes in the **Rectangle** command drop-down menu.

Now, we know how to sketch lines, rectangles, and squares. Next, we will develop our skills by addressing circles and arcs.

Sketching circles and arcs

In this section, we will sketch circles and arcs. First, let's break down what a circle and an arc are. The following figure shows a circle. Note that a circle is defined by its **Center** (a point) and its **Diameter**:

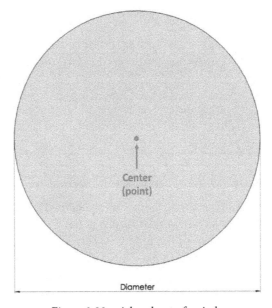

Figure 3.28 – A breakout of a circle

The following figure shows an arc, as well as the elements that define it. An arc can be defined by its **Center** (a point), as well as other points, which indicate the endpoints of the arc. This is in addition to its **Radius** and various distances:

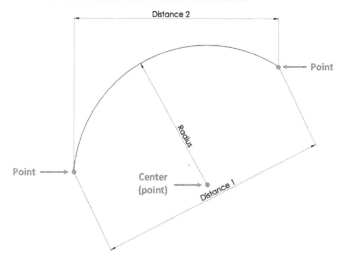

Figure 3.29 – A breakout of an arc

Each of these elements can be controlled with dimensions or relations. Each point can be understood as a standalone entity that we can use for relations or dimensions.

To illustrate these two commands, we will sketch the following shape:

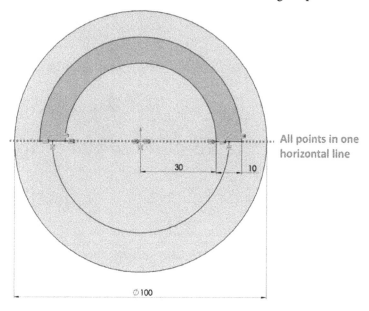

Figure 3.30 – The final result of this sketching exercise

Similar to what we did previously, we will do outlining and then defining. We will start with the outlining stage, as follows:

1. Open a new part file and make sure that the measurement system is set to **MMGS (millimeter, gram, second)**. Note that we have selected **MMGS** for practice purposes only.

2. Navigate to the **Sketch** mode using any of the default planes (for example, **Top Plane**).

3. On the command bar, select the **Circle** command. Click on the origin, and then move the mouse further to form a circle. Click again to finish drawing the circle:

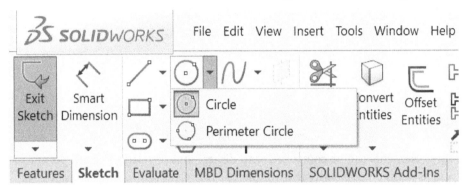

Figure 3.31 – The Circle sketching command

4. Click on the **Centerpoint Arc** command and follow the instructions shown in the small command image:

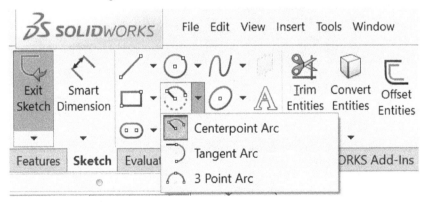

Figure 3.32 – The CenterPoint Arc sketching command

5. Connect the endpoints of the arcs using the **Line** sketch command. The result will be similar to the following figure. Note that we indicated the different points of the sketch using the letters **P1-P4**, which stand for point 1, point 2, and so on:

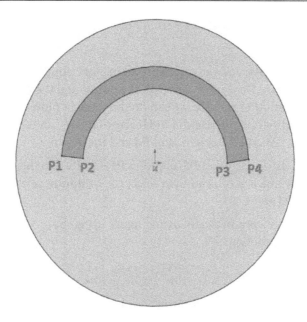

Figure 3.33 – The arcs endpoints can be connected using lines

6. Use the **Centerpoint Arc** to create the last lower arc, which links the two lines we created in *Step 5*, to create the following sketch:

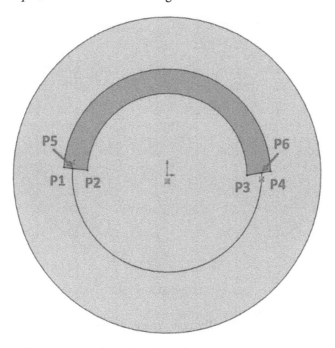

Figure 3.34 – All the sketch entitles required for our shape

Now that we've finished outlining, we can start defining the sketch elements in the defining stage. Let's get started:

1. Note that **P1**, **P2**, **P3**, **P4**, and the origin are all in a horizontal line in the initial image for this exercise. To do this, select all the points and the origin and select the **Horizontal** relation. Alternatively, we can do the same in more steps by selecting **P1** and the origin and setting the relation to **Horizontal**. We can then do the same with **P2** and the origin, **P3** and the origin, and **P4** and the origin.

2. Set a **Midpoint** relation between **P5** and lines **P1** and **P2**. Do this by selecting the point, **P5**, and the line it is on, and then select the **Midpoint** relation. Do the same for **P6** and lines **P3** and **P4**.

3. Set the dimensions shown in the following figure using the **Smart Dimension** function:

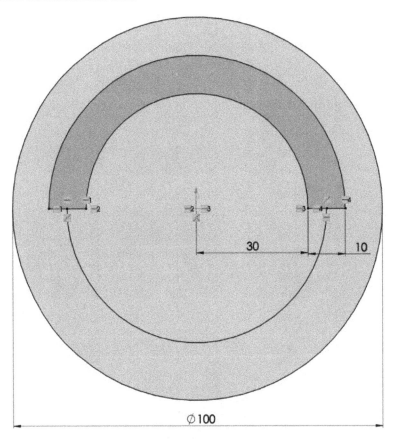

Figure 3.35 – The final sketch fully defined

This concludes this exercise of using circles and arcs. At this point, our sketch is fully defined. Before moving on, take some time to individually experiment with creating a **Perimeter Circle**, a **Tangent Arc**, and a **3 Point Arc**. These are some other ways we can create circles and arcs that we did not explore in this exercise. However, all these commands follow the same principles when it comes to making circles and arcs. Now that we've mastered how to create circles and arcs, we will address ellipses and construction lines.

Sketching ellipses and using construction lines

In this section, we will discuss what ellipses are, how to define them, and how to make them in SOLIDWORKS. We will also touch on the idea of construction lines. We can look at an ellipse as a combination of two axes and five points, as shown in the following figure:

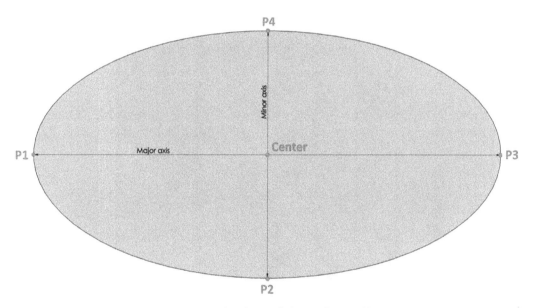

Figure 3.36 – A breakout of what makes an ellipse

When we define an ellipse in SOLIDWORKS, we can use the four points and the center as our defining factors. We can also define an ellipse with the help of construction lines, which we can use to define the size and the location of the ellipse. To illustrate this, we will sketch the following ellipse:

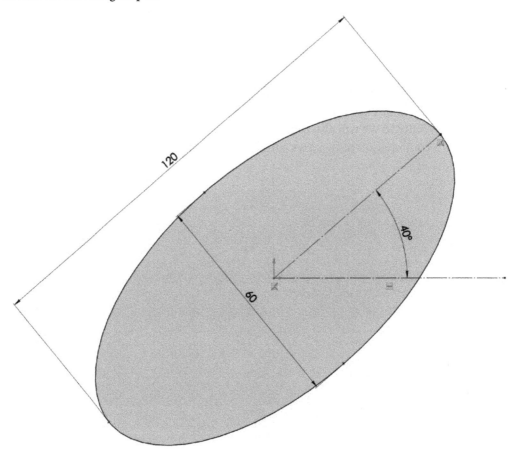

Figure 3.37 – The final result of this sketching exercise

Similar to all our other exercises, we will sketch the ellipse in two stages: outlining and then defining. Let's start with our outlining stage:

1. Start a new part file and set the measurement system to **MMGS (millimeter, gram, second)**.

2. Start the Sketch mode using the **Right Plane** (or any other plane).

3. Select the **Ellipse** command from the command bar:

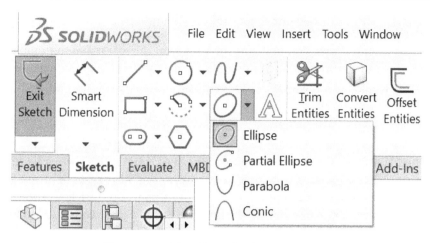

Figure 3.38 – The Ellipse sketching command

4. We will need to click three times to create an ellipse. First, click on the origin.
    Then, move the mouse to create the major axis and then left-click to confirm
    this. After that, move the mouse once more to create the minor axis and left-click
    again to confirm this. Make sure that the ellipse is tilted a bit to avoid unnecessary
    automated relations being made by SOLIDWORKS. We should have a shape similar
    to the one shown in the following figure:

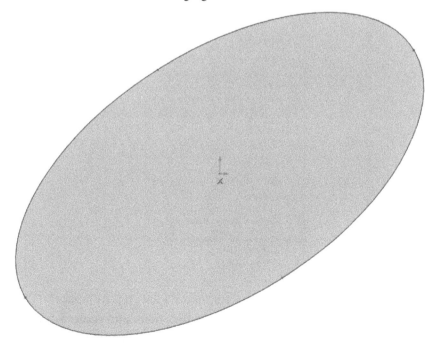

Figure 3.39 – An arbitrary starting sketch for an ellipse

As usual, now that we've finished our shape outline, we can start our definition stage:

1. First, use a smart dimension to define the lengths of the major and minor axes. We will set them to 60 mm for the minor axis and 120 mm for the major axis. We can define the lengths of the axes by defining the distance between the points on the perimeter of the ellipse.

2. To set up the angle of the ellipse, we can use construction lines. Construction lines are dotted lines that are used for the purpose of supporting the definition of our sketches. However, they are not accounted for by SOLIDWORKS when building features. To set up construction lines, select the normal **Line** command and then check the **For construction** option in the **Options** panel on the left, as shown in the following screenshot. After that, all the lines we sketch will be construction lines:

Figure 3.40 – The construction line setup from the line property manager

Alternatively, we can sketch normal lines and turn them into construction lines. Do that by clicking on the normal line or any other sketch entity and checking **For construction** from the **Options** tab that appears on the left.

3. Draw the two construction lines that are shown in the following figure:

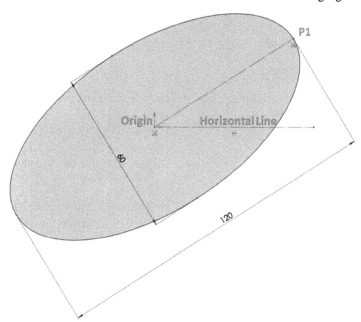

Figure 3.41 – Construction lines can play a role in defining our sketches

4. Using the smart dimension, set the angle between the two construction lines to 40 degrees.

This concludes this exercise on creating an ellipse. In this exercise, we covered how to draw and define an ellipse, and what construction lines are and how to utilize them to define our sketches.

In the exercise, we defined construction lines using the **Line Properties**. However, you can also sketch one directly using the drop-down menu for the **Line** command and selecting **Centerline** as shown in the following figure.

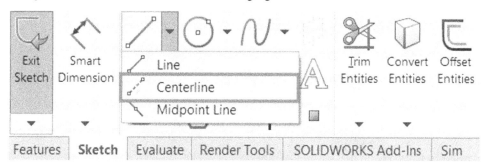

Figure 3.42 – The Centerline command can directly sketch construction lines

Both ellipses and construction lines will be very useful as we advance our SOLIDWORKS skills. Now, we will look at the fillets and chamfers commands so that we can improve our sketching skillset further.

Fillets and chamfers

In this section, we will discuss making fillets and chamfers for our sketches. Fillets and chamfers can be applied between two sketch entities, usually between two lines. They are defined as follows:

- **Fillets**: Fillets can be viewed as a type of arc. Thus, they are defined in the same way, that is, with a center and a radius.

- **Chamfers**: Chamfers can be defined in different ways. These include **two equal distances**, **two different distances**, or a **distance and an angle**.

The following figure illustrates the shapes of fillets and chamfers, as well as how they are defined:

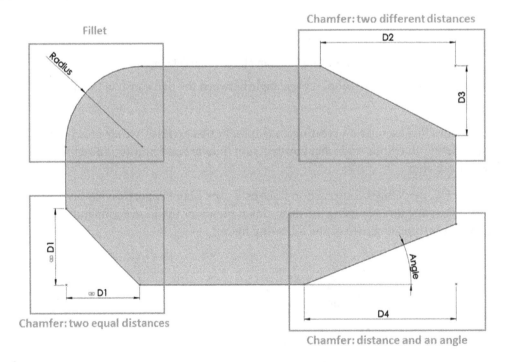

Figure 3.43 – The sketch fillets and the different types of chamfers

To illustrate how to create fillets and chamfers, we will sketch what's shown in the following figure:

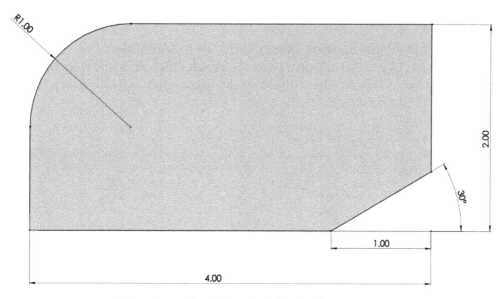

Figure 3.44 – The final result of this sketching exercise

Fillets and chamfers are different than other sketching commands in that we will apply and define them at the same time. Thus, they won't follow our typical procedure of outlining and then defining. To create the sketch, we will use the IPS measurement system. To achieve the preceding sketch, we will need to sketch a 4 x 2-inch rectangle, which will be our starting point. Note that we marked the different lines with the letter **L** and marked the different vertices with the letter **V** for ease of reference:

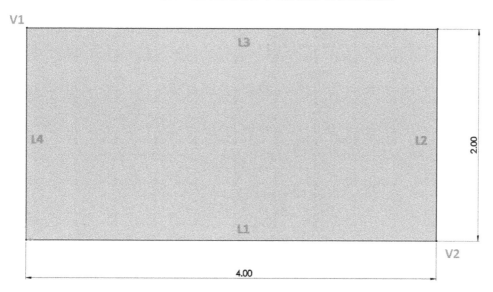

Figure 3.45 – A fully defined rectangle as a starting point to apply the fillet and chamfer to

Now, we can start creating the fillet, as follows:

1. A sketch is positioned based on a vertex or the two lines around a vertex. Select lines **3** and **4** (**L3** and **L4**) and then select the **Sketch Fillet** command from the command bar, as follows. Alternatively, we can select vertex 1 (**V1**) and then select the **Sketch Fillet** command:

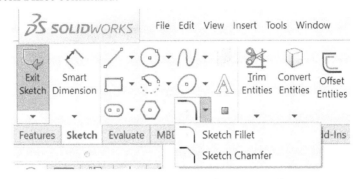

Figure 3.46 – The Sketch Fillet command

2. This will bring up a preview of the fillet on the canvas. You can find more options on the left-hand side, in place of the design tree, as shown in the following screenshot. Under **Fillet Parameters**, fill in the radius of the fillet as **1** inch. Then, click on the green checkmark. This concludes making the fillet. Now that we've made the fillet, we can press *Esc* on the keyboard to exit the fillet sketching mode:

Figure 3.47 – The Sketch Fillet property manager showing the fillet parameters

> **Tip**
> Press *Ctrl + Z* to undo the fillet you've applied.

Now that we've sketched the fillet, we will move on to sketching the chamfer:

1. The chamfer that we had in our final sketch is defined by an angle and a distance. Note that the angle displayed indicates the angle measurement between **L2** and the chamfer itself. Thus, hold down *Ctrl*, select **L1**, and then select **L2**. After that, select the **Sketch Chamfer** command. Similar to the **Fillet** command, we will get a preview of the chamfer. Also, on the left-hand side, we will find more options that we can use to define our chamfer.

2. Select **Angle-distance**. Then, set the angle to 30 degrees and the distance to 1 inch.

3. Click on the green checkmark afterward:

Figure 3.48 – The Sketch Chamfer property manager showing the chamfer parameters

> **Tip**
> If you apply the wrong chamfer, you can click on the **Undo** button.
> Alternatively, you can press *Ctrl + Z* on your keyboard.

This concludes how to create fillets and chamfers. In this exercise, we learned what fillets and chamfers are and what defines them, and how to create fillets and chamfers in SOLIDWORKS sketching mode.

Now we know how to use all the major basic sketching commands. You will use these commands over and over again when working with the software. Now, we will dig deeper into what the different types of definition statuses mean, that is, under defined, fully defined, and over defined.

Under defined, fully defined, and over defined sketches

SOLIDWORKS sketches can fall under three status categories according to how they are defined. They can be under defined, fully defined, or over defined. These terms have already been mentioned briefly, but in this section, we will explore what those statuses are, as well as some ways to deal with them. We will explore these different statuses by drawing and defining a triangle that's under defined so that it becomes over defined.

Under defined sketches

Usually, the starting point of a sketch is under defined. Under defined sketches have parts of them that are loose or lack proper definition; for example, a line without a specific length. To find out more about under defined sketches, we'll examine the following sketch. We will use the MMGS measurement system for this exercise. We have indicated the lines and points with the letters **L** and **P** for reference:

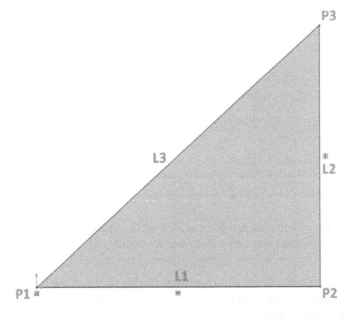

Figure 3.49 – An under defined sketch consisting of three unrestrained lines

Take some time to sketch the preceding sketch and examine it. From this sketch, we can see the following:

- Note that, in your SOLIDWORKS canvas, **P1** and **L1** are black in color. Black colors indicate that those parts are fully defined. To test this further, click and hold **P1** or **L1** and try to move the mouse. You will notice that the sketch doesn't move. This is an indication that that part of the sketch is fully defined.

- The other parts of the sketch (**P2, P3, L2**, and **L3**) are blue in color. This indicates that those parts haven't been defined yet. To test this, click and hold any of those parts and move the mouse. You will notice that those sketch parts move around the canvas. If any part of the sketch is blue, the sketch will be labeled as **Under Defined** at the bottom of the SOLIDWORKS interface, as shown in the following screenshot:

Under Defined   Editing Sketch1   MMGS   ▲

Figure 3.50 – The sketch indicated as under defined

We will always try to make our sketches fully defined. To fully define a sketch, we can simply add more dimensions and/or relations to turn the blue parts black.

To find out which parts of the sketch need definition, we can click and hold on any of the blue parts and move them. The resulting movement tells us which parts need definition. For example, if we hold and move **P2** left and right, we will notice that **L1** changes in length. This indicates that we can define **P2** by setting a dimension for **L1** (or between **P1** and **P2**), as shown in the following figure. After defining the length, we will notice more lines turned black. Now, if we click on **P2** and try to move it, it will be fixed:

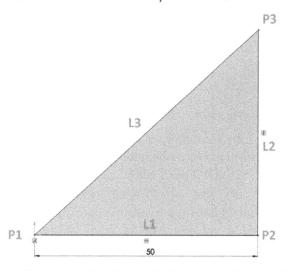

Figure 3.51 – A sketch that is partially defined, with the interface showing blue and black lines

We can do the same movement test for **L3** and **P3** and decide what elements of the sketch we can define further.

Fully defined sketches

Fully defined sketches are where all of the parts of the sketch are fully fixed. In other words, no part of the sketch can be moved from its current position. To illustrate this, we will take another look at the sketch we started and fully define it. We will do this by adding an angle of **45** between **L1** and **L3**. The fully defined result is as follows:

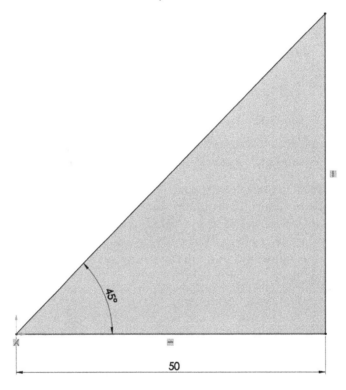

Figure 3.52 – A fully defined triangle

Note that the sketch is now fully black on your screen. Also, if you hold and try to move any of the sketch parts, they will not move because they are fully restrained. When the sketch is **Fully Defined**, SOLIDWORKS will take note of this at the bottom of the interface. Remember that we fully defined a sketch by adding relations and dimensions until all the sketch elements were fixed:

Fully Defined | Editing Sketch1   MMGS   ▴

Figure 3.53 – The sketch indicated as Fully Defined

When sketching with SOLIDWORKS, we will mostly try to make our sketches fully defined in order to capture our full design intent. Now, let's look at over defined sketches.

Over defined sketches

Over defined sketches are those with more relations and dimensions than are needed for the sketch elements to be fully fixed. This is not a recommended status to have for a sketch. Over defined sketches occur when we apply contradicting relations to define a particular part of a sketch. To illustrate this, we can add an extra dimension to **L2**. Once we do that, we will get the following message:

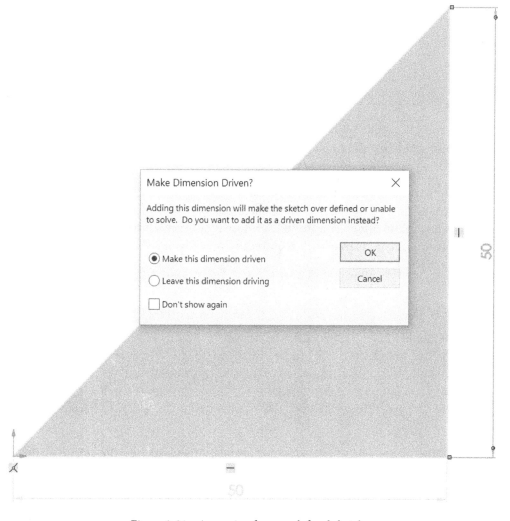

Figure 3.54 – A warning for over defined sketches

Let's examine these two options and their results:

- **Make this dimension driven**: This will make the dimension so that it's just for show. This option will only tell us the length of the line. If the lengths of the other lines change, the displayed length here will change accordingly. Choosing this option will not cause any issues.

- **Leave this dimension driving**: This will give the dimension driving power, just like the other dimensions. However, since the sketch is fully defined already, selecting this option will leave our sketch with conflicting items and a warning message that the sketch is **Over Defined**, as shown in the following screenshot. Once we are in an over defined sketch situation, we will need to delete some existing relations or dimensions to get rid of the over defined status. In SOLIDWORKS, we shouldn't set more than one relation or dimension that governs the same sketch entity, even if they are geometrically non-conflicting:

Figure 3.55 – A sketch indicated as Over Defined

Keep in mind that the best practice is to have our sketches fully defined to ensure that we are fully capturing the design intent of our sketch.

All the definition statuses are related to how many relations and dimensions we add to our sketch. A fully defined sketch has all the sketch elements fully fixed with the proper number of relations and dimensions to fully capture the design's intent. Under defined sketches have fewer relations than they require, and over defined sketches have more relations than needed.

Summary

In this chapter, we learned about the different aspects of SOLIDWORKS sketching that form our sketching foundations. We learned what sketching is and how to sketch different sketching elements, including lines, rectangles, circles, arcs, ellipses, fillets, and chamfers. We also covered using dimensions and relations to define sketches, as well as the meaning of the different sketch definition statuses, that is, under defined, fully defined, and over defined.

All of this information is part of our sketching foundation, which we will use every time we build a 3D model with SOLIDWORKS. In the next chapter, we will address additional sketching commands that can greatly enhance our sketching performance and speed, such as patterns and mirrors.

Questions

Answer the following questions to test your knowledge of this chapter:

1. What is the position of SOLIDWORKS sketching in modeling?

2. What are the two stages we commonly follow when sketching?

3. Sketch the following shape using the MMGS measurement system:

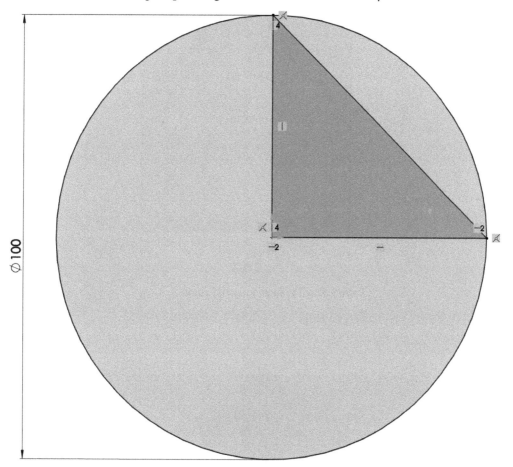

Figure 3.56 – The final output of question 3

4.   Sketch the following shape using the IPS measurement system:

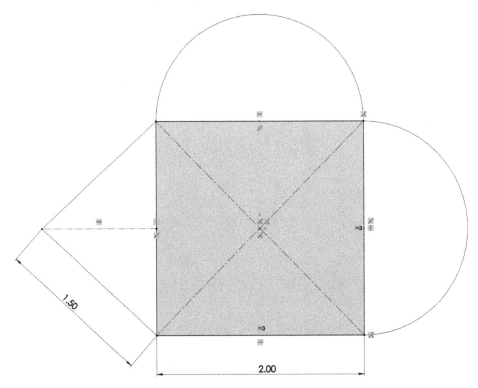

Figure 3.57 – The final output of question 4

5.   Sketch the following shape using the CGS measurement system:

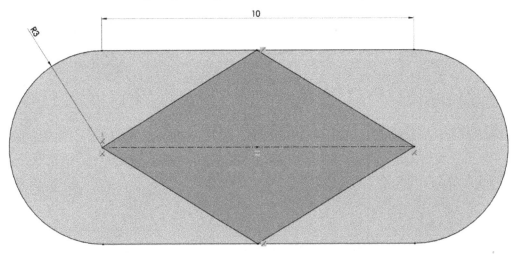

Figure 3.58 – The final output for question 5

6.   Sketch the following shape using the MMGS measurement system:

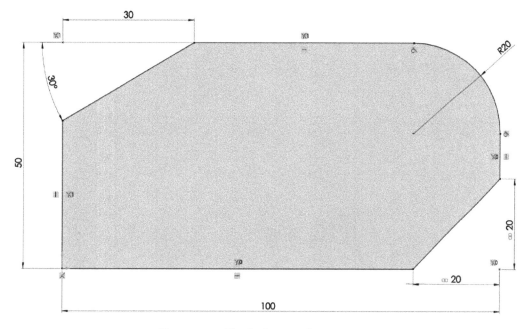

Figure 3.59 – The final output for question 6

7.   What are under defined, fully defined, and over defined sketches?

Important Note

The answers to the preceding questions can be found at the end of this book.

4
Special Sketching Commands

Mastering SOLIDWORKS sketching is not only about sketching shapes such as rectangles and ellipses but also depends on other special commands that will greatly enhance our ability to sketch complex shapes faster.

In this chapter, we will introduce sketching commands such as mirroring, offsets, patterns, and trimming. We will also cover examples where we will use multiple shapes and commands to create relatively complex sketches. Even though we can continue without using these commands, they will greatly enhance the efficiency of our sketching creation process.

In this chapter, we will cover the following topics:

- Mirroring and offsetting sketches
- Creating sketch patterns
- Trimming in SOLIDWORKS sketching

By the end of this chapter, you will be able to use the mentioned sketching commands to both optimize and speed up your sketching process.

Technical requirements

In this chapter, you will require access to the SOLIDWORKS software.

Check out the following video to see the code in action: `https://bit.ly/3oT1L4I`

Mirroring and offsetting sketches

Some of the sketching commands in SOLIDWORKS allow us to easily create more sketch entities based on ones we already have, including circles, rectangles, lines, or any combination of sketch entities. Examples of such sketching commands are mirroring and offsetting. Using these commands will help us avoid creating similar sketch entities more than once. Here, we will start by exploring the mirroring and offsetting sketching commands. We will learn about what these commands do and how we can use them.

Mirroring a sketch

As the name suggests, mirroring a sketch means to reflect one or more sketch entities around a mirroring line. It is very similar to reflecting an image in a mirror. The following figure illustrates the components of mirroring in SOLIDWORKS:

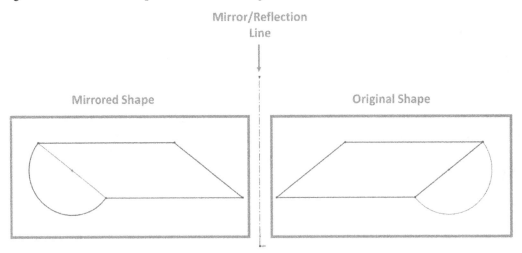

Figure 4.1 – Mirroring a sketch creates a reflection of it around a mirror line

In the preceding diagram, we can see that there are two parts that we need in order to use mirroring:

- Sketch entities to mirror

- A mirroring or reflection line

Since the two shapes are mirror sketch entities of each other, any changes that happen to one shape will automatically be reflected on the other.

To highlight how we can use mirroring, we will create the following shape. Note that the shape is a right-angle triangle with a mirrored reflection and that we will use the **Millimeter, Gram, Second** (**MMGS**) measurement system for this exercise:

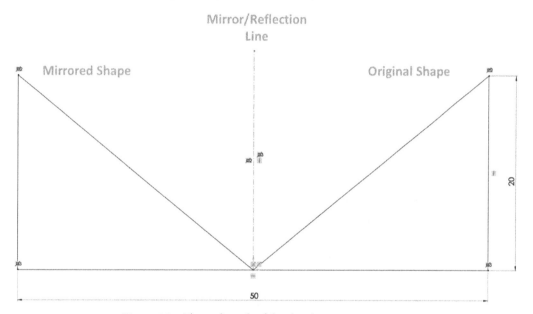

Figure 4.2 – The end result of the sketch mirroring exercise

To create this shape, we need to complete the outlining and defining stages. Here, we will create the base/original sketch and then mirror it. Then, we will define the resulting sketch.

Let's start by outlining our general shape and applying the **Mirror Entities** command. Follow these steps:

1. Sketch the outline of the first triangle and the mirroring line, using a centerline or construction line for the latter:

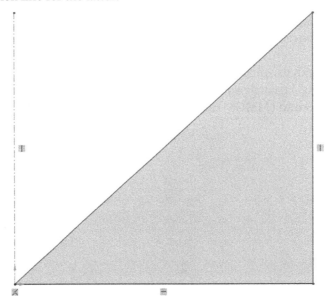

Figure 4.3 – An underdefined outline of a right-angle triangle

2. Select the **Mirror Entities** command, as shown in the following screenshot:

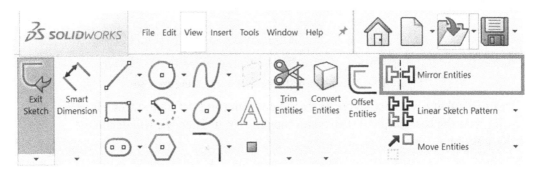

Figure 4.4 – The Mirror Entities command

3. On the left-hand side, we will see the available **Mirror** options. For **Entities to mirror:**, select the three lines that make up the right-angle triangle (**L2**, **L3**, and **L4**). For **Mirror about:**, select the centerline (**L1**). Make sure that the checkbox for **Copy** is also checked. Now, we will see a preview of the triangular mirrored shape on the left. Click on the green check mark to approve the mirroring:

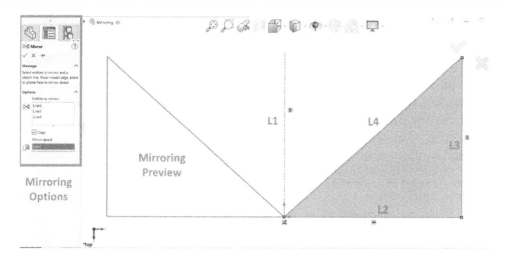

Figure 4.5 – Mirror PropertyManager and preview

To find out more about the mirroring effect, click and hold any of the blue parts of the sketch and move them. You will notice that the mirrored entity will reflect the same movement as the original shape.

Defining mirrored entities

Now that we have our shapes outlined, we can start defining our whole sketch. We can define the sketch by adding the dimensions and relations shown in the following figure. Note that, when defining one line, the mirrored line will also be defined in the same way, as shown in the **20** mm dimension. Also, note that we can add dimensions between the mirrored entities, as shown in the **50** mm dimension. After adding those dimensions, we will notice that it becomes fully defined:

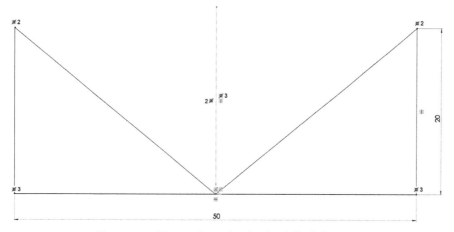

Figure 4.6 – The resulting sketch after fully defining it

In this exercise, we defined the whole sketch after applying the mirror. An alternative approach is to fully define the first triangle before using the **Mirror Entities** command. In this scenario, the mirrored triangle will also be fully defined, directly after mirroring.

The **Copy** option in mirroring is used if we wish to keep the original shape. If the **Copy** option is checked, we will keep the original shape and create a mirrored shape. If it is unchecked, the original shape will be deleted, and we'll be left with the mirrored shape only. In addition, whatever dimensions and relations are on the original shape will be removed.

> Tip
> Centerline/construction lines are not explicitly required to mirror sketch entities. An alternative to construction lines is to use an existing line in the sketch as a reflection line for mirroring.

This concludes the **Mirror Entities** sketching command. Note that, in practice, whichever order we outline and define our sketches in can vary. As you gain more experience with sketching, you will have your own approach to doing things. Now, we can start learning about the **Offset Entities** sketch command.

Offsetting a sketch

Offsetting a sketch allows us to generate sketch entities that are similar to existing ones by shifting the original entities by a certain distance while maintaining all their features. The following figure shows an example of a sketch and its offset. Note that the original sketch is the one we sketch first; after that, the **Offset sketch** is created by applying the **Offset Entities** sketch command. The **Offset sketch** is defined by inputting an **Offset Distance**:

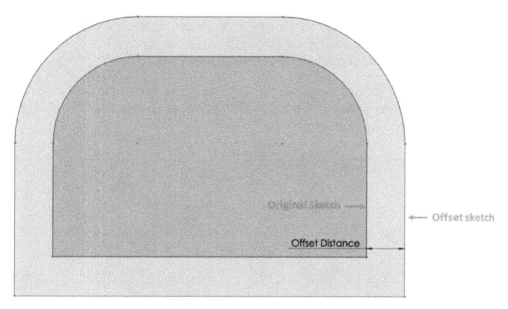

Figure 4.7 – A sketch and its offset

To illustrate how we can create an **Offset Distance**, we will create the sketch shown in the following figure:

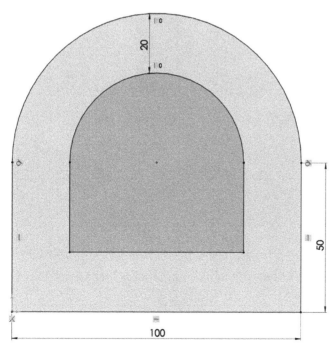

Figure 4.8 – The end result after the offset entities exercise

We will use the MMGS measurement system for this exercise. To sketch the given shape, we will create and fully define the original sketch. After that, we will create the offset sketch using the **Offset Entities** command. Follow these steps to do this:

1. Sketch and define the original sketch using the sketching commands we covered in the previous chapter (*Chapter 3, SOLIDWORKS 2D Sketching Basics*). The result is shown in the following figure:

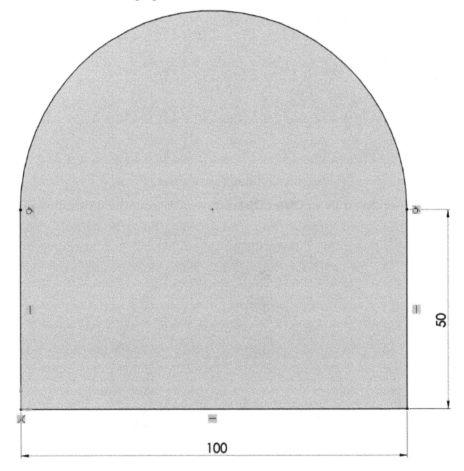

Figure 4.9 – The original sketch before applying an offset

2. Select the **Offset Entities** sketching command, as highlighted in the following screenshot:

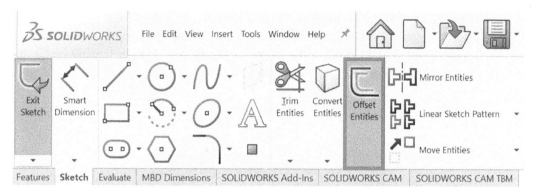

Figure 4.10 – The Offset Entities command

3. An options panel for the command will appear on the left-hand side of the screen. From there, we can customize our way of defining our sketch. Set the options that are shown in the following screenshot, which will give us our desired result:

Figure 4.11 – The Offset Entities PropertyManager options

4. Select any of the lines in the original sketch we created. A preview will then be shown in the sketch canvas, illustrating the result of the offset. The preview will be shown in yellow, as highlighted in the following figure. Any changes that are made to the options will be reflected in the preview as well:

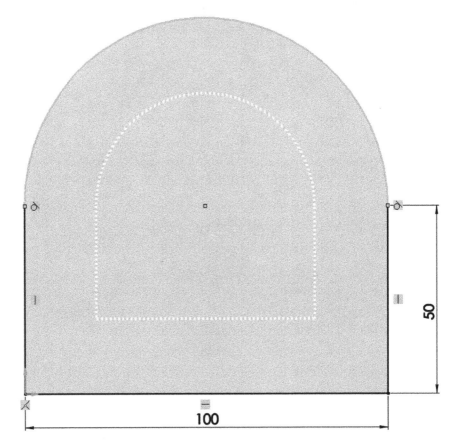

Figure 4.12 – An offset preview

5. Click on the green check mark in the options panel to apply the offset. This will result in the following figure, which matches the final required sketch. Note that the **Offset Sketch** is already fully defined since our **Original Sketch** was fully defined as well:

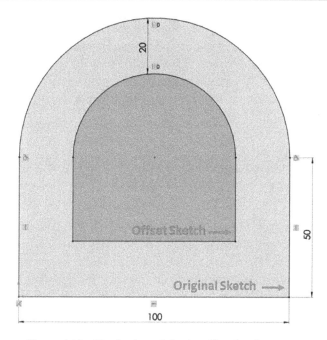

Figure 4.13 – The final result in the offset sketch exercise

This concludes how we can use the **Offset Entities** sketching command. However, in addition to using the **Offset Entities** command, we need to learn how to delete an offset and learn about the customization options associated with offsets. Let's explore those.

Deleting an offset

To delete an offset entity, we can select that entity and press *Delete* on the keyboard. Alternatively, we can right-click on the entity and select **Delete**. Deleting an offset is just like deleting a normal sketch entity.

Customization options

The following are some of the different options that accompany the **Offset Entities** command. Those options are displayed on the command's PropertyManager, as shown in *Figure 4.11*. They allow us to customize the command to best fit our needs:

- **Add dimensions**: This option makes the offset fully defined upon implementation, as in our previous example. Unchecking this option will create the offset; however, it will not add the offset dimension, thus making it undefined.

- **Reverse**: This changes the direction of the offset. For example, the default offset of an enclosed circle will be outward, forming a bigger circle. Checking this option makes the offset go onward.

- **Select chain**: This command helps us select all the sketch entities that are linked together. For example, with this option, selecting one line of a rectangle will automatically select the whole rectangle since the four lines that make up the rectangle are connected. We should uncheck this option if we only want to offset a part of the shape, for example, to offset only one line of a rectangle.

- **Bi-directional**: This will apply two offsets in two different directions, outward and inward.

- **Caps ends**: This command is applied when the offset is not an enclosed shape, such as an enclosed rectangle or a circle. The **Caps ends** option allows us to easily enclose open loops between the original and the offset sketch entities.

- **Construction geometry**: This allows us to make the original or the offset sketch entities construction entities (for example, construction lines, arcs, and more). Checking **Base geometry** will switch the original sketch entities into construction entities. Checking **Offset geometry** will do the same for the offset entities.

This concludes the **Offset Entities** sketching command. We have learned what the **Offset Entities** command does and how to use it. We also covered the **Mirror Entities** sketching command.

To recap, the **Mirror Entities** command allows us to quickly reflect a copy of a sketch around a reflection line; this means both the original and the mirrored entities will keep imitating each other. The **Offset Entities** command allows us to create a copy of a sketch entity by offsetting it by a certain distance.

Now, we can start learning about how to create patterns.

Creating sketch patterns

Sketch patterns allow us to easily copy a sketch entity multiple times in a pattern formation. Such sketch patterns can be created in a **linear** or **circular** formation. In this section, we will cover creating patterns in both formations.

Defining patterns

Patterns are repeated formations that can be commonly found in consumer products, architecture, fabrics, and more. In patterns, we often have a base shape, sometimes called a **base cell** or **patterned entity**, which is created from scratch. Then, the basic shape is repeated multiple times to form a bigger piece. There are two common types of patterns, **linear patterns** and **rotational/circular patterns**. Examples of both types of patterns will be shown in this section.

Linear patterns are ones in which you have a base shape (patterned entity) that is repeated linearly in different directions. Linear patterns are commonly found in curtains, carpets, building tiles, floors, and architecture. The following diagram is an example of a linear pattern. The **Patterned Entity** shape is highlighted with a red square:

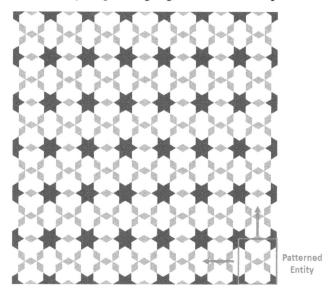

Figure 4.14 – An example of a linear pattern

Rotational patterns are ones in which we have a base shape that is repeated as we rotate it. A common application for the rotational pattern is in car rims, as highlighted in the following figure. Within the SOLIDWORKS interface, rotational patterns are known as circular patterns:

Figure 4.15 – An example of a rotational/circular pattern

SOLIDWORKS' pattern tools make it easier for us to create similar linear and circular patterns within sketching. Now that we know what these two types of patterns are, we can start exploring them within SOLIDWORKS, starting with linear patterns.

Linear sketch patterns

Linear sketch patterns allow us to pattern sketch entities in a linear direction. The following sketch shows us how we can define linear patterns in SOLIDWORKS sketching:

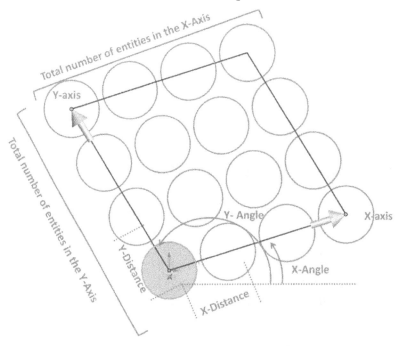

Figure 4.16 – Elements needed to define a linear pattern in SOLIDWORKS

In the preceding sketch, the shaded circle is the base circle, while the other ones are additions to be made with the pattern command. The annotations in writing are the parameters that we need in order to communicate a pattern with SOLIDWORKS sketching. Each of the annotations is repeated twice, one for the **X-axis** and one for the **Y-axis**. They are as follows:

- **Axes**: The **X-axis** and **Y-axis** represent the direction in which our pattern is implemented.

- **Total number of entities**: The number of times we want the entity to be sketched, including the base entity. In the preceding sketch, the number of entities is four for the *X* and *Y* directions.

- **Distance**: This specifies the distance that divides every two entities from each other in a certain direction. In the preceding sketch, they are highlighted as **X-Distance** and **Y-Distance**.

- **Angle**: This specifies how tilted our axes should be since the axes determine the direction of the patterns. Therefore, the whole pattern will shift as we change the direction of an axis. In the preceding sketch, the angles are highlighted as **X-Angle** and **Y-Angle**. Note that the X and Y angles start from the same baseline.

Now that we know what elements define a linear pattern, we can start creating one in SOLIDWORKS. To highlight this, we will sketch the following figure. We will use the MMGS measurement system for this exercise:

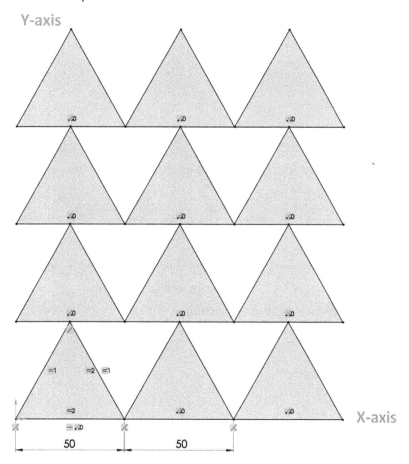

Figure 4.17 – The final result of the linear pattern exercise

To sketch the preceding diagram, follow these steps:

1. Sketch and define the base equilateral triangle, as follows:

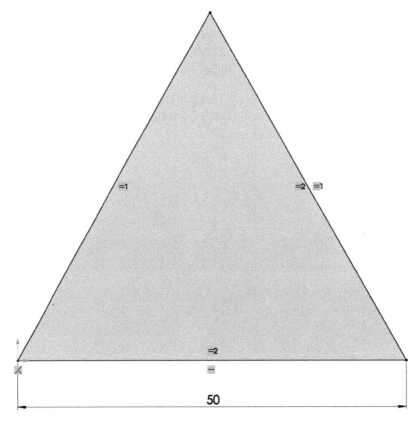

Figure 4.18 – The base triangle used for the pattern

2. Select the **Linear Sketch Pattern** command:

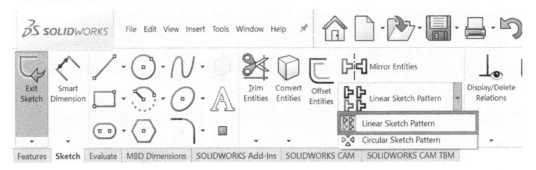

Figure 4.19 – The Linear Sketch Pattern command

3. Select the three lines that form our base triangle. Now, we can set up our linear sketch pattern using the options that appear on the left-hand side of SOLIDWORKS. Set the following options:

Figure 4.20 – The options for setting a linear sketch pattern

As we are adjusting those options, a preview of the final shape will appear on the canvas. From top to bottom, the options we are using are as follows:

- **Direction**: For **Direction 1**, we can see the direction of the **X-axis**. If we click on the two black and gray arrows next to **X-axis**, the direction of the pattern will change by 180 degrees. For **Direction 2**, the direction of the pattern will go along the **Y-axis**.

- **Distance**: This determines the distance between every two entities that are adjacent to each other.

- **Dimension X spacing**: This makes the dimension set a fixed driving dimension. Unchecking this option will not add a dimension to the pattern; instead, the listed dimension in **Distance** will be the starting point. The **Dimension Y spacing** option does the same in the second direction.

- **Number of instances**: This indicates how many times we want an entity to be drawn, including the base sketch.

- **Display instance count**: Checking this option will show the number of instances on the drawing canvas.

- **Angle**: This is used to set the direction of **X-axis** and **Y-axis**, which governs the pattern's direction.

- **Fix X-axis direction**: Checking this option will make the angle a driving dimension. Unchecking this option will not add the angle direction to the pattern; therefore, we need to identify it separately.

- **Dimension angle between axis**: Checking this option will make the angle between the x and y axes a driving dimension. This option is used to define the **Y-axis** direction in our canvas.

- **Entities to Pattern**: This lists all the entities that will be patterned with the command. We can delete entities by deleting them from the list. We can add entities by selecting them in the sketch canvas.

Our preview may look something like the following figure:

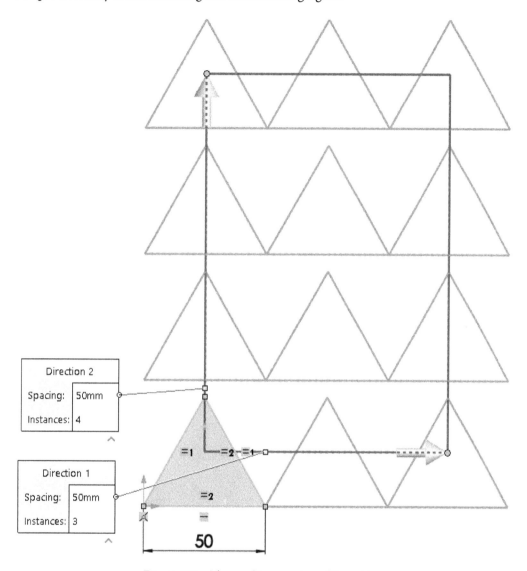

Figure 4.21 – The resulting preview of the pattern

4. After adjusting these options, we can click on the green check mark. We will see the following shape. Note that the first row of triangles in the *x* direction is fully defined, while the other triangles extending in the *y* direction are not. To understand how such patterns work, we can click and drag the blue parts around our screen. We will see that all the patterned shapes move together:

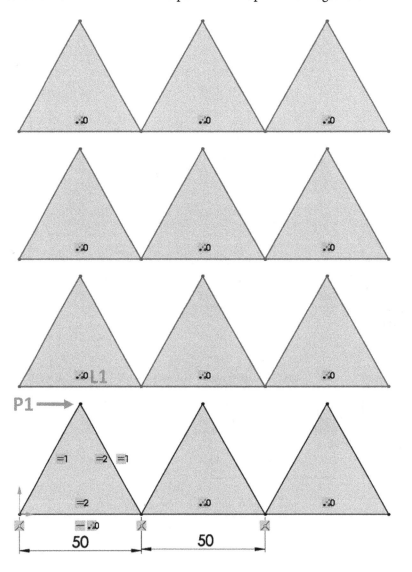

Figure 4.22 – The pattern might not be fully defined after application

5. To fully define the pattern, we can define the shapes in the *y* direction. We will need to add a midpoint relation between **L1** and **P1**. This will make the sketch fully defined, as shown in the following diagram.

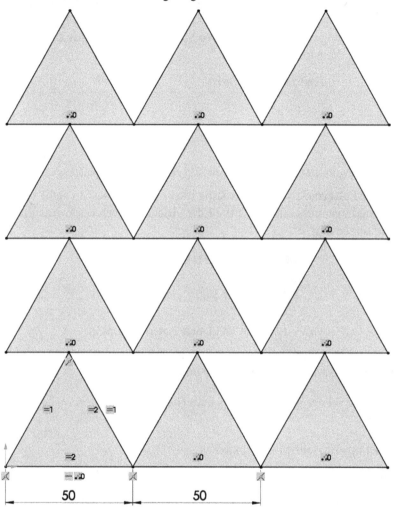

Figure 4.23 – The final result of the linear sketch pattern exercise

> **Tip**
>
> We can also define a dimension for the *y*-direction pattern using the **Linear Pattern** command PropertyManager, as shown in *Figure 4.20*, by checking the **Dimension Y spacing** option. However, in the exercise, we used a relation to define the **Y-axis** pattern.

Let's have a look at some related commands:

- **Instances to Skip**: At the bottom of the **Linear Pattern** PropertyManager options, we will find the **Instances to Skip** option. We can use this to skip instances of the pattern. For example, in the preceding exercise, we can remove the two middle triangles by adding them to **Instances to Skip**. This will exclude the middle triangles from the pattern:

Figure 4.24 – The Instances to Skip option in sketch patterns

- **Edit Linear Patterns**: To edit an existing linear pattern, we can right-click on any of the patterned instances and select the **Edit Linear Pattern** option, as shown in the following screenshot:

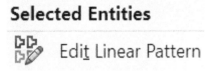

Figure 4.25 – The Edit Linear Pattern command

- **Delete Linear Patterns**: To delete the pattern (or parts of it), we can simply select or highlight those entities and press *Delete* on the keyboard.

This concludes this exercise on using linear patterns. In the exercise, we have covered the following topics:

- How to set up and define linear patterns
- The different options that we can use to define a linear pattern
- How the linear pattern entities interact as under defined entities

Now that we have mastered linear patterns, let's move on and look at circular patterns.

Circular sketch patterns

Circular sketch patterns allow us to pattern sketch entities in a circular direction. The following figure highlights how we can define a circular pattern in SOLIDWORKS sketching:

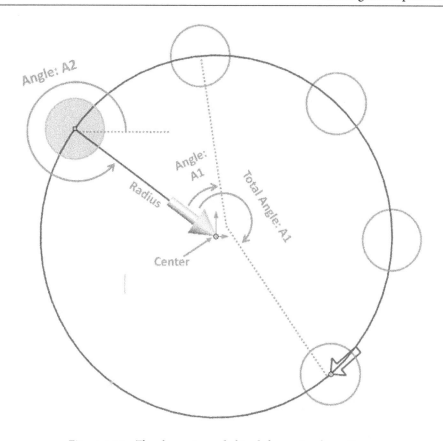

Figure 4.26 – The elements needed to define a circular pattern

In the preceding sketch, the shaded circle is the base circle, while the others are additions to be made with the pattern command. The annotations in red are the parameters that we need in order to communicate a pattern to SOLIDWORKS sketching. The following is a small description of the different annotations in the preceding figure:

- **Center**: This represents the center of rotation for the circular pattern. This can be determined with specific x and y coordinates or by relating it to another point.

- **Radius**: This is the distance between the original entity and the center of the pattern.

- **Angle: A1**: This is the angle between two adjacent patterned entities.

- **Total Angle: A1**: This is the angle between the original and the last patterned entity.

- **Angle: A2**: This is the angle between the original patterned entity and the **center**.

- **Number of patterned entities**: This shows the total number of patterned entities, including the base sketch.

To illustrate how to use circular sketch patterns, we will create the following sketch, using the **Inch, Pound, Second (IPS)** measurement system in this exercise:

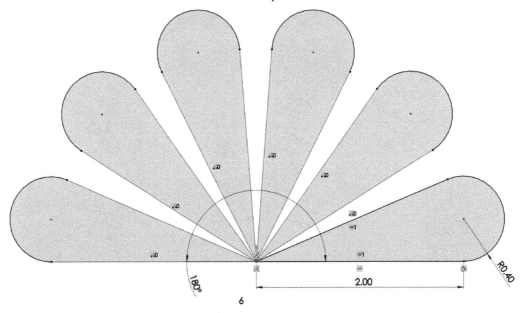

Figure 4.27 – The final result of the circular pattern exercise

To sketch the preceding diagram, follow these steps:

1. Sketch and fully define the base entity, as shown in the following figure:

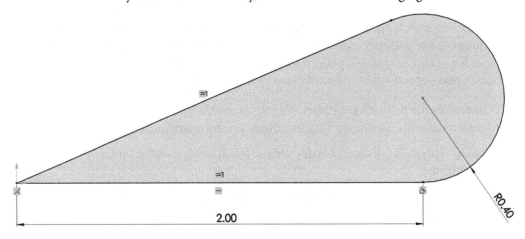

Figure 4.28 – The base sketch for the circular pattern

2. Select the **Circular Sketch Pattern** command from the **Sketch** command bar:

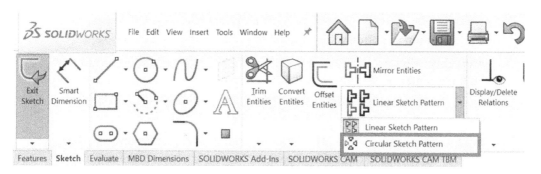

Figure 4.29 – The Circular Sketch Pattern command

3. Select the three lines that form our base sketch. Now, we can set up our circular pattern using the options that appear on the left-hand side of SOLIDWORKS. Set the options that are shown in the following screenshot:

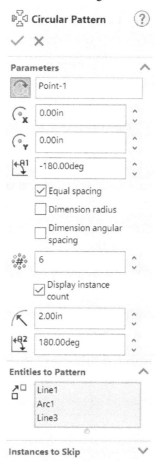

Figure 4.30 – The Circular Pattern options relating to this exercise

As we are adjusting those options, a preview of the final shape will appear on the canvas. From top to bottom, the options we will be using are as follows:

- **Center**: This field starts as **Point-1** and represents the center of the pattern. By default, the center of the pattern will be the origin; however, we can change it by selecting other points. Rotational direction, which is to the left of the **Point-1** selection, is a button that will flip the direction of the rotation.

- **X and Y center locations**: The two fields marked with **X** and **Y** represent the location of the center of our circular pattern. Since our center is the origin, the location is marked as 0.00 inches for the *x* and *y* directions. We can use these fields to set up an exact center in the coordinate system.

- **A1 angle**: This defines the angle that will govern the locations of the circular pattern. In our example, it is set as -180 since the pattern goes counterclockwise by 180 degrees. Note that this does not define the dimension; instead, it helps us approximate the location and the look of the pattern. We can fully define the pattern after its implementation.

- **Equal spacing**: This will ensure that all the patterns are equally distributed in the angle range.

- **Dimension radius**: This will add a driving dimension to the radius of the pattern. Note that this is not needed if we merge the center with a fixed point, such as the origin.

- **Dimension angular spacing**: Checking this option will allow us to dimension the angle between the adjacent patterned instances, instead of the angle between the base sketch and the last patterned entity.

- **Number of patterned instances**: This indicates how many times we want the entity to be drawn or patterned, including the base sketch.

- **Display instance count**: Checking this option will show the number of instances on the drawing canvas.

- **Radius**: This allows us to linearly increase or decrease the radius between the patterned entities and the center.

- **A2 angle**: This will shift the center of the pattern so that it's at a certain angle.

4. After adjusting these options, we can click on the *green check mark*. This will give us the following shape:

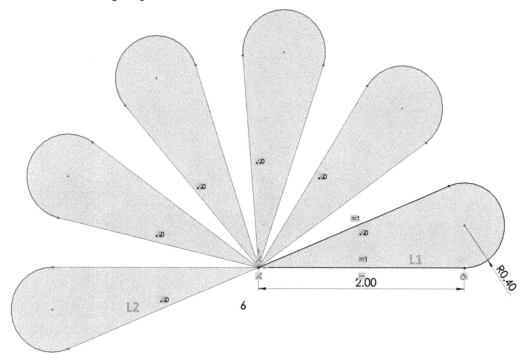

Figure 4.31 – The resulting sketch after applying the pattern might not be fully defined

Important Note

The pattered entities are not fully defined. To understand how circular patterns work, we can click and drag the blue parts of the sketch around. We will see that the patterned shapes move together.

5. Set the angle between **L1** and **L2** so that it's equal to 180 degrees. Doing this will fully define the sketch, as shown in the following figure:

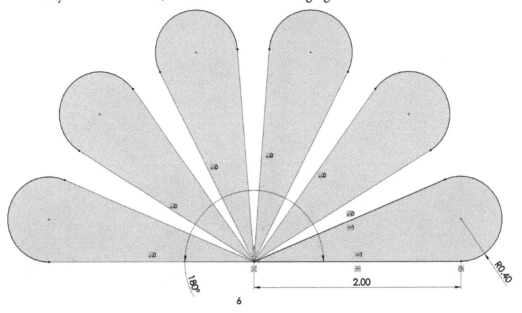

Figure 4.32 – The final resulting sketch from the exercise

Similar to linear patterns, we have the option of skipping instances, editing the circular patterns, and deleting instances. To do these things, we can follow the same procedure that we followed for linear patterns.

This concludes this exercise of using circular patterns. In the exercise, we have covered the following topics:

- How to set up and define circular patterns
- The different options that we can use to define a circular pattern
- How the circular pattern entities interact with each other as under defined entities

In this section, we talked about linear and circular sketch sketching patterns. Both allow us to quickly repeat a selection of sketch entities multiple times, saving a lot of time and energy. A linear pattern is allowed to pattern entities linearly while a circular pattern allowed us to pattern entities rotationally. We can also use both linear and circular patterns to build one sketch. However, we have to apply them separately. Now, we can start looking at another special sketching command – trimming.

Trimming in SOLIDWORKS sketching

Trimming in SOLIDWORKS allows us to easily remove unwanted sketch entities or unwanted parts of sketch entities. This makes it easier for us to create complex sketches. In this section, we will cover what trimming is, why we use trimming, and how to use trimming within SOLIDWORKS.

Understanding trimming

Trimming allows us to delete parts of sketches that are unwanted. This makes it easier to create complex sketches that go beyond the standard sketch commands. This is because it makes it easier to utilize segments from different sketching commands. To explore this further, let's examine the following sketch and how the trimming command can help create it:

Figure 4.33 – Trimming allows us an easy or alternative way to sketch this

We can start this simple sketch by sketching two circles, as shown in the following figure. After that, we can trim/remove the interfering parts to get our desired shape:

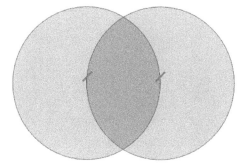

Figure 4.34 – Trimming unwanted sketch entities can get us new shapes

Also, depending on the sketch we want, trimming could make it easier for us to define the sketch according to our specific design intent. Now that we know what trimming is, we can start using the command.

Using power trimming

To show you how we can use the trimming tool in SOLIDWORKS, we will create the following sketch, which consists of a circle and a rectangle. We will use the MMGS measurement system for this exercise:

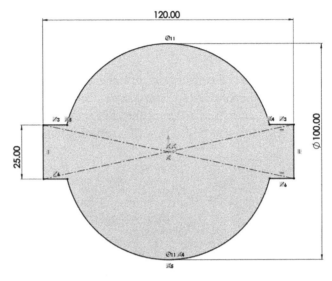

Figure 4.35 – The end result of the power trim tool exercise

To create the given sketch, follow these steps:

1. Sketch and fully define the base shapes of a circle and a rectangle, as follows:

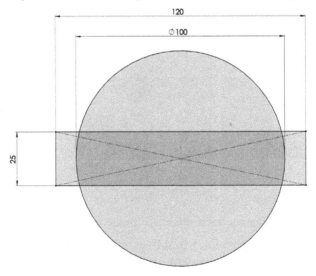

Figure 4.36 – Original sketches with different intersecting enclosures

2. Select the **Trim Entities** command from the command bar, as shown in the following screenshot:

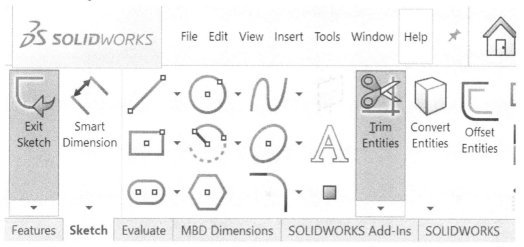

Figure 4.37 – The Trim Entities command

3. After selecting this command, the PropertyManager for it will appear on the left of the interface. This will show different types of trimming tools. Here is a brief description of the different trimming tools that are available:

*   **Power trim**: This is the easiest method for trimming entities. **Power trim** is a multipurpose trimming tool that allows us to cut entities by going over them in the canvas. Power trim is what we will use in this section.

*   **Corner, Trim away inside, Trim away outside**, and **Trim to closest**: These are the different and more specific ways we can trim. A **Corner** trim makes it easier to trim two entities till they intersect at a projected corner. **Trim away inside** allows us to trim an entity that lies inside bounding entities, while **Trim away outside** allows us to trim entities that lie outside bounding entities. **Trip to closest** trims an entity to its closest two boundaries. We will utilize these specific trim options in this book.

*   **Keep trimmed entities as construction geometry**: When checked, trimming will not remove entities from the canvas. Instead, they will be converted to construction lines.

*   **Ignore trimming of construction geometry**: When checked, trimming will not function with construction lines.

As shown in *Figure 4.38*, we have picked **Power trim** and, as we didn't want to revoke the construction lines that define our rectangle, we checked the **Ignore trimming of construction geometry** option:

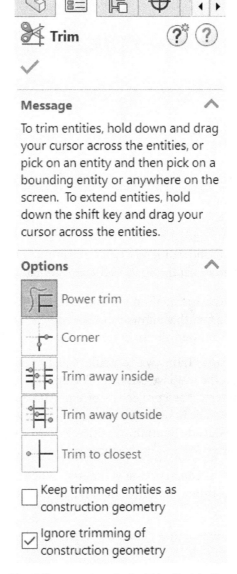

Figure 4.38 – The different options in the Trim Entities PropertyManager

4. Go back to the canvas and start trimming. To trim unwanted parts, we can click, hold, and move the mouse, as illustrated by the red lines in the following diagram:

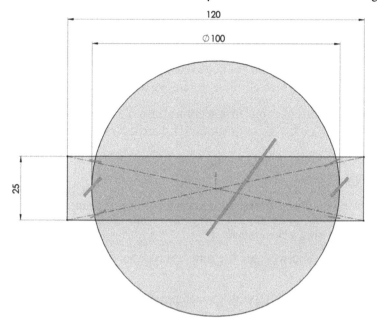

Figure 4.39 – The lines indicating the needed movement of the power trim to trim unwanted entities

As we cross each of these lines, we will notice that the line will disappear. By doing this, we will only remove the lines that are not desirable. If we trim the wrong line, we can simply undo this by pressing *Ctrl + Z*. After doing this, we will end up with the final shape, which was shown at the beginning of this section.

> **Tip**
>
> After trimming an entity using the power trim command, a small red square will appear after each trim. Moving the mouse back to that red rectangle will undo that trim.

This concludes this exercise on trimming entities using SOLIDWORKS. In this exercise, we have covered the following topics:

- How to set up the **Trim Entities** command
- How to use trimming to get our desired result

The trimming tool can be used as an enhancement tool; we can always create a certain sketch out of it. However, the tool can greatly enhance our sketching abilities by allowing us to create certain sketches faster.

Summary

In this chapter, we covered different sketching commands that can enhance our sketching capabilities. These included sketch mirrors, which allow us to generate mirrored sketch entities around a mirror line. We also covered offsetting, which allows us to generate duplicated sketch entities at an offset. Then, we learned about linear and circular patterns, which allow us to create many instances of a sketch entity in a specified pattern. Finally, we covered trimming tools, which allow us to remove unwanted parts of our sketches.

In the next chapter, we will cover our first set of features. So far, we've only learned about 2D sketches. Features will allow us to turn our 2D sketches into 3D models.

Questions

Answer the following questions to test your knowledge of this chapter:

1. What does mirroring a sketch do?

2. What are patterns, and what are the different types of patterns we can make in SOLIDWORKS?

3. What is trimming in SOLIDWORKS sketching?

4. Sketch the following diagram using the MMGS measurement system:

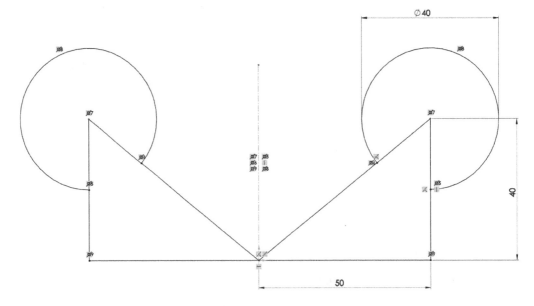

Figure 4.40 – The resulting sketch from question 4

5. Sketch the following diagram using the IPS measurement system:

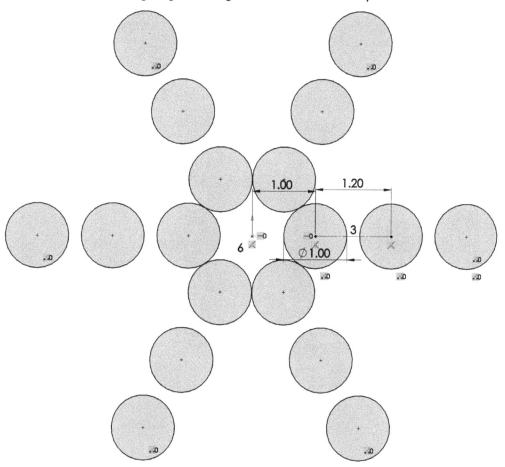

Figure 4.41 – The resulting sketch from question 5

6. Sketch the following diagram using the IPS measurement system (tip – use ellipses):

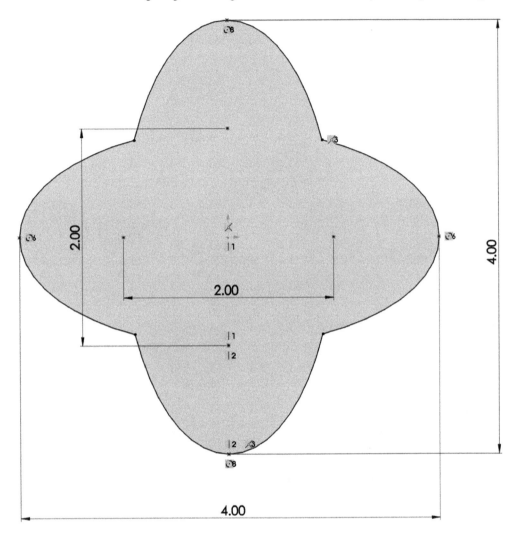

Figure 4.42 – The resulting sketch from question 6

7. Sketch the following diagram using the MMGS measurement system:

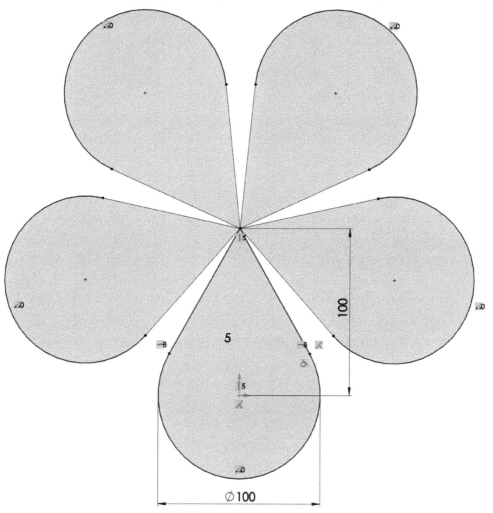

Figure 4.43 – The resulting sketch from question 7

Important Note

The answers to the preceding questions can be found at the end of this book.

Section 3 – Basic Mechanical Core Features – Associate Level

The associate level is the first level of proficiency for SOLIDWORKS users. This section will cover all the 3D modeling features expected of that level. These include extruded boss and cut, fillets, chamfers, revolved boss and cut, swept boss and cut, loft boss and cut, and reference geometries.

This section comprises the following chapters:

- *Chapter 5, Basic Primary One-Sketch Features*
- *Chapter 6, Basic Secondary Multi-Sketch Features*

5
Basic Primary One-Sketch Features

In SOLIDWORKS, features are what can turn a 2D sketch into a 3D model. In this chapter, we will move on from 2D sketches and start creating 3D models. We will explore the most basic features, such as extruded boss and extruded cut, fillets, chamfers, and revolved boss and revolved cut. We will study how to apply, modify, and delete features. We will also start creating more complex models by applying multiple features. Each feature that's covered in this chapter requires only one sketch to apply or no sketch at all.

In this chapter, we will cover the following topics:

- Understanding features in SOLIDWORKS
- Understanding and applying extruded boss and cut
- Understanding and applying fillets and chamfers
- Understanding and applying revolved boss and revolved cut

By the end of this chapter, we will be able to create 3D models using the most common features in SOLIDWORKS. Even though the features covered in this chapter are simple, they will enable us to create complex-looking 3D models.

Technical requirements

In this chapter, you will need to have access to SOLIDWORKS.

Check out the following video to see the code in action: `https://bit.ly/3oUIyjl`

Understanding features in SOLIDWORKS

SOLIDWORKS features are our way of moving from 2D to 3D. Similar to sketches, SOLIDWORKS provides many features that can help us create simple shapes. For more complex shapes, we will have to use more features. In this section, we will discuss SOLIDWORKS features, simple versus complex models, and additional sketch planes.

Understanding SOLIDWORKS features and their role in 3D modeling

Features is the term we use to refer to the tools that allow us to construct 3D models based on sketches. We usually use features directly after sketching to go from two-dimensional sketches to 3D models, which are mostly built based on sketches.

For example, if we were to model a cube, we would follow these steps:

1. First, we would create a sketch of a square.

2. Then, we would apply a feature in order to make the square into a cube. We do this by extruding it.

The following screenshot illustrates these two steps:

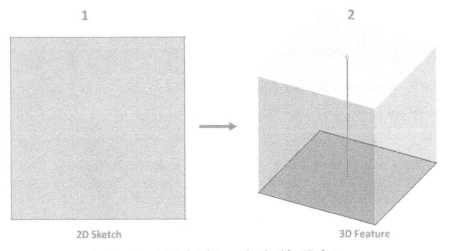

Figure 5.1 – A 2D sketch is used to build a 3D feature

Now that we know what features are, as well as their purpose, we can address how features differ in terms of simple and complex models.

Simple models versus complex models

SOLIDWORKS offers a wide variety of features that can help us easily create different shapes. Most of these features are for simple shapes such as simple hexahedrons (cubes), rotational shapes such as spheres and tubes, and much more. Thus, we will apply more features that build on top of each other to create more complex models.

When we were sketching, we applied and mixed multiple sketch commands to create more complex sketches. This is similar to what we do with features. The more complex the model is, the more features it may require.

To highlight this, take a look at the models shown in the following figure. The one on the left is a simple model of a cube. We only used one feature to create this cube. The model on the right is a turbine rotor. It is a more complex model, and we had to use 11 features to build it:

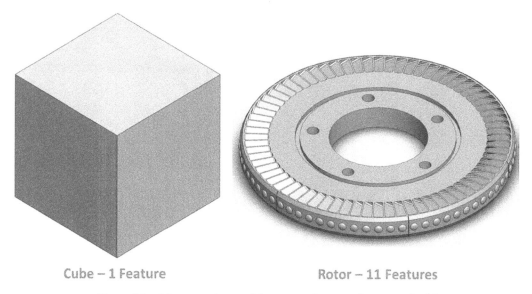

Cube – 1 Feature   Rotor – 11 Features

Figure 5.2 – More complex models can require more features to build

As we continue using SOLIDWORKS, we will be able to create more complex models. Now, let's learn about one fundamental aspect of all features – **planes**.

Sketching planes for features

By default, SOLIDWORKS provides three default planes: the front plane, the top plane, and the right plane. We will use one of these planes to create our first sketch and feature. As we start applying features, these three basic planes may not fullfill our needs for further sketches and features.

Thus, by creating more features, the resulting straight surfaces can also be used as sketching planes. We can use these to create even more sketches and features.

For example, for a new file, we will only have three sketch planes – the default base ones. If we create a cube, each face of the cube will also be a possible sketch plane. Thus, after creating the cube, we will be adding five potential sketch planes for the five new faces of the cube. The following image shows the three base planes, as well as some of the new planes that were created with the new cube. Note that some of the sketch planes may coincide with each other:

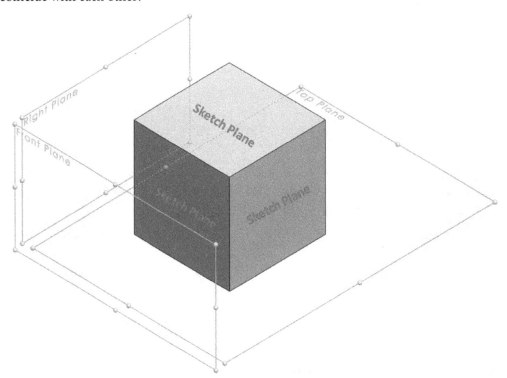

Figure 5.3 – A visual representation of the base planes and a cube's surfaces

> **Tip**
> Any straight surface can also be used as a sketch plane.

We have just learned what features are, which features are used in complex and simple models, and how sketch planes relate to features. Now, let's explore our first set of features – extruded boss and cut.

Understanding and applying extruded boss and cut

Extruded boss and extruded cut are the most basic and easiest features to apply. They are direct extensions of a sketch and push it into the third dimension. In this section, we will cover what extruded boss and extruded cut are, how to apply them, how to edit them, and how to delete them.

What are extruded boss and extruded cut?

Extruded boss and extruded cut are two of the most basic features we'll use when modeling with SOLIDWORKS. Let's look at them in more detail:

- **Extruded boss**: This is a direct extension of a sketch that pushes it into the third dimension, resulting in adding materials.

- **Extruded cut**: This is a direct extension of a sketch that pushes it into the third dimension, resulting in removing/subtracting materials.

From these definitions, you can see that extruded boss and extruded cut are quite similar, but they have opposite effects. Extruded boss adds materials, while extruded cut removes material. The following figure illustrates the effect of the extruded boss feature. Note that we were able to go from a 2D sketch to adding materials to form a cube:

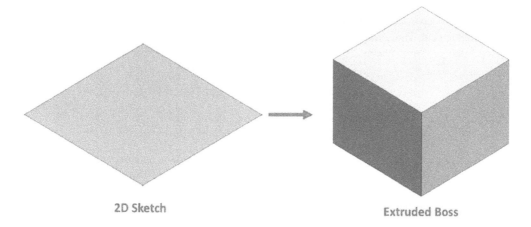

2D Sketch   Extruded Boss

Figure 5.4 – A representation of what the extruded boss feature does

The following figure illustrates the effect of extruded cut. Note that we were able to use a sketch to remove materials:

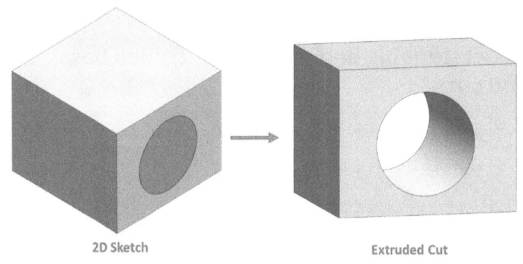

2D Sketch   Extruded Cut

Figure 5.5 – A representation of what the extruded cut feature does

Let's learn how to apply the extruded boss feature.

Applying extruded boss

In this section, we will discuss how to apply the extruded boss feature. To show this, we will create the model shown in the following figure. We have added annotations for each view, including information about the view's type and its dimensions:

> **Important Note**
> We will keep building with the same model as we explore extruded boss and extruded cut. Thus, keep saving the model as we go along.

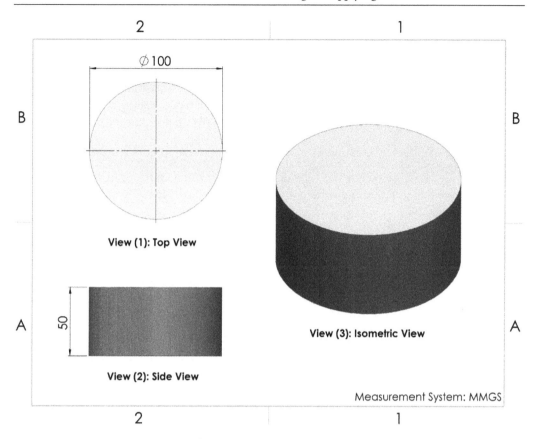

Figure 5.6 – The end shape of the extruded boss exercise

When applying the extruded boss and extruded cut features, we will always start with a 2D sketch and then apply that feature based on that 2D sketch. Thus, for this exercise, we will split each feature application into two stages – the sketching stage and applying the feature stage.

One important aspect to keep in mind is that, as we continue modeling, we will need to plan a strategy when it comes to how to model the targeted object. There is no right or wrong way to create a model. Thus, different people will have different plans for making the same model. It is always good to plan ahead when it comes to creating a model. We can do that by either sketching or writing down our ideas. Since we are taking our first steps toward 3D modeling, we will need to have a brief written plan before we start modeling:

- **Planning**: We start by creating a circle and then extruding that into a cylinder.

- **Sketching**: Next, we sketch and fully define a circle with a diameter of 100 mm. We can see this in *Figure 5.6*, where it says **View (1): Top View**. The circle will look as follows:

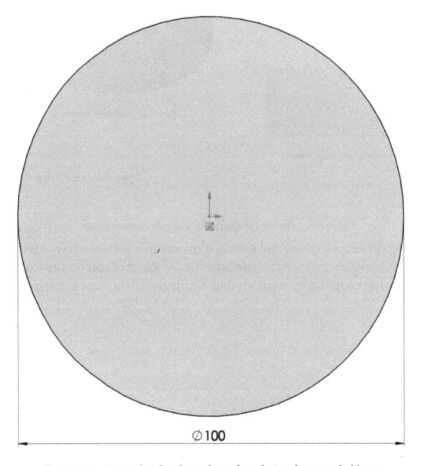

Figure 5.7 – A circular sketch used as a foundation for extruded boss

- **Applying the feature**: In this example, we will apply the extruded boss feature.

To apply the features, follow these steps:

1. Click on the **Features** tab and select the **Extruded Boss/Base** command, as shown in the following screenshot. We don't need to exit sketch mode. As soon as we select this command, SOLIDWORKS will understand that we want to apply this feature to the active sketch:

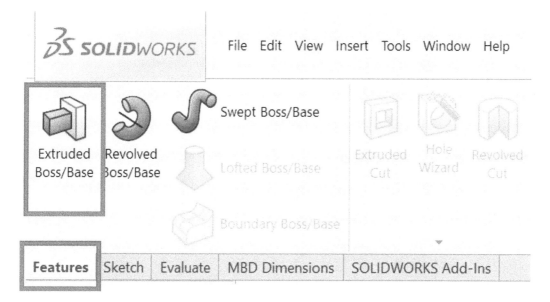

Figure 5.8 – The location of the extruded boss feature

2. Once we click on the **Extruded Boss/Base** command, we will notice that an options panel appears on the left-hand side. The extrusion preview will also appear on the canvas:

> **Important Note**
> If we exit the sketch mode before selecting the **Extruded Boss/Base** command, we can simply select the command and then select the sketch on the canvas.

3. Fill out the options in the **PropertyManager**, as shown in the following screenshot. Fill in the height as 50 mm. The PropertyManager will appear on the left-hand side of the interface:

Figure 5.9 – The extruded boss feature setting for this exercise

4. Once we've filled in these options, we will see the following preview:

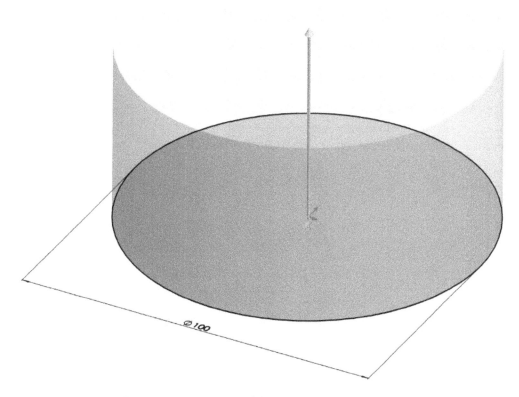

Figure 5.10 – A preview of the extruded boss application

5. After adjusting the options for our extrusion, we can click on the green check mark at the top of the **PropertyManager** panel to apply the extrusion:

Figure 5.11 – The green check mark on top of the PropertyManager to apply the feature

The result will be the following model:

Figure 5.12 – The resulting 3D model after the extrusion

Before we finish looking at the extruded boss feature, let's take a look at options in the **PropertyManager**. We will look at them based on their listing order, that is, from top to bottom, as shown in the following screenshot. We will start with the options we used in this exercise and then move on and look at the options we didn't use:

Figure 5.13 – The different options available for the extruded boss feature

The following options are available for the extruded boss feature:

- **From**: This determines where the extrusion features should start. Since we are still beginners, we will mostly use **Sketch Plane**. This means that the extrusion will start from the sketch that was used to create it. Other options include starting from **Surface**, **Vertex**, and **Offset**. The first two can't be used in this case because our model doesn't have multiple surfaces and vertices. The last option, **Offset**, can be used to offset the whole extrusion by a certain distance. You can see the **From** option on top of the PropertyManager feature, as shown in *Figure 5.13*.

- **Direction 1**: This is active by default. Under this heading, we can customize the previewed extrusion that's shown on the canvas. We can hover the mouse over the options to see their full names. The options under **Direction 1** are as follows:

  a) **End Condition**: This determines how the extrusion stops. In this exercise, we will only be using the **Blind** option, which is selected by default. This means that the extrusion will be extended by the dimensions that we indicate. We will explore other end conditions later in the book.

  b) **Reverse Direction**: This is the arrow to the left of **End Condition**. This can easily reverse the direction of the extrusion from up to down and vice versa.

  c) **Depth (D1)**: This determines the depth of the extrusion. In our case, we want the extrusion to be 50 mm deep, so we will input 50 mm.

  d) **Draft**: The icon below **Depth** is used to draft the extrusion. We can activate drafting by clicking on the icon. We will cover this option at a more advanced level later in this book.

- **Direction 2**: This is very similar to **Direction 1**; however, it applies the extrusion in the opposite direction as well. We can use this if we ever want to have different length extrusions in two directions on the sketch. We can simply check this box if we require the second direction. This wasn't needed in our example.

- **Thin Feature**: This applies an extrusion based on the thin borders of the sketch rather than the enclosed shape. It can be activated by checking the box next to it. If we apply **Thin Feature** to our circle, we will get a result similar to the one shown in the following figure. Take some time to draw another circle and experiment with **Thin Feature**:

Figure 5.14 – A sketch of one circle can translate to a ring using the thin feature

- **Selected Contours**: This can be used if we have more than one enclosed area. Then, we can select which ones we want to apply the extrusion to. An example of a sketch with more than one enclosed area is highlighted in the following figure, which has three enclosed areas:

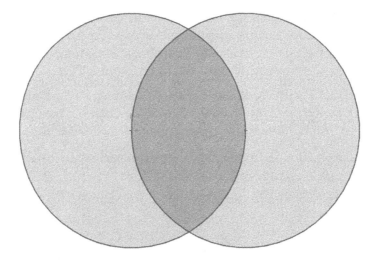

Figure 5.15 – A sketch with three enclosed areas

This concludes the section on creating the requested cylinder using the extruded boss feature. We learned about the following topics:

- The sequence we follow to create a model using features, that is, planning, sketching, and applying features

- How to relate a sketch to the extruded boss feature

- How to set up the extruded boss feature and apply the extrusion

Now that we know about the extruded boss feature, we will look at the extruded cut feature.

Applying extruded cut and building on existing features

In this section, we will discuss how to apply the extruded cut feature. The extruded cut feature is very similar to the extruded boss feature in terms of the options that are available to us. Due to this, we will explain them in less detail. We will build upon the model we created previously with extruded boss so that we can learn how to build upon existing features. To demonstrate this, we will create the following model:

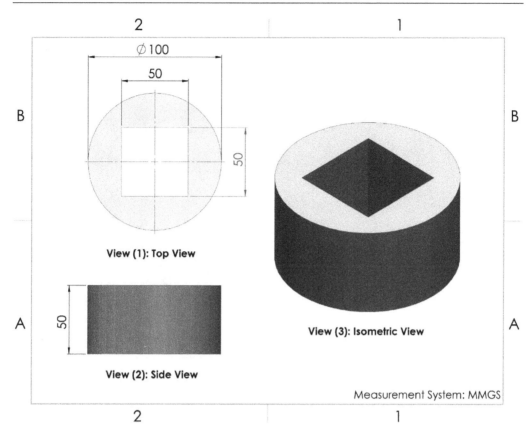

Figure 5.16 – The end model of the extruded cut exercise

Note that, in the preceding model, we are only applying an extra cut over the cylinder that we created with the extruded boss feature. Thus, we will start from the cylinder we created earlier and create an extra extruded cut. We will go through the following phases to do so:

- **Planning**: We will draw a square on top of the cylinder and apply it using the extruded cut feature.

- **Sketching**: The cutoff shape is a square, and so we will sketch a square on the top surface of the cylinder. Note that the top surface is not a default sketch plane. However, it is a straight surface, which means we can use it as a sketch plane.

- **Applying the feature**: When we have our sketch, we can apply our feature; in this case, this is extruded cut.

Follow these steps to create the sketch:

1. Select the top surface of the cylinder and click on the **Sketch** command, as shown in the following screenshot. We can also do this the other way around, that is, select the **Sketch** command first and then select the surface we want to sketch on:

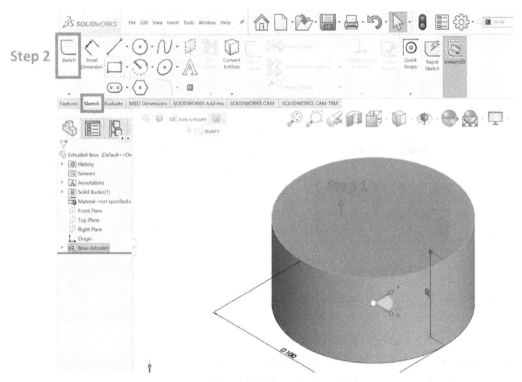

Figure 5.17 – A visual demo of steps 1 and 2

Now we need to start sketching. However, we may have a tilted view of our new sketch surface, which will make it harder for us to sketch. We can adjust our view so that it's normal to the sketch surface to make it easier to sketch. To do that, we can *right-click* on the new sketch at the bottom of the design tree and select **Normal To**, as shown in the following screenshot:

Figure 5.18 – The Normal To command changes our view to facing the sketch plane

This will change our view of the canvas so that it's facing the sketch surface. If we select **Normal To** again, the model will flip 180 degrees.

2. Sketch and fully define the required square, as shown in the following figure. The side of the square is 50 mm in length, as shown by **View (1): Top View** in *Figure 5.16* at the beginning of this section:

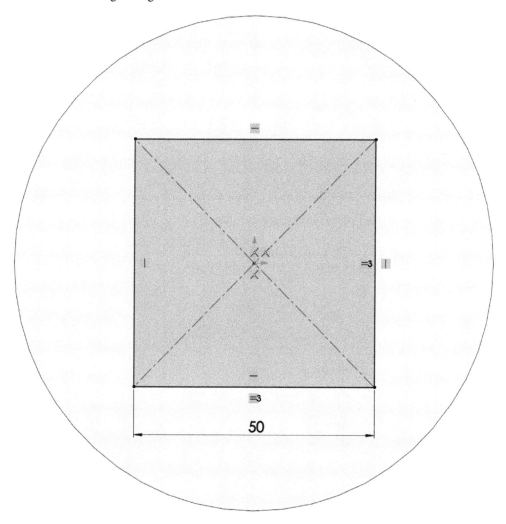

Figure 5.19 – The fully defined sketch for the extruded cut feature

Now that we have our sketch, we can apply our extruded cut feature.

3. Select the **Features** tab and select the **Extruded Cut** command, as shown in the following screenshot. (As with the extruded boss feature, we don't need to exit sketch mode.) As soon as we select the command, SOLIDWORKS will interpret that we want to apply the feature to the active sketch:

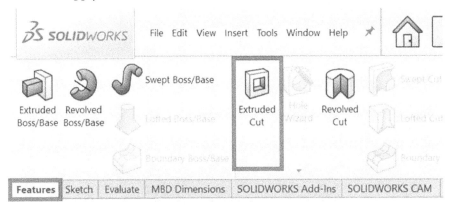

Figure 5.20 – The location of the extruded cut feature

4. Set the options in the PropertyManager as follows:

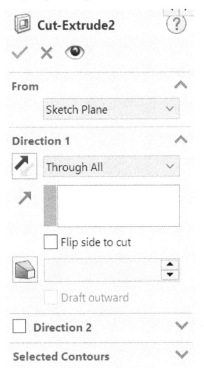

Figure 5.21 – The PropertyManager setting for the extruded cut feature

5. As with the extruded boss feature, we will also see a preview of the cut appear on the canvas. This preview is shown in the following figure:

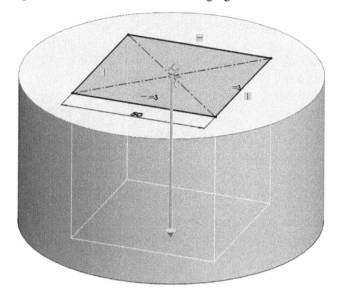

Figure 5.22 – A preview of the extruded cut

Note that the options that are shown in the PropertyManager for the extruded cut feature are almost the same as the options for the extruded boss feature. We will only elaborate on those that are highlighted in the following screenshot:

Figure 5.23 – The end condition as displayed in the PropertyManager

Let's take a look at these options in more detail:

- **End Condition**: Many end conditions are the same as for the extruded boss feature. By default, the end condition is usually **Blind**. However, in our case, we changed it to **Through All**. The **Through All** end condition means that the cut will extend to the end of the model, which is what we want. We can also get the same result by using the **Blind** end condition and setting the depth of the cut to 50 mm or more. To change the end condition, click on the drop-down menu and select the desired condition.

- **Flip side to cut**: Checking this option will turn the cut part around. In our example, if we have this option checked, we will keep the contained square and delete everything else. Having this option unchecked will delete the contained square and keep everything else. Experiment with this option to understand what it does.

6. Click the green check mark at the top of the PropertyManager to apply the extruded cut feature. We will end up with the following model:

Figure 5.24 – The final model after applying the extruded cut feature

Recall that after entering a sketch, we switched the view to normal for that sketch to make it easier to deal with. We can have the software auto-rotate the view to normal to sketch a plane on sketch creation and edits by enabling that from the system options. You can find the option by going to **Tools** and then **Options**. Under the **System Options**, click on **Sketch**, and then find the **Auto-rotate view normal to sketch plane on sketch creation and sketch edit** option, as shown in the following figure:

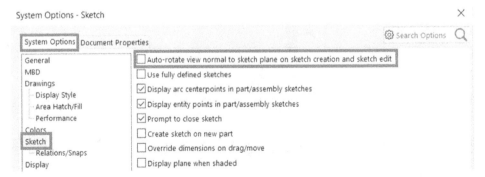

Figure 5.25 – Auto-rotate to a normal view

This concludes using the extruded cut feature. In this section, we covered the following topics:

- How to sketch over existing surfaces

- How to apply the extruded cut feature

Now that we know how to apply the extruded boss and cut features, we will learn how to modify and delete them.

Modifying and deleting extruded boss and extruded cut

Often, we apply a feature and then need to edit it or delete it. In this section, we will address how to edit and delete features. To illustrate this, we will apply the modifications shown in the following figure to the model we created earlier:

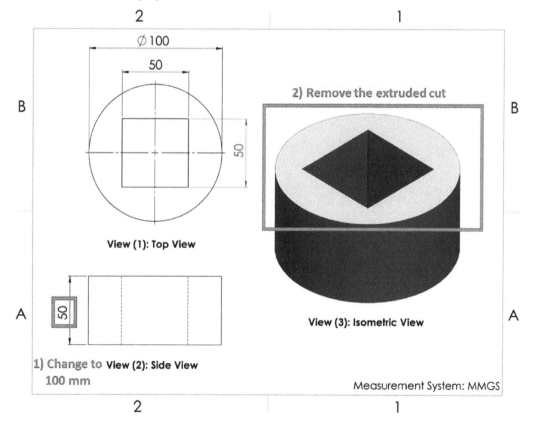

Figure 5.26 – The changes to the model to be applied in this exercise

As shown in the preceding figure, the changes that we are going to make are as follows:

- Changing the height of the cylinder from 50 mm to 100 mm
- Removing the extruded cut that goes through the cylinder

Let's go ahead and apply these changes.

Editing a feature – changing the height of the cylinder from 50 mm to 100 mm

To edit an implemented feature, we can right-click (or left-click) on it on the design tree and select the **Edit Feature** option. In this case, we want to edit the boss extrude feature we applied. Thus, we can right-click on the first **Boss-Extrude** feature and select **Edit Feature**, as shown in the following screenshot:

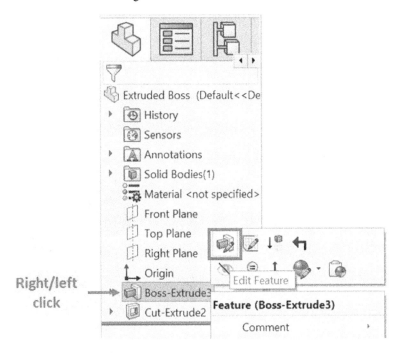

Figure 5.27 – The Edit Feature command as shown in the SOLIDWORKS interface

Once we select **Edit** Feature, the extruded boss features will be shown on the left, while a preview of the extruded boss feature will be shown on the right. Note that the preview doesn't show the extruded cut feature because it is located lower in the design tree. From the available options, we can change **D1** from 50 to 100, as shown in the following screenshot. After making this change, we can click on the green check mark to implement it:

Figure 5.28 – A feature can be edited by directly editing its PropertyManager

Now our cylinder will become longer, as shown in the following figure. If the shape of the model failed to update, try clicking on the traffic light (Rebuild) icon at the top of the SOLIDWORKS interface or use the *Ctrl + B* shortcut:

Figure 5.29 – The shape of our model after editing the extruded boss feature

This concludes how to edit a feature. Now, let's look at deleting a feature.

Deleting a feature – removing the extruded cut feature that goes through the cylinder

To delete a feature, right-click on a feature in the design tree and select the **Delete…** option. In this case, we want to delete the extruded cut feature. Thus, we can right-click on it and select **Delete…**, as shown in the following screenshot:

Figure 5.30 – Features can be deleted with the Delete… command

Once we've selected **Delete…**, we will get the following message, asking us to confirm that we want to delete the feature. The message will specify the item to be deleted, which in our case is **Cut-Extrude2 (Feature)**. It will also specify any dependent items that will be deleted with the feature; there are none in this case. We will cover dependent items at a more advanced level later in this book. We can click **Yes** to confirm that we want to delete the feature:

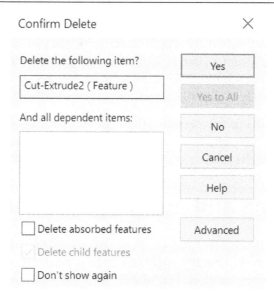

Figure 5.31 – A conformation box appears when deleting a feature

Now, we will end up with the following model. Note that we only deleted the feature, so the sketch will still remain in the model for us to use for any other purpose, as shown in the following figure:

Figure 5.32 – The sketch still remains even though its feature was deleted

If we want to delete both the feature and its sketch, we can check the following **Delete absorbed features** option, as shown in *Figure 5.31*, when deleting the feature.

Tip

We can delete the feature by directly selecting it on the design tree and pressing *Delete* on the keyboard.

This concludes the sections on editing and deleting features. Every feature can be edited and deleted in the same way.

As designers and 3D modelers, we will be faced with many situations where we receive models from other individuals and are asked to edit them. Also, we ourselves will modify our models as part of improvement cycles. Thus, it is very important for us to know how to modify models. As our SOLIDWORKS skills grow, we will pay special attention to modifying models, especially pre-existing ones.

We have just learned about our first set of features, that is, extruded boss and extruded cut. We learned how to apply them and how to modify them. Now, we will move on to another set of features – fillets and chamfers.

Understanding and applying fillets and chamfers

Fillets and chamfers are used to modify edges and vertices on our models by making them less sharp. If we look at everyday objects around us, such as phones, laptops, and furniture, we will notice the common use of small fillets and chamfers on the edges. In this section, we will discuss what fillets and chamfers are, how to apply them, and how to modify them.

Understanding fillets and chamfers

Fillets and chamfers are modifications that are made to the edges and vertices of our models. A **fillet** is a curved surface defined by a radius, while a **chamfer** is a transitional straight surface defined by lengths and angles. They help remove sharp edges and turn them into softer ones in order to provide a safer product or a better user experience. They are similar to fillets and chamfers in sketching. The following figure illustrates the effect of the fillet feature:

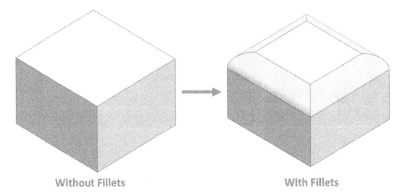

Without Fillets   With Fillets

Figure 5.33 – Edges with and without fillets

The following figure illustrates the effect of the chamfer feature:

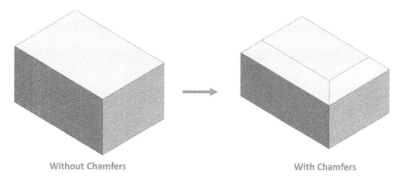

Figure 5.34 – Edges with and without chamfers

Note that fillets and chamfers can only be applied to existing features. In the preceding figure, we explored the extruded boss feature first, and then we were able to apply fillets or chamfers. Also, to apply fillets and chamfers, we don't need to start with a 2D sketch.

Now that we know what fillets and chamfers are, we can start using them in SOLIDWORKS.

Applying fillets

In this section, we will discuss how to apply the fillet feature. To show this, we will create the model shown as follows:

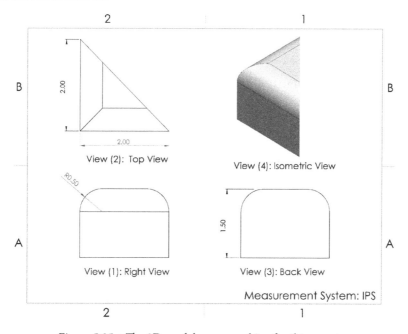

Figure 5.35 – The 3D model we are making for this exercise

As usual, we will start with planning, sketching, and then applying a feature:

- **Planning**: Here, we will create a triangular prism and then apply fillets to the top edges.

- **Sketch**: We will sketch and fully define a triangle, as shown by the **Top View** view in the preceding figure.

- **Feature**: We will apply the extruded boss feature by 1.5 inches, as shown by the **Back View** view in the preceding figure. This will result in a triangular prism that looks as follows:

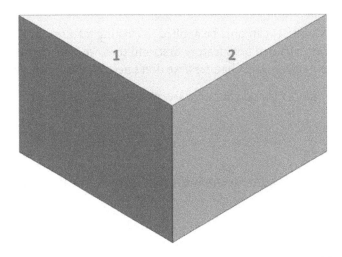

Figure 5.36 – The base triangular prism we will use to apply the fillet

To apply the fillet feature, select the **Fillet** command, as shown in the following screenshot:

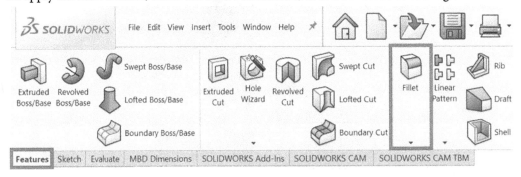

Figure 5.37 – The location of the Fillet command

Select **Edge<1>** and **Edge<2>**, which have fillets applied to them. Adjust the **Fillet** options, as shown in the following screenshot showing the PropertyManager:

Figure 5.38 – The fillet feature PropertyManager

The following is a brief explanation of the options that we used in this exercise:

- **Fillet Type**: This specifies the type of fillet we're using. The first type is the constant size fillet, which we are using in this exercise. As we are only beginners, we will only use this type.

- **Items to Fillet**: This highlights where the fillet will be applied. In our example, we're applying it to **Edge <1>** and **Edge <2>**. We can use this window to remove edges from the selection if we've added one by mistake. For the selection, we can select individual edges or faces. Selecting a face will apply the fillet to all the edges related to that face.

- **Show selection toolbar**: If this option is checked, then we can select an edge; a small selection toolbar will appear with some shortcuts that we can use to select where we want to apply the fillet. The toolbar looks as follows. Since the selection toolbar only contains selection shortcuts, we can disregard it for now and start using it later, once we are comfortable with the different selections it provides:

Figure 5.39 – The selection toolbar is an easy way to select the edges to apply the fillet to

- **Tangent propagation**: This only applies if there's more than one edge and they are tangential to each other, for example, two curved edges that have a tangent relationship with each other. In the case of our triangular prism, none of the edges are tangent to each other. Thus, selecting or deselecting this option will not make any difference.

- **Full preview**, **Partial preview**, and **No preview**: These options decide on the type of fillet preview we can see when we are making the shape. Selecting **Full preview** will show us what the fillet looks like in its entirety, whereas selecting **No preview** won't show it at all.

- **Fillet Parameters**: Here, we have two options, **Symmetric** and **Asymmetric**. A **Symmetric** fillet is a uniform fillet in which we only need to define one radius. An **Asymmetric** fillet, on the other hand, has different curvatures on each side and requires us to define more than one dimension.

- **Radius**: Here, we can input a numerical value for the fillet radius. In this exercise, it is 0.5 inches, as shown by **View (1): Right View** in the figure at the beginning of this section.

- **Multi Radius Fillet**: Clicking this option will allow us to determine a different radius for different edges. For example, if we want the fillet radius for **Edge <1>** to be different from the fillet radius for **Edge <2>**, we can use this option.

- **Profile**: This determines the profile of the fillet. The **Circular** profile option projects the fillet profile as a quarter of a circle. Other options include **Conic Rho**, **Conic Radius**, and **Curvature continuous**.

We won't be using the **Setback Parameter**, **Partial Edge Parameter**, and **Fillet Options** just yet. While we are here, take some time to combine some of the preceding options and look at the result in the canvas preview. Once done, click on the green check mark at the top of the **Options** tab to apply the fillet. We should get the following shape:

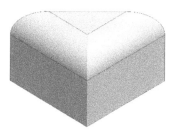

Figure 5.40 – The final 3D model after applying the fillet

This concludes how to apply fillets. In this section, we covered the following topics:

- How to apply fillets in SOLIDWORKS
- The different options we can use to define fillets in SOLIDWORKS

Now that we know how to apply fillets, let's learn how to apply chamfers.

Applying chamfers

In this section, we will discuss how to apply the chamfer feature. To do this, we will create the model shown in the following figure:

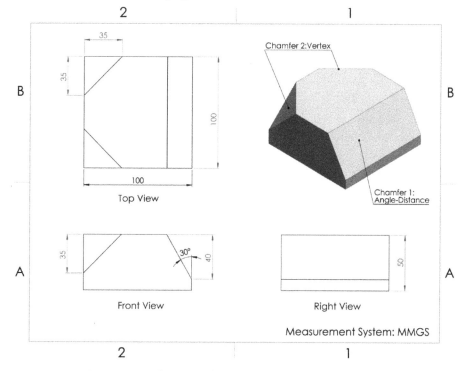

Figure 5.41 – The 3D model we are making with this exercise

The preceding figure specifies two different types of chamfers, each with different specifications. In SOLIDWORKS 2022, there are five different types of chamfers, based on how they are defined. The following table illustrates the angle-distance, distance-distance, and vertex chamfer types:

Chamfer Type	Description	Illustration
Angle-distance	This is defined by specifying an angle and the distance that makes up the chamfer. The location of the chamfer is identified by selecting an edge, a face, or multiples of both.	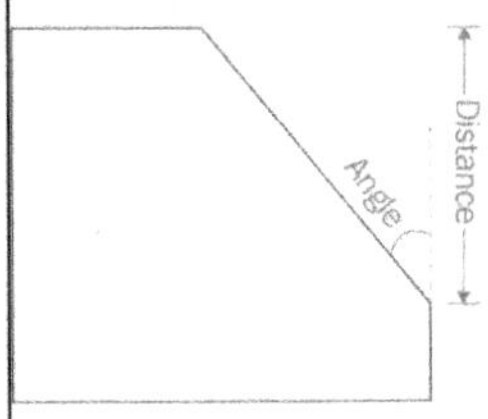
Distance – distance	This is defined by two distances that make up the chamfer. The location of the chamfer is identified by selecting an edge, a face, or multiples of both.	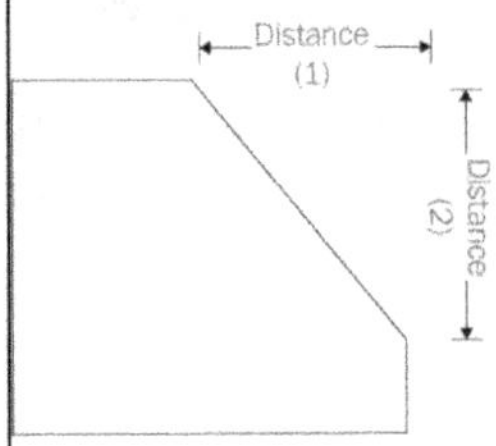
Vertex	This is defined by three distances that move away from a vertex. The location of the chamfer is identified by selecting a vertex.	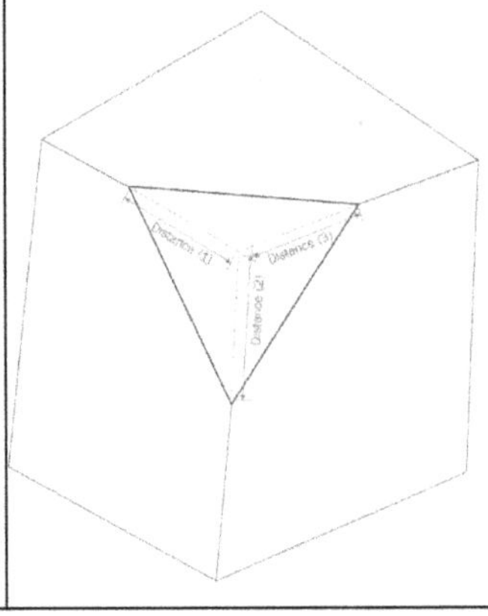

Figure 5.42 – Illustrations of the angle-distance, distance-distance, and vertex chamfer types

The following table illustrates the offset face and face-face chamfer types. Each chamfer type is defined differently giving us different options to help maintain our required design intent:

Chamfer Type	Description	Illustration
Offset face	This is defined by an offset distance for the two faces that are adjacent to the chamfer's edge. This gives the chamfer a special definition if the two faces are not perpendicular to each other. The location of the chamfer is identified by selecting one or more edges.	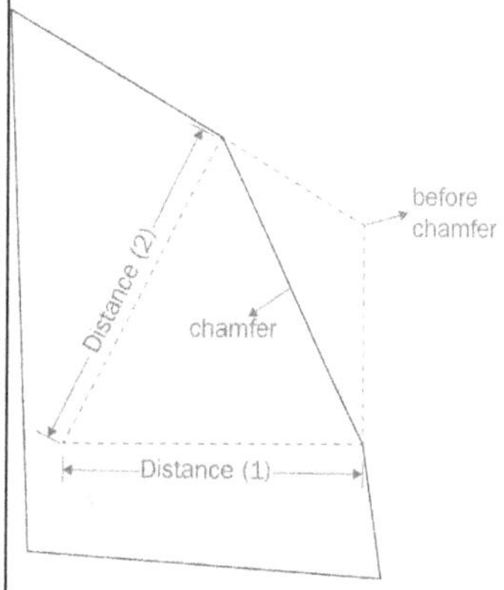
Face—face	This is defined by the distance between two or more faces. The location of the chamfer is defined by selecting the faces that surround the chamfer.	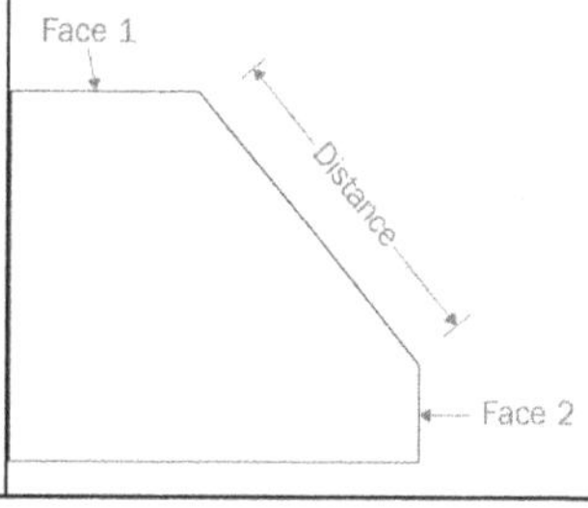

Figure 5.43 – Illustrations of the offset face and face-face chamfer types

The model we are going to create uses two types of chamfers, angle-distance and vertex. To create the chamfer, we will follow our usual procedure, where we start by planning, then creating sketches, and then applying a feature:

- **Planning**: Here, we will create a rectangular prism and then create an angle-distance chamfer, followed by two vertex chamfers.

- **Sketch**: We will sketch and fully define a 100 x 100 mm square, as shown in the top view of the diagram at the beginning of this section.

- **Feature**: We will be applying the extruded boss feature and extruding the square by 50 mm, as shown in the right view of the diagram at the beginning of this section. At this point, we will have the following rectangular prism:

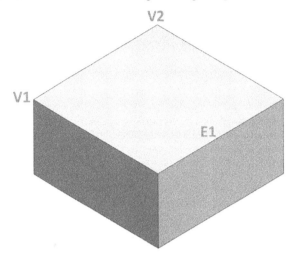

Figure 5.44 – The base rectangular prism we will use to apply the chamfers to

To apply angle-distance chamfers, follow these steps:

1. To apply the chamfer feature, select the **Chamfer** command, as shown in the following screenshot. The command can also be accessed from the drop-down menu, under the **Fillet** command:

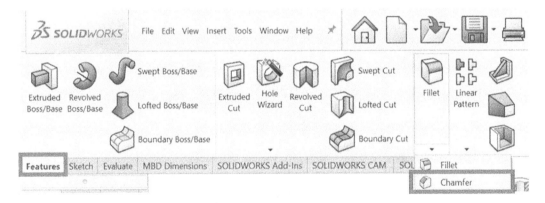

Figure 5.45 – The location of the Chamfer command

2. Set up the chamfer's options in the PropertyManager as follows. Some options that are available for the chamfer feature are the same as for the fillet feature, so we won't go over them in as much detail here:

Figure 5.46 – Chamfer's PropertyManager

The following is a brief description of the options that we used in this exercise:

- **Chamfer Type**: From here, we can select one of the five types of chamfers, as specified in the preceding table. We can hover over each one to see their names. For our first chamfer, we will use the angle-distance chamfer.

- **Items To Chamfer**: This specifies the location of the chamfer. We can apply the angle-distance chamfer by selecting edge 1 (**E1**), as highlighted in the following rectangular prism. If we select another edge or face by mistake, we can remove it by deleting it from the list under **Items to Chamfer**, as shown in the PropertyManager:

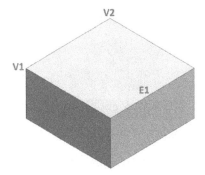

Figure 5.47 – Select edge 1 (E1) to apply the angle-distance chamfer to

- **Tangent propagation**: This works the same as for fillets. Checking this option will propagate the chamfer across any edges that are tangent to each other. In our case, this option will not make a difference, as there are no other edges that have a tangent relationship to our selected edge.

- **Full preview**, **partial preview**, and **no preview**: This is similar to the option for fillets. This can determine what our chamfer's preview looks like on the canvas.

- **Flip direction**: This will flip the location of the defining angle and distance.

- **Distance (D)**: This value will determine the distance value of our chamfer. In this exercise, the distance is 40 mm.

- **Angle (A)**: This value will determine the angle value of our chamfer. In this exercise, the angle is 30 degrees.

> **Tip**
> While you are here, play around and mix these different options to understand their effects.

3. Click on the green check mark to apply the chamfer. After applying the chamfer, we should have the following model:

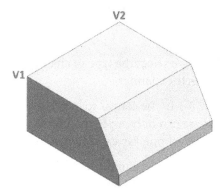

Figure 5.48 – The shape of the 3D model after applying the angle-distance chamfer to E1

At this point, our model has one chamfer of the angle-distance type. Now, we will apply the other two chamfers, which are vertex types. To apply a vertex chamfer, follow these steps:

1. Select the **Chamfer** command from the command bar to apply a second chamfer.

2. Set up the chamfer's options as follows in the **PropertyManager**:

Figure 5.49 – The chamfer PropertyManager for the vertex chamfer type

Here is a brief description of some of the special features we can apply to the vertex chamfer feature:

- **Chamfer Type**: From here, we select the type of chamfer we want to create; in this case, we want to create a vertex chamfer.

- **Items To Chamfer**: Select **V1** in the canvas to locate the chamfer. Note that, when using the vertex chamfer, we can only apply chamfers to one vertex at a time. In other words, we can only select one vertex in **Items To Chamfer**.

- **Equal distance**: The vertex chamfer can be defined with three distances that extend away from the vertex into the edges it consists of (refer to the preceding table regarding chamfer types). Checking this option will make all three distances equal, so we only need to input one dimension. If we uncheck this option, we will need to enter three different distances.

- **Distance (D)**: This value will determine the distance of the chamfer. In this exercise, the distance is 3 5 mm in all three directions.

3. Click on the green check mark to apply the chamfer. After applying the chamfer, we should have the following model:

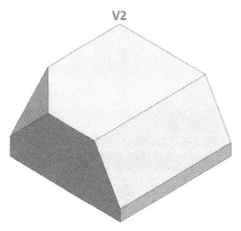

Figure 5.50 – The 3D model after applying the vertex chamfer

4. Now, we can repeat *steps 1–3* for **V2** to apply the second vertex chamfer. After repeating those steps, we will have the final model. This will look as follows:

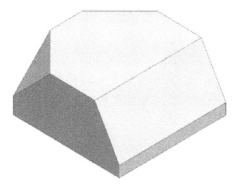

Figure 5.51 – The final model after applying the two vertex chamfers

This concludes how to apply chamfers. In this section, we discussed the following topics:

- The five different types of chamfers, as well as how are they defined and located

- How to apply different types of chamfers in SOLIDWORKS

- The different options we can use to define chamfers in SOLIDWORKS

At this point, we know how to apply fillets and chamfers. Now, let's learn how to modify them.

Modifying fillets and chamfers

To edit or delete a fillet or chamfer, we can follow the same procedure that we followed to edit and delete the extruded boss and cut features. In fact, every feature can be modified in the same way. Here, we *right-click on a feature* in the design tree and then select **Edit** to edit it or **Delete** to delete it.

One special aspect when it comes to modifying chamfers is that there are some limitations between switching from one chamfer type to another. For example, if we modify an angle-distance chamfer, we won't have the option to change the type in order to offset the face. If we are faced with such a limitation, we can simply delete the chamfer and start again with a new one.

Applying partial fillets and chamfers

So far, we talked about making fillets and chamfers that extend to a full edge. However, we can also apply partial fillets and chamfers that are only applied to a specific section of an edge. Let's talk about those here. To do this, we are going to create the following 3D model:

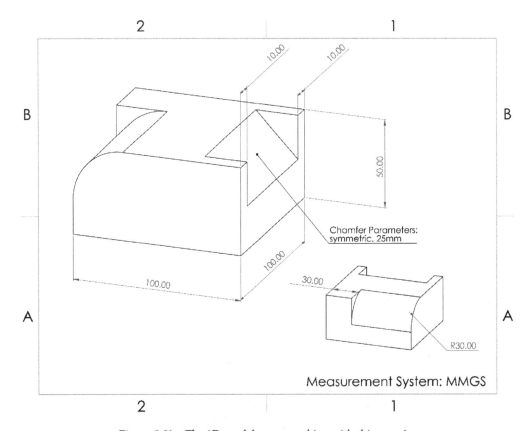

Figure 5.52 – The 3D model we are making with this exercise

We are going to create it in three phases, as follows:

1. Creating the rectangular prism

2. Applying the partial fillet

3. Applying the partial chamfer

The first stage is simple; you can start by sketching a **100** mm square, then extrude it by **50** mm using the **extruded boss** feature. We are already familiar with applying extruded boss, as shown in the following figure:

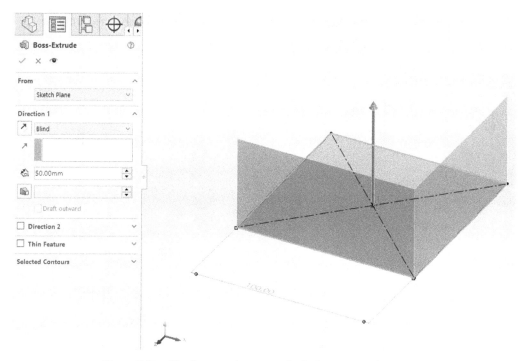

Figure 5.53 – The first step is to extrude the base square by 50 mm

Now that we have our base rectangular prism, we can start applying the partial fillet and chamfers. We will start by applying the fillet.

Applying a partial fillet

To apply a partial fillet, we can follow these steps:

1. Select the **Fillet** command. Then, under **Fillet Type**, select **Constant Size Fillet**, select an edge to apply the fillet to, and set the parameters to **Symmetric** with the radius being 30 mm, as shown in the following screenshot:

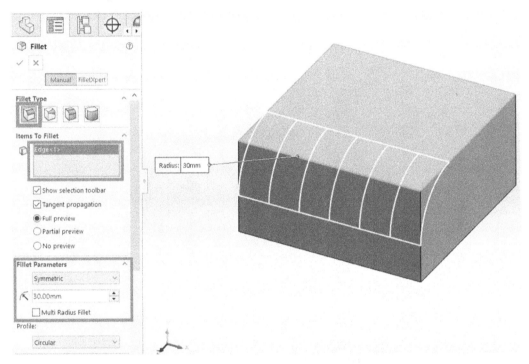

Figure 5.54 – The fillet preview and the PropertyManager applying a full fillet

> **Important Note**
> Partial fillets can only be applied to the **Contact Size Fillet** type.

2. Scroll to the bottom of the fillet **PropertyManager**, and check the **Partial Edge Parameters** option.

3. Set the start condition to **Distance Offset** and input 30 mm, as shown in the following screenshot:

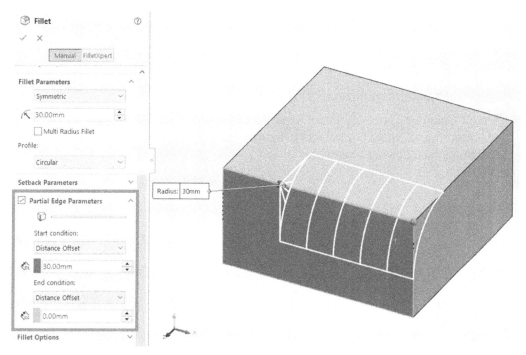

Figure 5.55 – The preview and the PropertyManager applying the fillet to a partial edge

4. Click on the green check mark on top of the **PropertyManager** to apply the partial fillet.

Apart from the **Distance Offset** start and end conditions, there are another two conditions that can be used. Let's explore them:

- **Percentage offset**: This allows you to offset the fillet by a certain percentage in relation to the edge.

- **Reference offset**: This allows you to offset the fillet in relation to an added reference. For example, you can sketch a point on the edge and then use it as a reference to the start or end of the partial fillet.

> **Tip**
> While you are here, take some time to experiment with the different types of start and end conditions.

Now that we know how to apply a partial fillet, let's move to apply a partial chamfer to our model.

Applying a partial chamfer

With a few exceptions, applying a partial chamfer follows a similar procedure to applying a partial fillet. We can follow these steps:

1. Select the **Chamfer** command.

2. Under **Chamfer Type**, select the **Offset Face** type, select an edge to apply the chamfer to, and set the parameters to **Symmetric** with the distance being 25 mm, as shown in the following screenshot:

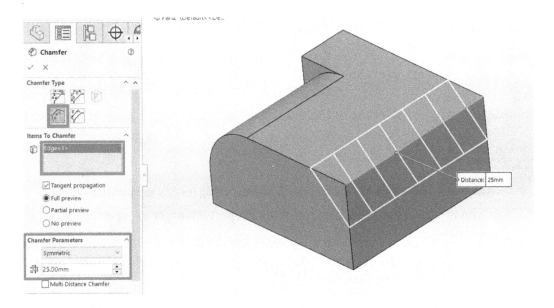

Figure 5.56 – The full chamfer preview and its PropertyManager

> **Important Note**
> Partial chamfers can only be applied to the **Offset Face** type.

3. Scroll to the bottom of the fillet PropertyManager and check the **Partial Edge Parameters** option.

4. Set the start condition and end condition to **Distance Offset** and input 10 mm for both, as shown in the following screenshot:

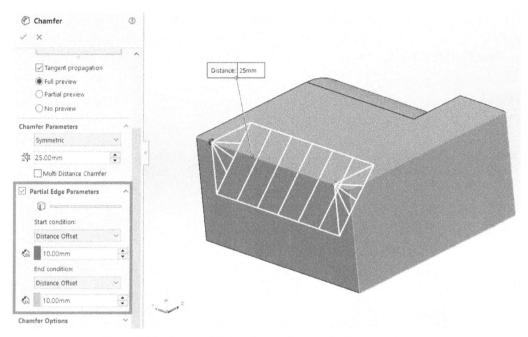

Figure 5.57 – The partial chamfer preview and its PropertyManager

5. Click on the green check mart to apply the partial chamfer.

This concludes this section on fillets and chamfers. In this section, we have learned how to apply them and how to modify them. Now, we can start learning about another feature set – revolved boss and revolved cut.

Understanding and applying revolved boss and revolved cut

Revolved boss and revolved cut are two of the most common features in SOLIDWORKS and are also easy to apply. They capitalize on rotational movements to add or remove materials. In this section, we will discuss what revolved boss and revolved cut are, how to apply them, and how to modify them.

What are revolved boss and revolved cut?

Revolved boss and revolved cut are among the most basic features in SOLIDWORKS. Let's explain them in more detail:

- **Revolved boss**: This adds materials by rotating a sketched shape around an axis.

- **Revolved cut**: This removes materials by rotating a sketched shape around an axis.

From these definitions, we can see that revolved boss and cut are similar. However, they have the opposite effect. Revolved boss adds materials, while revolved cut removes materials. The following figure illustrates the effect of the revolved boss feature:

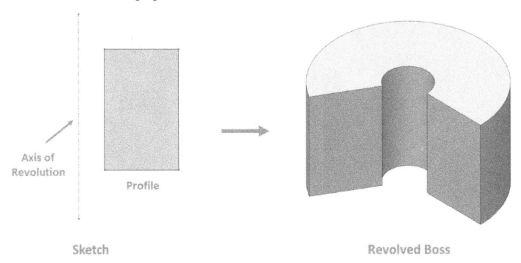

Figure 5.58 – An illustration of what the revolved boss feature can do

The following figure illustrates the effect of the **Revolved Cut** feature:

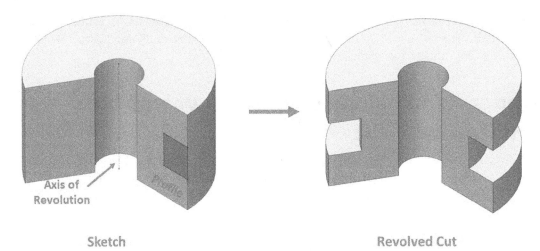

Figure 5.59 – An illustration of what the revolved cut feature can do

As we can see, two elements are required if we want to apply revolved boss and revolved cut – a profile and an axis of revolution. We have to sketch both of them in sketch mode before we can apply our features.

> **Important Note**
> The axis of revolution can also be part of the profile.

Now that we know what the revolved boss and revolved cut features are, we will learn how to apply them.

Applying revolved boss

The revolved boss feature adds materials by revolving a sketch around an axis of revolution. To show you how to apply the revolved boss feature, we will create the following model. Now that we are using more and more features, we will start to notice that we use the same options repeatedly. For example, when applying the revolved boss, we notice that most options are the same ones that we use while applying extruded boss. Therefore, we won't explain these again here:

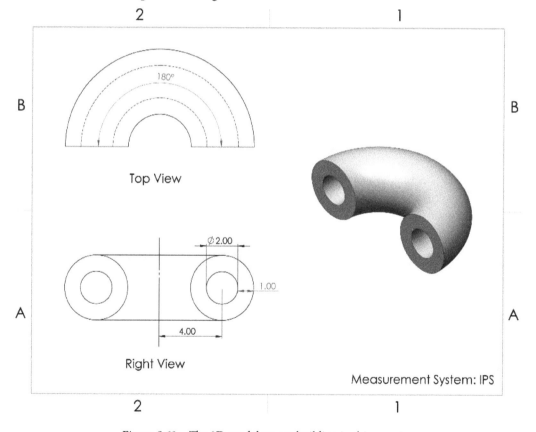

Figure 5.60 – The 3D model we are building in this exercise

As usual, we will follow our standard procedure planning, sketching, and applying features:

- **Planning**: Here, we'll draw the profile shown in the right view of the preceding drawing, and then we'll rotate that by **180** degrees.

- **Sketching**: Here, we're going to sketch the profile highlighted in **Right View** using the right plane. This includes using the axis of revolution. Our sketch should look as follows:

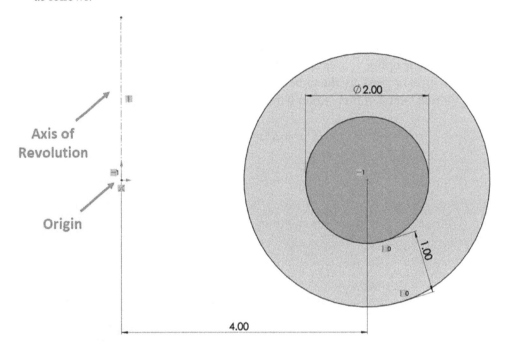

Figure 5.61 – The sketch used to apply the revolved boss feature

- **Applying the feature**: Here, we are going to apply our revolved boss feature.

To apply the revolved boss feature, follow these steps:

1. Select the **Revolved Boss/Base** command from the **Features** command bar. We don't need to exit the sketch to select the command. However, if we do exit the sketch for whatever reason, we can select the **Revolved Boss/Base** command and then select the sketch we want to apply it to:

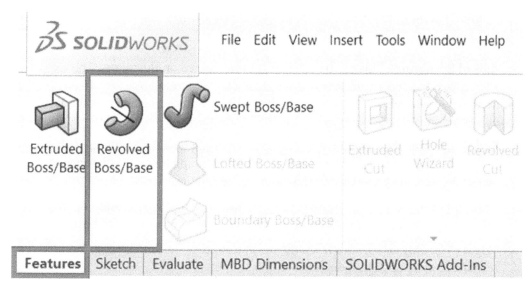

Figure 5.62 – The location of the revolved boss feature

2. Adjust the options in the PropertyManager, as shown in the following screenshot:

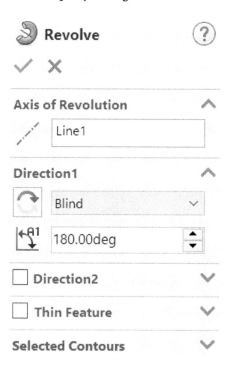

Figure 5.63 – The revolved boss PropertyManager

Here is a brief description of the unique options that can be used with the revolved feature:

- **Axis of Revolution**: This can be any straight line located on the canvas. This axis will be used to apply the feature by rotating the sketch around that axis. In our exercise, we will select the centerline indicated in the preceding figure as the axis of revolution.

- **End Condition**: This provides a few options regarding how the revolution will stop. In this exercise, we will use the **Blind** condition. By doing this, we will decide on the end of the revolution by determining the angle of rotation.

- **Reverse Direction**: This is the two curved arrows next to the end condition field. Clicking on this icon will reverse the rotation direction of the revolved boss.

- **Angle (A1)**: This determines when the revolution stops. In this exercise, the revolution will stop at 180 degrees.

3. Since we're setting the feature's options, we will be able to see a preview of the feature on the canvas. This will look as follows:

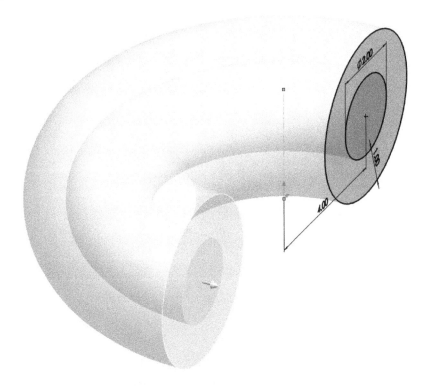

Figure 5.64 – A preview of the revolved boss feature

The rest of the options (**Direction 2**, **Thin Feature**, and **Selected Contours**) were explained when we look at the extruded boss feature. They have the same functionality.

4. To apply the revolved boss feature, we can click on the green check mark at the top of the **Options** panel. The resulting model should look as follows:

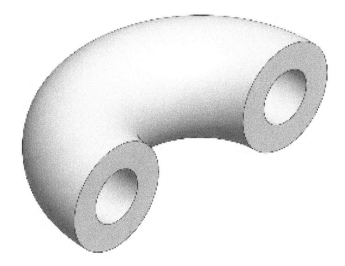

Figure 5.65 – The final shape after applying revolved boss

This concludes how to use the revolved boss feature. In this section, we discussed the following topics:

*   How to apply the revolved boss/base feature
*   The different options we can use to define the revolved boss/base feature

Now, we can start learning about the revolved cut feature.

Applying revolved cut

The revolved cut feature removes materials from existing bodies by revolving a sketch around an axis of revolution. To show you how to apply the revolved cut feature, we will create the following model. Note that the final body is a continuation of the body we just created when we applied the revolved boss feature. Thus, we will use and build upon that model:

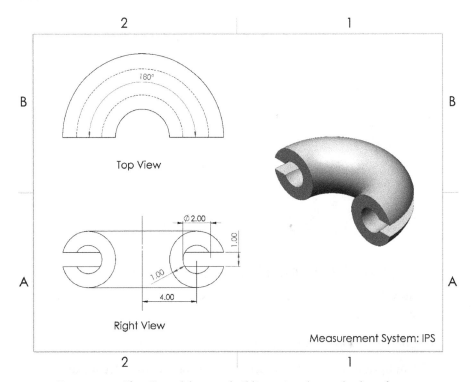

Figure 5.66 – The 3D model we are building using the revolved cut feature

Let's go ahead and complete the model:

- **Planning**: We'll use the model we created earlier and add the revolved extruded cut feature to create the desired final shape in the drawing.

- **Sketching**: Here, we will sketch and fully define a rectangle to make the new cut that's shown in the following figure. The height of the cut is shown in **Right View** as **1** inch and cuts through the thickness of the wall. You will notice that we picked **1.2** inches for the width of our rectangle since this will be enough to cut through the whole wall. There are more advanced ways to define the cut, but we won't be covering them here. The axis of revolution goes through the center of the model, which also goes through the origin that we provided in the previous exercise:

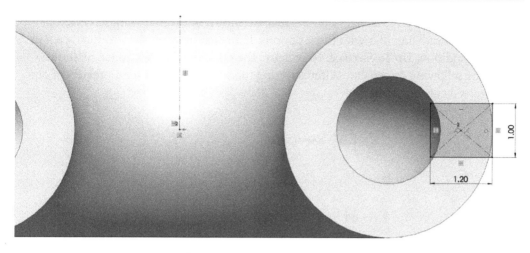

Figure 5.67 – The sketch used to apply the revolved cut feature

- **Applying the feature**: Here, we will be applying the revolved cut feature.

To apply the revolved cut feature, follow these steps:

1. Select the **Revolved Cut** feature from the command bar, as follows:

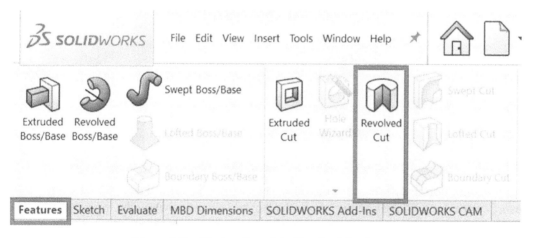

Figure 5.68 – The location of the Revolved Cut feature

2. Set up the PropertyManager options that are shown in the following screenshot. We are already familiar with these options. However, here, we're introducing a new end condition, **Up To Surface**. Now, select the other end of the surface, as shown in the following screenshot. After selecting the surface, we will see a preview of our revolved cut:

Figure 5.69 – The revolved cut PropertyManager

This end condition means that the cut will start from our sketch and end at a selected surface. Since our revolved cut goes through all of our shapes, it would be more convenient to end it by selecting the end surface rather than writing a numerical value for an angle. Another advantage of selecting this end condition is that it preserves our design intentions. For example, if we wanted to modify the angle of the revolved boss so that it's 270 degrees instead of 180, the revolved cut will update automatically and go through all of our shapes. However, if we use 180 degrees in a **Blind** end condition for our revolved cut and then update the revolved boss to 270 degrees, the cut will stay at 180 degrees:

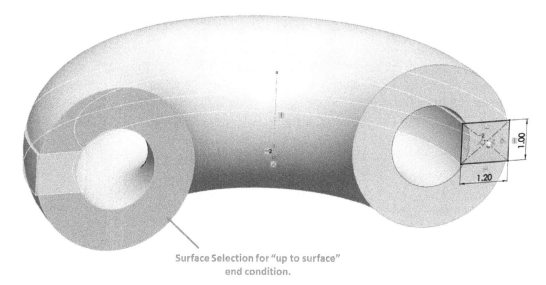

Surface Selection for "up to surface"
end condition.

Figure 5.70 – The preview of the revolved cut feature

3. Click on the green check mark to apply the cut. This will result in the
    following model:

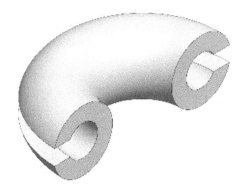

Figure 5.71 – The final 3D model after applying the revolved cut feature

This concludes how to apply the revolved cut feature. In this section, we discussed the
following topics:

- How to apply the revolved cut feature

- How to use the **Up To Surface** end condition

Now that we've learned how to apply the revolved boss and revolved cut features, we will
learn how to modify them.

Modifying revolved boss and revolved cut

The same procedure that we follow to modify the extruded boss and extruded cut features applies when we modify the revolved boss and revolved cut features. To recap, to edit or delete a revolved boss or revolved cut, we can right-click on the feature from the design tree and select **Edit** or **Delete...**, as required. The following screenshot highlights the **Edit** and **Delete...** commands that appear after right-clicking on a feature from the design tree:

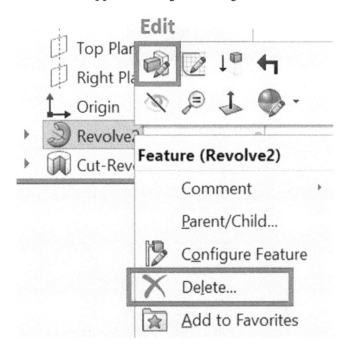

Figure 5.72 – The locations of the Edit and Delete... commands

This concludes this section on revolved boss and revolved cut. In this section, we learned how to apply these features, how to set them up, and how to modify them. Those two features will be important when we deal with rounded objects such as shafts and cylinders.

Summary

In this chapter, we have learned about our first set of SOLIDWORKS features, all of which allow us to go from creating 2D sketches to creating 3D models. Here, we learned about the extruded boss/extruded cut and revolved boss/revolved cut features. Each of these feature sets has an additive feature and a subtractive feature. We also learned about the fillet and chamfer features, which mainly aim to remove sharp edges for our 3D models. For each feature, we learned about what it was, how to apply it, and how to modify it.

In the next chapter, we will explore more features that are considered more advanced than the features we explored in this chapter. This includes swept boss/swept cut and lofted boss/lofted cut, both of which require more than a single sketch to be applied. We'll also address adding more reference geometries, such as new planes and coordinate systems.

Questions

Answer the following questions to test your knowledge of this chapter:

1. What are the features in SOLIDWORKS?
2. What are the extruded boss and extruded cut features?
3. What are the fillet and chamfer features?
4. What are the revolved boss and revolved cut features?
5. Create the model shown in the following figure:

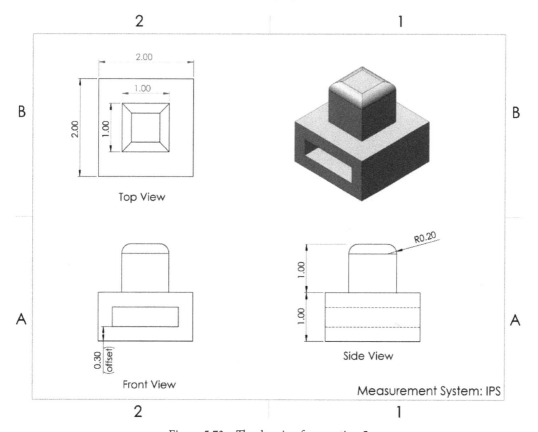

Figure 5.73 – The drawing for question 5

6.   Create the model shown in the following figure:

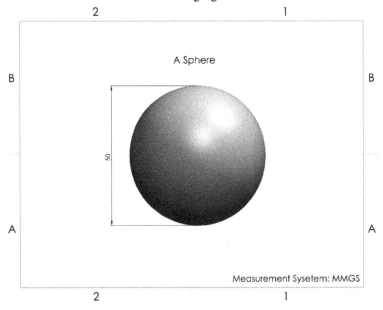

Figure 5.74 – The drawing for question 6

7.   Create the model shown in the following figure:

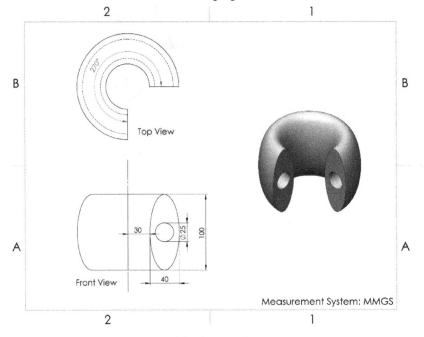

Figure 5.75 – The drawing for question 7

8. Create the model shown in the following figure:

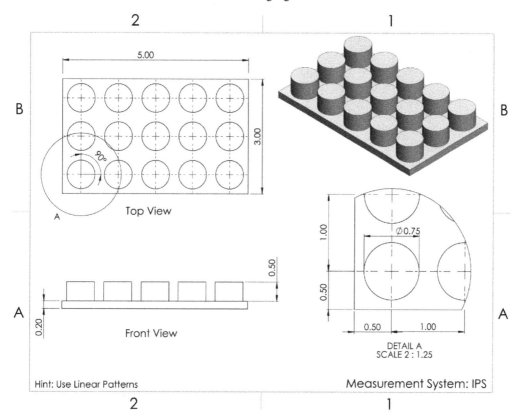

Figure 5.76 – The drawing for question 8

Important Note

The answers to the preceding questions can be found at the end of this book.

6
Basic Secondary Multi-Sketch Features

While the most basic SOLIDWORKS features require that only one sketch is made, others require more than that so that more complex shapes can be modeled. In this chapter, we will cover essential multi-sketch features, such as swept boss and swept cut, and lofted boss and lofted cut. In addition to that, we will define some new planes that are completely different than the default ones.

In this chapter, we will cover the following topics:

- Reference geometries – additional planes
- Understanding and applying swept boss and swept cut
- Understanding and applying lofted boss and lofted cut

By the end of this chapter, we will be able to create complex-looking 3D models compared to what we have built already. They will also enable us to create irregular shapes compared to the previous chapter.

Technical requirements

In this chapter, you will need to have access to SOLIDWORKS software.

Check out the following video to see the code in action: `https://bit.ly/3pZxvo2`

Reference geometries – additional planes

By default, SOLIDWORKS provides us with three planes that we can start sketching on. In addition, we can use any other straight surface as a plane. However, sometimes, we need planes that are different. In this case, we need to introduce our own planes. In this section, we will discuss how we can create additional planes in our 3D space. We will also introduce reference geometries.

Understanding planes, reference geometries, and why we need them

Reference geometries are like the origin points we need for the different planes that we use for sketching. They are used as a base for sketches, features, and coordinate locations. In SOLIDWORKS, reference geometries include planes, coordinate systems, axes, and points. In this section, we will focus on planes.

Whenever we create a sketch, we start by selecting a sketch plane to base our sketch on. Previously, we used the default planes and the surfaces resulting from features. However, in some cases, we may need additional planes that do not exist. To illustrate this, take a look at the following example model.

The following model consists of a cube and a cylinder that intersect. The cube is a normal cube, just like the ones we've made in the previous chapters. However, the cylindrical part has been created with an angle. In the following diagram, we used a new plane (shown as **Plane 1**) to sketch and create that cylinder. Note that the plane is different than the default planes and different than all of the other surfaces.

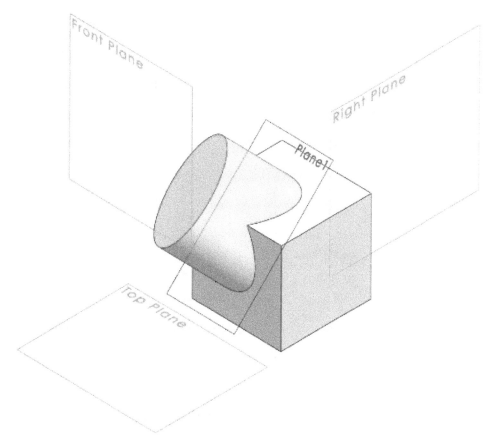

Figure 6.1 – An illustration of the default planes and an additional one

In this section, we will learn how to introduce new planes, such as the one shown in the preceding diagram. But first, we will learn about some of the geometrical principles that define a plane.

Defining planes in geometry

When defining planes in a three-dimensional space, we need to take geometrical principles into account. Hence, before we get practical with SOLIDWORKS, we need to review some of the basic geometrical principles that define a plane. As you may recall, a plane can be understood as infinity and can extend surfaces in all directions. To define a plane, we only need to define a piece of it. There are eight common ways that a plane can be defined in space. The following are the ways in which they can be defined, along with an example of each:

1. **Three points**: Any three points in a space can be connected to define a plane, as shown in the image below:

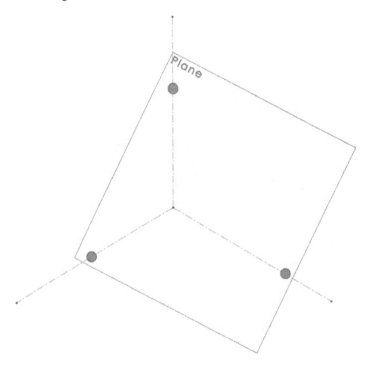

Figure 6.2 – Defining a plane using three points

2. **One point and a line**: Linking a line to a point in space will define a plane, as shown in the image below:

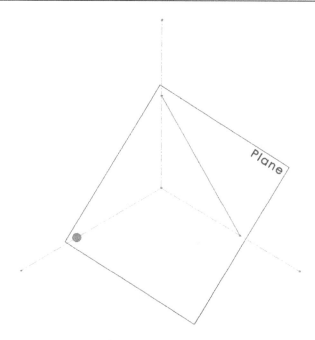

Figure 6.3 – Defining a plane using a point and a line

3. **Two parallel lines**: Any two parallel lines in a space can define a plane, as shown in the image below:

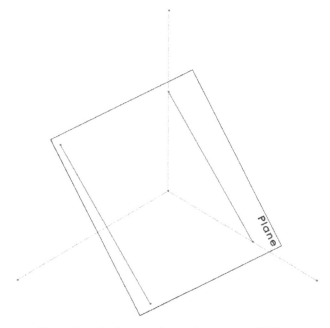

Figure 6.4 – Defining a plane using two parallel lines

4. **Two intersecting lines**: Any two intersecting lines can define a plane, as shown in the image below:

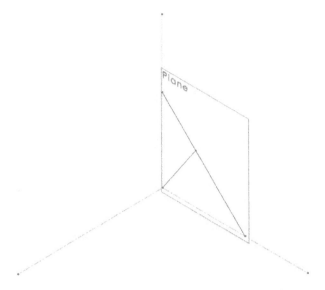

Figure 6.5 – Defining a plane using two intersecting lines

5. **Other planes**: Any existing plane or flat surface can be used to define new planes if we offset it by a certain distance. We can also make the new plane related to the other two. For example, the new plane can be midway between two parallel planes or flat surfaces. The image below highlights a base plane and another plane defined by an offset:

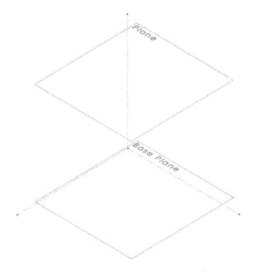

Figure 6.6 – Defining a plane using another plane

6. **A plane and a line**: Mixing an existing plane and a line can result in a new plane. This is done by defining a relationship between the existing plane and a line. These relations can be parallel, perpendicular, or coincident. We can also place an angle or distance between them, as shown in the image below:

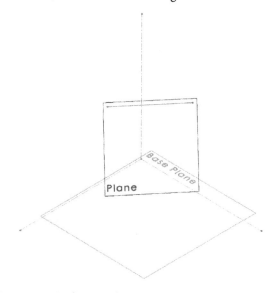

Figure 6.7 – Defining a plane using another plane and a line

7. **A plane and a point**: A plane can define another plane if we offset it. A point can be used to define the offset, as shown in the image below:

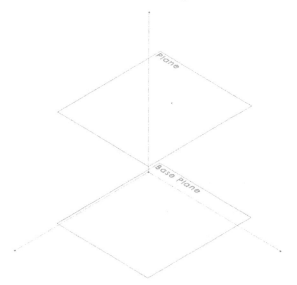

Figure 6.8 – Defining a plane using another plane and a point

8. **A plane and a curve**: Curve tangents can be used to define planes when we wish to relate the new plane to an existing one, as shown in the image below:

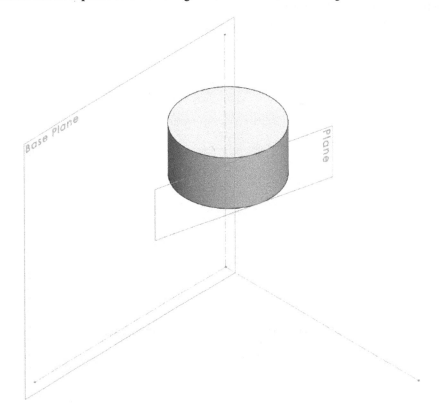

Figure 6.9 – Defining a plane using another plane and a curved surface

Note that these are just the basic ways of defining planes in a space. In most cases, we can manipulate angles, distance, and geometric relations further to generate more diverse planes.

Once we are familiar with how to define a plane from a geometrical perspective, we can easily define a plane in SOLIDWORKS by selecting the different components to define it with. Now, we will learn how to define new planes in SOLIDWORKS.

Defining a new plane in SOLIDWORKS

Now that we know how to define planes in different ways in terms of geometry, we can start learning how to define new planes in SOLIDWORKS. To illustrate this, we will create the following model:

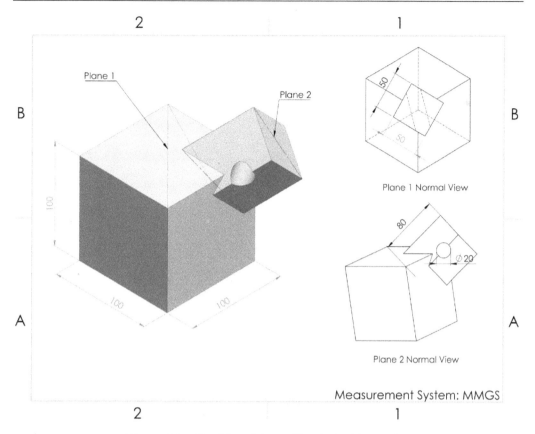

Figure 6.10 – The 3D model we will make in this exercise

Note that, in the preceding model, we have used two additional planes: **Plane 1**, which is in orange, and **Plane 2**, which is in purple. They are defined as follows:

- **Plane 1**: Three points (vertexes)
- **Plane 2**: Two parallel lines (edges) that come from the side extrusion

To create this model, we need to plan, sketch, and apply features. However, we will also create additional reference planes.

Our plan is to create the base cube first and then create a new reference plane. After that, we'll create the side extrusion using the new plane. Then, we'll create the second reference plane. Finally, we'll create a circular hole. Let's start applying this plan by following these steps:

1. Create the base 100 mm cube. We will get the following model by doing so:

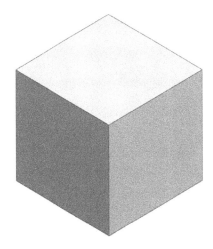

Figure 6.11 – The first step will be to create a 100 cube

2. Now, we can start creating the first reference plane for our model. With the **Features** tab selected, click on the **Reference Geometry** tab and select **Plane**, as follows:

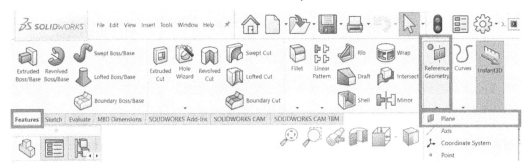

Figure 6.12 – The location of the Reference Geometry command

3. The **PropertyManager** panel will appear on the left-hand side so that we can select our references. Select three points (vertexes), as shown in the following screenshot. Also, you may need to check the last box under **Options**, that is, **Flip normal**, since we want the normal dimension to be outward, as indicated by the blue arrow. We will get the following preview:

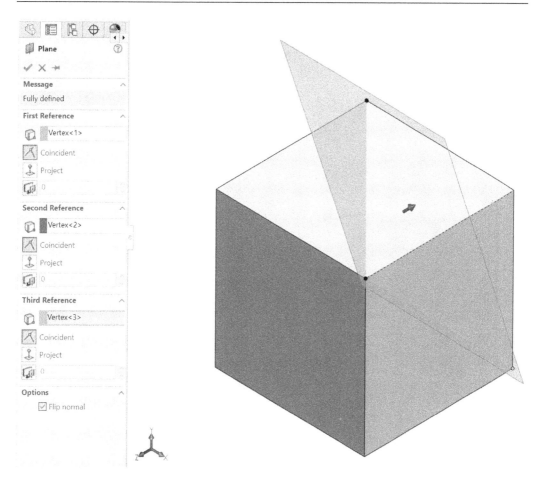

Figure 6.13 – PropertyManager and the plane preview

The options for reference planes are simple. First, we need to select a reference. The reference could be a point, a plane, or a straight or curved surface. Then, we can select the relation of our new plane to that reference.

In the case of the plane we are creating, the first, second, and third references are all points. Also, the relation of these points to the new plane is that they are all coincident.

> **Note**
>
> We do not always need three references; the number of references depends on how we define the plane, as we explained in the *Defining planes in geometry* section. On top of these options, we will be told whether the plane is fully defined.

4. Click on the green checkmark to approve the plane.

5. We can now start sketching on top of the new plane we just created. In terms of sketching, we can treat the new plane just like the other default planes we have. Select the new plane from the design tree and sketch a 50 mm square that's centered on the edge of the cube, as follows:

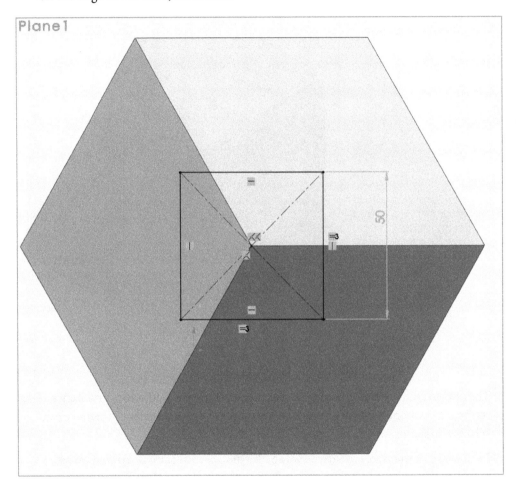

Figure 6.14 – Using the new plane to create new sketches

6. Apply a feature based on the new sketch. We can apply features to new sketches in the same way we apply features to other sketches: based on default or surface planes.

7. Apply the extruded boss tool by extruding the shape by 80 mm. We will get the following shape:

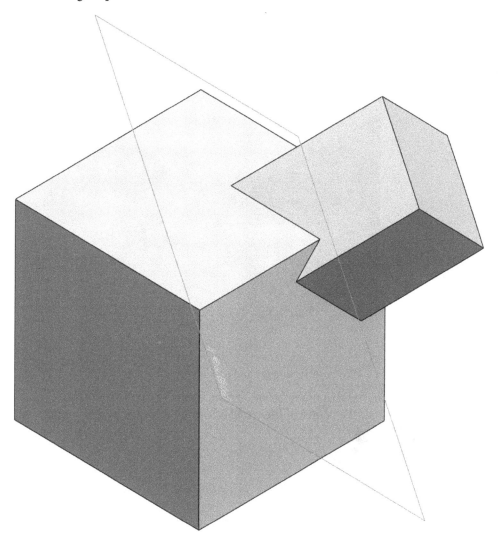

Figure 6.15 – Features can be applied based on sketches from the new planes

8. Now, we will create the second reference plane, which is based on two parallel lines. Select the geometries and select the two parallel edges, as shown in the following screenshot. Then, click on the green checkmark to generate the new plane.

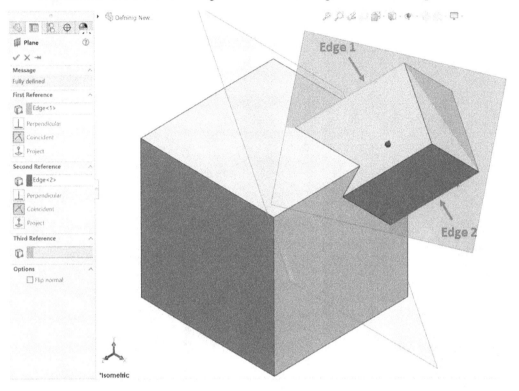

Figure 6.16 – The PropertyManager and preview for our second plane

9. Sketch the new circle with a diameter of 20 mm, as shown in the following screenshot. Note that the center of the circle is located mid-point on the edge.

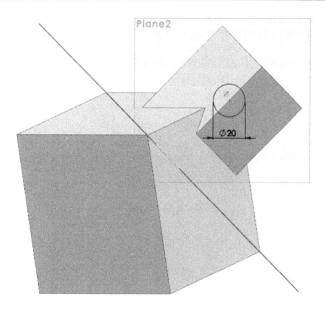

Figure 6.17 – A new circle sketch applied on a new plane

10. Apply the extruded cut feature by setting the end condition to be **Through all-both**. We are doing this since the hole goes through the whole model and in both directions. We will get the following shape:

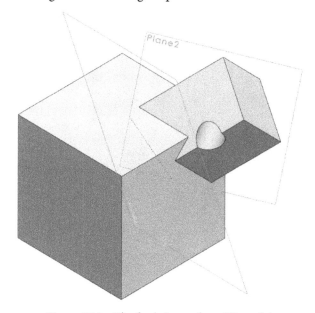

Figure 6.18 – The final shape of our 3D model

This concludes the process of creating the model.

Note that, in our resulting model, we can see that the two planes we introduced are visible. To make the model look cleaner, we can hide the new planes after they have fulfilled our needs. We can do this by right or left-clicking on the plane listing in the design tree and selecting the eye-shaped option, as shown in the following screenshot:

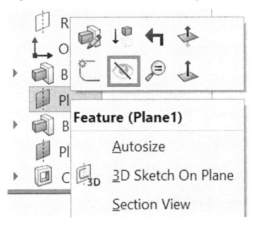

Figure 6.19 – The hide option to hide planes from view

Once we've hidden the visible planes, our cleaner model will look as follows:

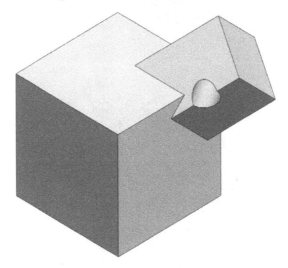

Figure 6.20 – A clean-looking 3D model after hiding the new planes

In this exercise, we created a reference plane with **three points** and **two parallel lines**. If we were to create a new plane based on any of the other six methods, we looked at earlier, for example, another plane, or two intersecting lines, we can simply follow the same steps. The only difference will be the reference selection and the available selected reference relations.

This concludes how to create planes as additional reference geometries in SOLIDWORKS. In this section, we covered the following topics:

- How to access the command for creating new reference planes
- How to create a reference plane based on three points
- How to create a reference plane based on two parallel lines
- How to use a new reference plane to create sketches and features

Now that we know how to generate new plane reference geometries, we can start learning about our next set of features: swept boss and swept cut.

Understanding and applying swept boss and swept cut

Swept boss and **swept cut** allow us to create shapes by sweeping a profile along a path. In this section, we will discuss swept boss and swept cut in detail. These features are opposites and require more than one sketch if we wish to apply them. We will learn about their definitions, how to apply them, and how to modify them.

What are swept boss and swept cut?

Swept boss and swept cut are opposing features; one adds materials, while the other removes materials. Let's talk about them in more detail:

- **Swept boss**: This adds materials by sweeping a profile shape on a designated path.
- **Swept cut**: This removes materials by sweeping a profile shape on a designated path.

The following diagrams highlight the effect of the swept boss and swept cut features in a better way:

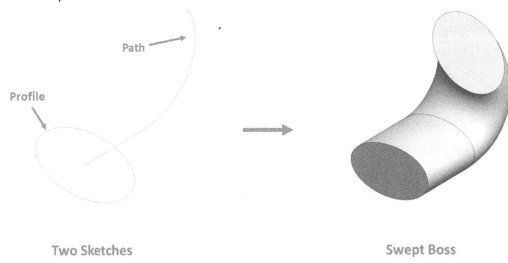

Figure 6.21 – An illustration of what the Swept Boss feature does

The preceding diagram shows **Swept Boss**, while the following diagram shows **Swept Cut**. Note that both features require two sketches before they can be applied:

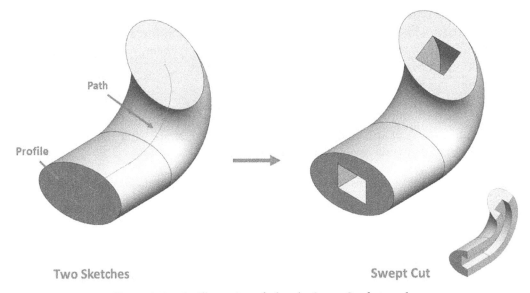

Figure 6.22 – An illustration of what the Swept Cut feature does

Compared to the extruded boss and extruded cut, swept boss and swept cut provide us with more flexibility when it comes to extruding a shape. While extruded boss and extruded cut add and remove materials directly perpendicular to the sketch plane, swept boss and swept cut allow us to guide the extrusion as we see fit. Let's start applying them, beginning with swept boss.

Applying swept boss

In this section, we will demonstrate how to apply the swept boss feature. To do this, we will create the model shown in the following diagram:

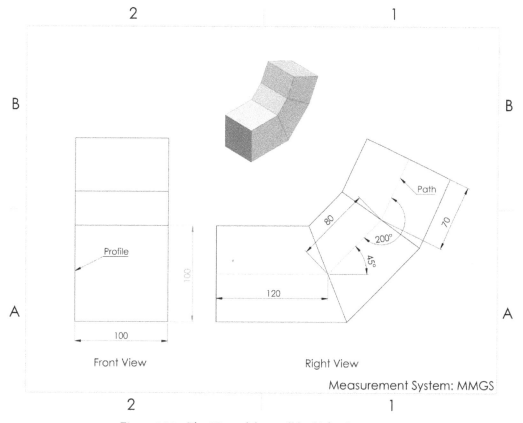

Figure 6.23 – The 3D model we will build for this exercise

To model this shape, we will go through the stages of planning, sketching, and applying features. Our plan will be to create a rectangular profile and then the path. After that, we will apply the swept boss feature. Follow these steps to do so:

1. Select **front plane** and draw a square centered at the origin. This will represent the profile. Set the side length to 100 mm, as shown in the following diagram. Then, exit sketch mode.

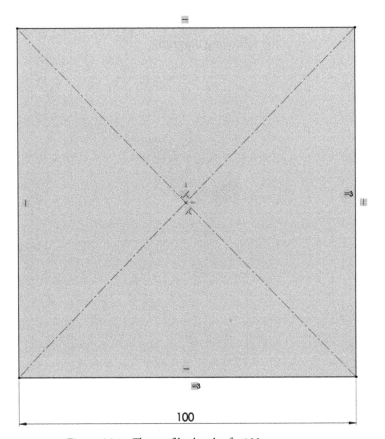

Figure 6.24 – The profile sketch of a 100 mm square

2. Select **right plane** and draw the path that's shown in the right-hand view of our diagram. We can make the start of the path the origin. Our sketch should look as follows. Exit sketch mode after that.

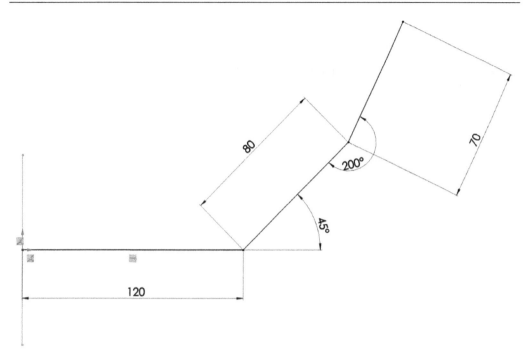

Figure 6.25 – The path sketch

3. Now, we will apply the swept boss feature. Go to the **Features** tab in the command bar and select the **Swept Boss/Base** command, as highlighted in the following screenshot:

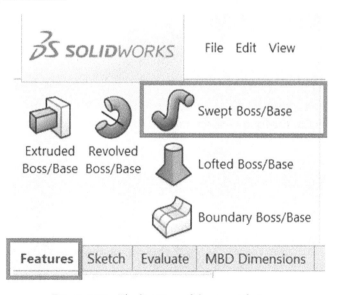

Figure 6.26 – The location of the swept features

4. **PropertyManager** will appear on the left-hand side so that we can select **Profile and Path**. We can then make the following selections:

    a) **Profile**: Select the first rectangle we drew in *step 1*. We can select it by clicking on it on the canvas.

    b) **Path**: Select the curve we sketched in *step 2*.

Once we've selected **Profile and Path**, we will see a preview of our shape, as shown in the following screenshot:

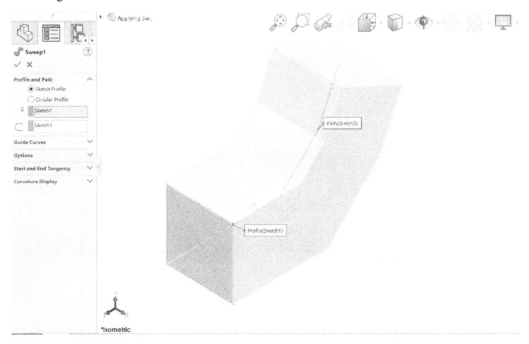

Figure 6.27 – PropertyManager and a preview of the swept boss feature

5. Click on the green checkmark to approve this feature. This will give us the final shape, as shown in the following screenshot:

Figure 6.28 – The resulting 3D model after applying swept boss

This concludes using the swept boss feature. Before we move forward with swept cut, let's talk about one of the aspects of the path and explain some of the other options that are available for the swept boss feature.

Note that for basic sweeps, the path must either intersect the profile itself or its extension so that it's captured by the feature. In the following screenshot, we have highlighted different types of acceptable and unacceptable paths. The unacceptable paths are indicated with an X sign.

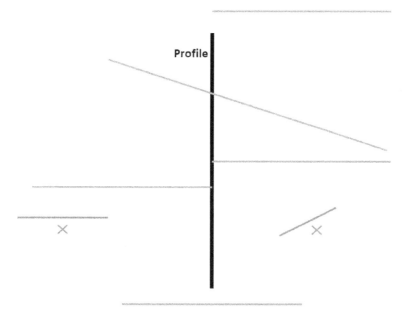

Figure 6.29 – An illustration of the acceptable paths location for the swept features

Now that we know how to create a basic swept boss, we can discuss the different options we can use to create more complex swept boss applications.

Swept boss feature options

Now, let's examine some of the other options that are available with the swept boss feature that we didn't utilize in the previous exercise. We will cover the options we missed from top to bottom, as shown in the following screenshot:

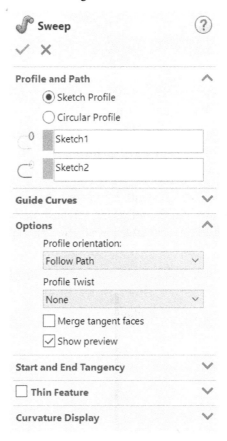

Figure 6.30 – The PropertyManager swept feature showing many options

The following is a brief explanation of these options:

- **Circular Profile**: Under **Profile and Path**, we can select the **Circular Profile** option. This makes it easier for us to create the swept feature when the profile is circular and goes through the middle of the path. When selecting the circular profile option, a profile sketch will not be needed.

- **Guide Curves**: With this option, we can add more sketches to help to guide our swept boss feature. This is helpful when the swept boss feature does not have a regular shape that matches the profile. However, it is an advanced feature that we will not use at this level.

- **Profile orientation**: This helps us to determine how the profile moves with the path. There are two options available here:

  a) **Follow Path**: This will make the profile tilt along with the path.

  b) **Keep Normal Constant**: This will keep the profile facing the same direction as it moves on the path.

- **Profile Twist**: This can be used to create shapes such as spiral springs or any type of twisted shape since it will allow the profile to twist around the path. The following screenshots are examples of using **Profile Twist** in two different ways. One was done by twisting a square by 90 degrees, as follows:

Figure 6.31 – A resulting sweep with the profile twist option

The other was done by twisting a circle by three revolutions around the path, as shown in the following screenshot:

Figure 6.32 – The Profile Twist option can also be used to create spirals

- **Start and End Tangency**: This ensures that the sweep is normal for the path toward the start or endpoint.

- **Thin Feature**: This will allow us to only sweep boss a thin layer around the profile instead of the whole enclosure. This is similar to the thin feature option in the extruded boss feature.

- **Curvature Display**: This option allows us to analyze the curvature of our swept feature better. This is done by adding visual elements such as mesh preview, zebra stripes, and curvature combs.

This concludes this exercise, which was all about applying the swept boss feature. We covered the following topics in this section:

- How to create sketches that can use swept boss

- How to apply the swept boss feature

- The acceptable paths that can be used with the swept boss feature

- The different options that can be used with the swept boss feature

Now, we can start learning about the opposite of swept boss, that is, swept cut.

Applying swept cut

In this section, we will demonstrate how to use the swept cut feature. To do this, we will create the model shown in the following diagram by adding a swept cut to the model we created earlier:

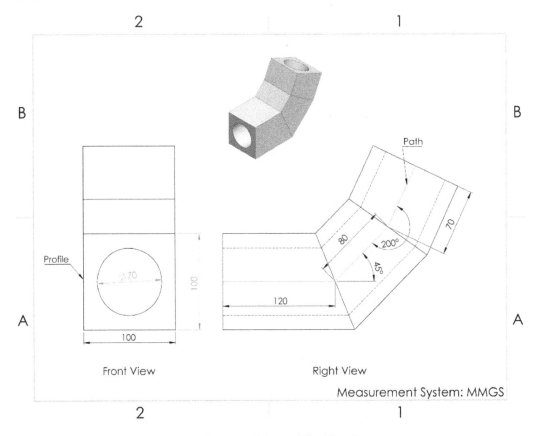

Figure 6.33 – The 3D model we will build in this exercise

To model this shape, we will go through the procedure of planning, sketching, and applying the feature. Our plan will be to create the profile on our existing swept boss. After that, we will apply the swept cut feature by following the same path we had for the swept boss. To act on this plan, we will follow these steps:

1. Select the face shown in the front view as a sketch surface. Then, sketch a circle with a diameter of 70 mm, as shown in the preceding diagram. The sketch will look as follows. Exit sketch mode afterward.

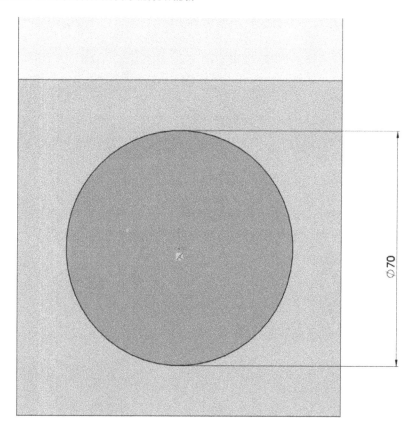

Figure 6.34 – The profile sketch

> **Note**
> The path for this cut is the same path that we used to apply the swept boss feature. Hence, we don't need to create another path. Instead, we can reuse the one we already have. A reused sketch will be indicated in the design tree with a small icon of an open hand.

2. Now, let's apply the swept cut feature. From the **Features** tab, select the **Swept Cut** command, as shown in the following screenshot:

Figure 6.35 – The location of the Swept Cut feature

3. Select the appropriate profile by selecting it from the canvas, just like we did with the swept boss feature.

4. We can select the path from the design tree that appears on the canvas. Expand that **Design Tree** and look for the **Sketch** we used for the **Path** under the **Sweep** feature. Then, select the sketch that corresponds to the path, as shown in the following screenshot. We will see a preview of the swept cut feature.

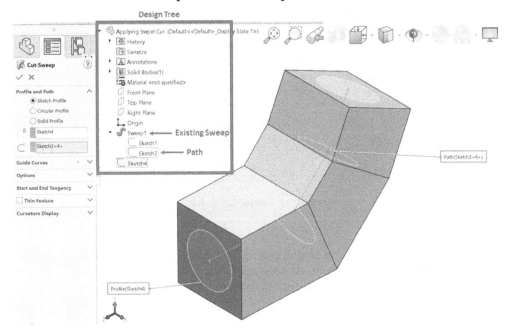

Figure 6.36 – The design tree can be used to select sketches for the path or profile

5. Click on the green checkmark to apply the swept cut feature. This will result in the following model:

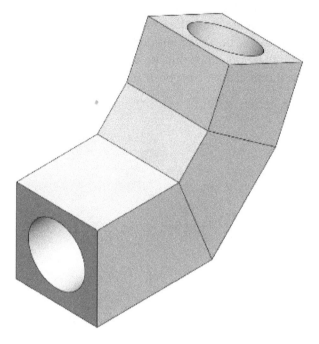

Figure 6.37 – The final 3D model after applying the swept cut

This concludes this exercise, which was all about applying the swept cut feature. The additional options that are available for the swept cut feature are the same as those for the swept boss festure. We covered the following topics in this section:

*   How to apply the swept cut feature

*   How to reuse a sketch in more than one feature

Now that we know how to apply the swept boss and swept cut features, we need to know how we can modify them.

Modifying swept boss and swept cut

Modifying features in SOLIDWORKS is done in the same way that it's done for all features; that is, by right- or left-clicking on the feature in the design tree and selecting **Edit Feature**. In addition to editing the feature, we can also edit the sketches that are guiding the feature. In the case of the swept boss and swept cut features, this includes the profile and path sketches. To demonstrate this, we will modify our previous model so that it looks as follows. These modifications have been annotated.

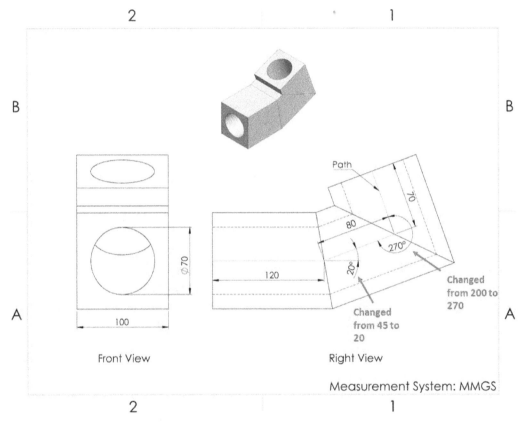

Figure 6.38 – The modifications we will apply for this exercise

Note that this model is very similar to the one we created in the previous exercise. The only difference is the path. Hence, we will only modify the sketch we used for the path. To do that, follow these steps:

1. Find the sketch path in the design tree by expanding the design tree listing of the sweep boss feature. Right-click on the path sketch and select **Edit Sketch**, as highlighted in the following screenshot. Note that if a sketch is being used in more than one sketch, a small hand icon will appear next to it.

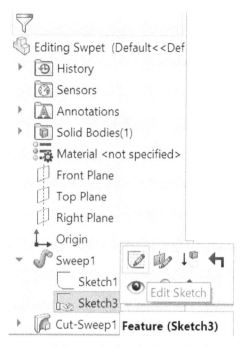

Figure 6.39 – We can edit sketches from the design tree

2. Adjust the sketch by double-clicking on the angle values and changing them so that they read as follows:

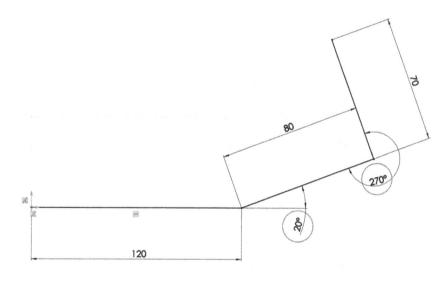

Figure 6.40 – We can adjust the sketches in the canvas once selected from the design tree

3.   Exit sketch mode. You will see that the model now looks as follows. Note that, by editing the sketch, we ended up editing the swept boss and swept cut features since both are utilizing the same sketch.

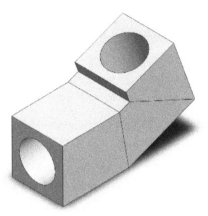

Figure 6.41 – The final 3D model after modifying the path sketch

> **Tip**
> A shortcut to edit the dimension is to double-click on the sketch from the design tree and then adjust the displayed dimensions directly without getting into edit mode.

This concludes our exercise on editing swept boss and swept cut. In this section, we covered the following topics:

- How to edit sketches that are being used for the swept paths
- How to identify sketches in the design tree that are being used in more than one feature

In this section, we covered how to apply and modify the swept boss and swept cut features. Next, we will cover another set of features: lofted boss and lofted cut.

Understanding and applying lofted boss and lofted cut

Lofted boss and **lofted cut** allow us to create a shape by sketching sections of it. In this section, we will discuss the features of the lofted boss and lofted cut. These features are opposites, and so we require more than one sketch if we wish to apply them. We will learn about their definitions, how to apply them, and how to modify them.

What are lofted boss and lofted cut?

With lofted boss and lofted cut, we are able to add or remove materials based on multiple cross sections. Let's talk about them in more detail:

- **Lofted boss**: This adds materials by linking different cross sections together. This includes the start and end of the shape. If we choose to, we can link these cross-sections with guide curves.

- **Lofted cut**: This removes materials by linking different cross sections together. This includes the start and end of the cut shape. If we choose to, we can link these cross sections with guide curves.

These features are polar opposites: one adds materials, while the other removes materials in the same way. The following diagrams illustrate these features:

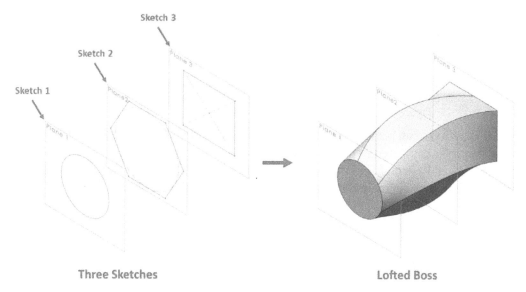

Figure 6.42 – An illustration highlighting the Lofted Boss feature

The preceding diagram shows **Lofted Boss**, while the following diagram shows **Lofted Cut**. Note that both features require at least two sketches if we wish to apply them. We can add as many sketches as we wish in order to define the sections:

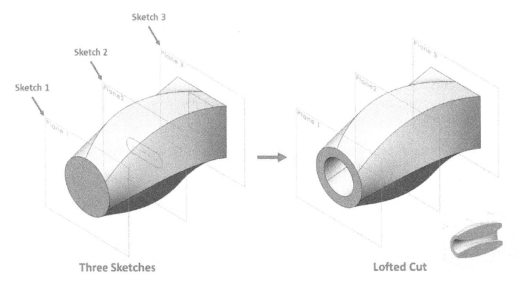

Figure 6.43 – An illustration highlighting the Lofted Cut feature

Lofted boss and cut can provide us with a unique way of controlling how we add or remove materials compared to other features, such as swept boss and swept cut and extruded boss and extruded cut. Now that we know what lofted boss and lofted cut are used for, we can start applying them. Let's start by applying a lofted boss.

Applying lofted boss

In this section, we will cover how to apply the lofted boss feature. To illustrate this, we will create the model shown in the following diagram:

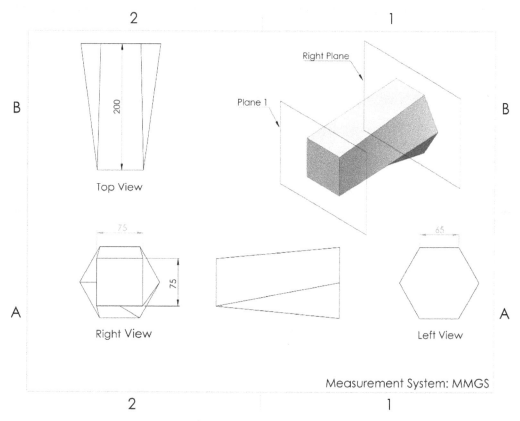

Figure 6.44 – The 3D model we will build in this exercise

To create this model, we will go through the steps of planning, sketching, and applying features. Our plan will be to utilize two planes to create two sketches and then apply the lofted boss feature. In this exercise, we'll have to create one additional reference plane based on the default right plane. Let's start by following these steps:

1. The first step is to define our new reference planes. Select the **Plane** sub-command, which can be found under **Reference Geometries**. Then, create **Plane 1** by offsetting a distance of 200 mm from **Right Plane**, as shown in the top view of the diagram. The option for plane creation is shown in the following screenshot, and can be found alongside the final shape of the two planes:

Figure 6.45 – PropertyManager for our new plane

After approving the new plane, it will appear parallel to **Right Plane**, as shown in the following diagram:

Figure 6.46 – The new plane after being defined

2. Since we are using the lofted boss feature, we will create two different sketches in the two planes. Select **Plane1**, which we created previously, and sketch a square whose sides are equal to 75 mm, as shown in the right-hand view of the preceding drawing. The resulting sketch will look as follows. Exit sketch mode after that.

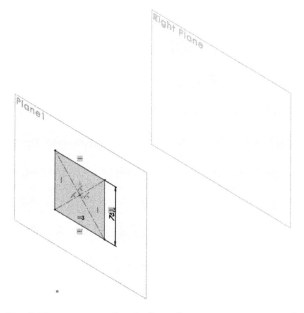

Figure 6.47 – A 75 mm square sketched on plane1 representing our first profile

> **Tip**
>
> You can show **Right Plane** in the canvas by right- or left-clicking on it in the design tree and selecting **Show**.

3. Select the right plane and sketch a hexagon whose sides are equal to 65 mm, as shown in the left view of the diagram. The resulting sketch will look as follows. Exit sketch mode after that.

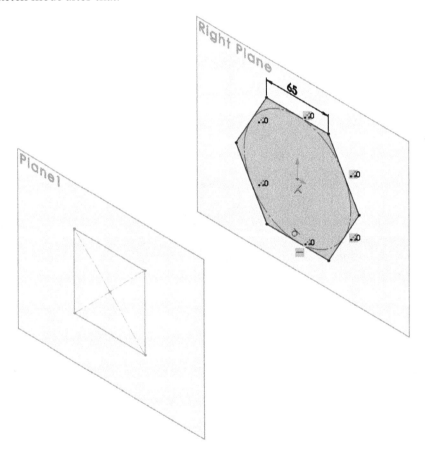

Figure 6.48 – A hexagon sketched on the right plane representing our second profile

4. Now, we will apply the lofted boss feature to connect the two sketches. Select the **Lofted Boss/Base** command from the **Features** tab, as shown in the following screenshot:

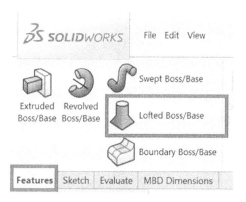

Figure 6.49 – The location of the Lofted Boss feature

5. After selecting the command, we will get the command's **PropertyManager** options on the left-hand side, as shown in the following screenshot:

Figure 6.50 – PropertyManager showing the profile selection

For the profiles, select the square first, then the hexagon. Note that the selection is order-sensitive. Once we do that, we will get the following review on the canvas. Note that there is one guideline in the preview to help us to define our loft. This line is controlled by the two green endpoints. To adjust it, we can drag the point to another location. This will change the shape of the loft. Take your time and adjust the guideline so that it looks as follows:

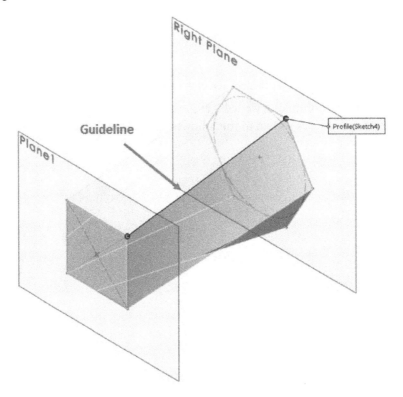

Figure 6.51 – A preview of the lofted boss showing a guideline

Note

The initial positioning of the guideline is determined by where you click on the sketch to select the profile.

6. Click on the green checkmark to apply the lofted boss feature. The result will be as follows:

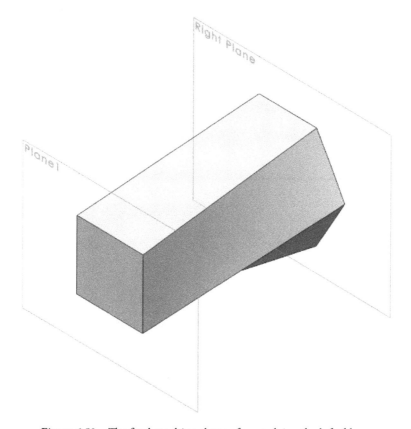

Figure 6.52 – The final resulting shape after applying the lofted boss

This concludes our exercise on the lofted boss feature. In this exercise, we only covered the most basic lofted shape. However, this feature has many advanced options that we did not get around to using. We will explore these options next.

Lofted boss feature options

In the feature's **PropertyManager**, we will be able to find all of the feature's options. The following screenshot highlights **PropertyManager** for the lofted boss feature:

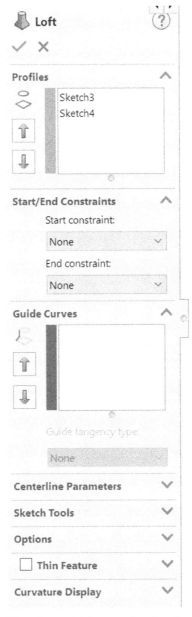

Figure 6.53 – The Lofted Boss PropertyManager showing the available feature options

The following is a brief explanation of these options:

- **Start/End Constraints**: This gives us more control over the areas that are close to the profile sketches. Under each, we have the following options:

  a) **Direction vector**: This pushes the loft toward a specific direction, as per an existing vector. To apply this, we may need to create additional lines to push the loft.

  b) **Normal to profile**: This pushes the loft in a direction that's normal/perpendicular to the existing profile we used to build the loft.

- **Guide Curves**: This gives much more flexibility in terms of how the loft is constructed. However, we still need to create more curves from different sides to guide the loft. The following screenshot shows our previous loft with multiple guiding curves:

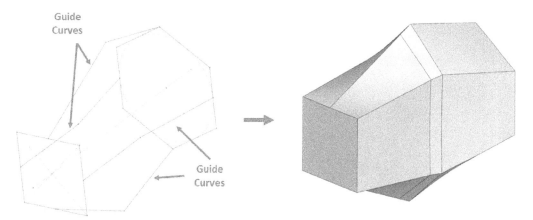

Figure 6.54 – An illustration of the guide curves function

- **Thin Feature**: This changes the loft so that it's only lofting the shell rather than the whole shape.

This concludes the various options of the lofted boss feature. In this section, we covered the following topics:

- How to create different profiles with the lofted boss feature
- How to apply the lofted boss feature
- The different options for the lofted boss feature

Now that we know how to apply the lofted boss feature, we can start learning how to apply the lofted cut feature.

Applying lofted cut

The lofted cut feature works the same as the lofted boss feature, except that it has the opposite effect. To show you how this feature works, we will build on the previous model and create the model shown in the following diagram:

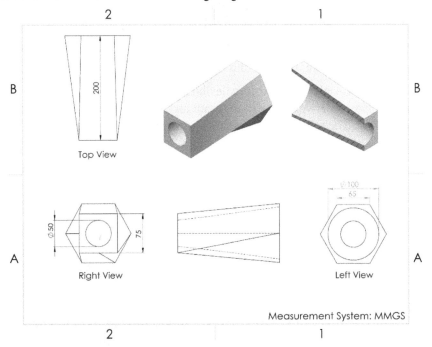

Figure 6.55 – The 3D model we will build in this exercise

Note that, for this model, we will only create an internal cut from the previous model. This cut is governed by two circles on each end. Like we did previously, we will create the model by planning, sketching, and then applying features. We will use the existing end faces for our shape to sketch two circles. Then, we will apply the lofted cut feature. Follow these steps to implement this plan:

> **Tip**
> Since both of our sketches are located on existing faces, we can hide the two visible planes and sketch on the faces that have been formed from the features instead. To hide a plane, we can right-click on it from the design tree and select the **Hide** option, which is the small eye icon.

1. Select the square face as a sketch plane and sketch a 50 mm circle, as shown in the following diagram. Use the origin as the center of the circle. Exit sketch mode after that.

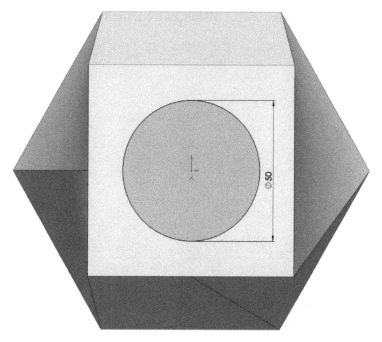

Figure 6.56 – The sketch of our first profile

2. Select the hexagonal face as a sketch plane and sketch a 100 mm circle, as shown in the following diagram. Use the origin as the center of the circle. Exit sketch mode after that.

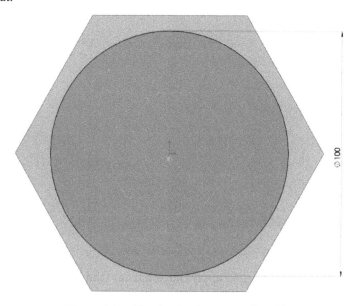

Figure 6.57 – The sketch of our second profile

3.   Now, let's apply the feature. Select **Lofted Cut** from the **Features** tab, as shown in the screenshot:

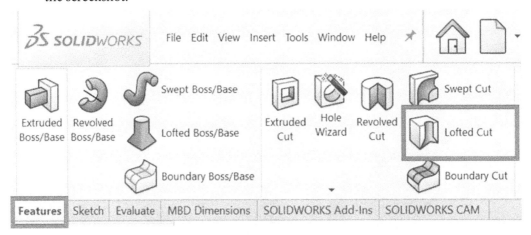

Figure 6.58 – The location of the Lofted Cut command

4.   The **PropertyManager** options for the lofted cut are the same as they are for the lofted boss feature. Under **Profiles**, select the two circles we sketched earlier. We will get the following preview. Note that since our two profile sketches are circles, the guideline provided will not have an effect.

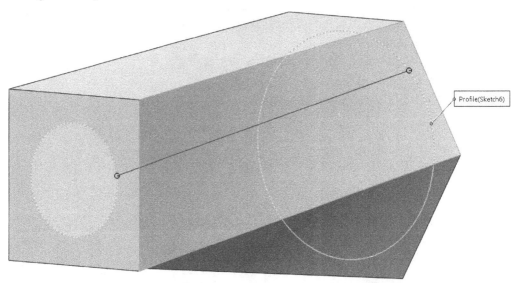

Figure 6.59 – The preview of the lofted cut

5.   Click on the green checkmark to apply the lofted cut feature. The final shape will be as follows:

Figure 6.60 – The final 3D model after applying the lofted cut

We can use the cross-section viewing feature at the top of the canvas to view the shape from the inside as well. The result of doing this is shown in the following screenshot:

Figure 6.61 – A cross section of the model showing the cut

6. To create a section view, click on the icon shown in the following screenshot. Then, adjust the **Section View** cross in **PropertyManager**.

Figure 6.62 – The Section View command on top of the canvas

This concludes our exercise of applying the lofted cut feature. In this section, we covered the following topics:

- How to apply the lofted cut feature
- The effects that the provided guidelines have on circular profiles
- How to make a section view

Now that we know how to apply these two features, we need to know how we can modify them.

Modifying lofted boss and cut

Modifying the lofted boss and lofted cut features follows the same procedure that we follow to modify any other feature. We can right-click on the feature to modify its options. We can also modify the sketches of the profile to adjust the shape of the loft.

Before we conclude this section, let's cover a key aspect of the lofted boss and lofted cut features: guide curves.

Guide curves

Guide curves provide us with much more flexibility when it comes to lofted features that we lack when applying the feature without them. Because of that, we will create the shape shown in the following diagram to learn about one way of using **Guide Curves**:

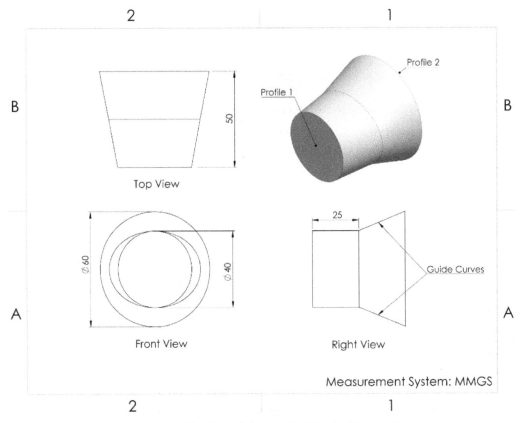

Figure 6.63 – The 3D model we are building in this exercise

Let's go through our usual procedure of creating models: planning, sketching, and applying features. Our plan will be to create the two profiles and then create the two guidelines. After that, we will apply the lofted boss feature. Follow these steps to implement this plan:

1. Create an extra plane by offsetting the front plane by 50 mm. Then, create the two circle profiles that are shown in the following diagram. The smaller diameter is 40 mm, while the larger one is 60 mm.

> **Tip**
>
> We can hide **Front Plane** and **Plane 1** to make our canvas clearer and less crowded.

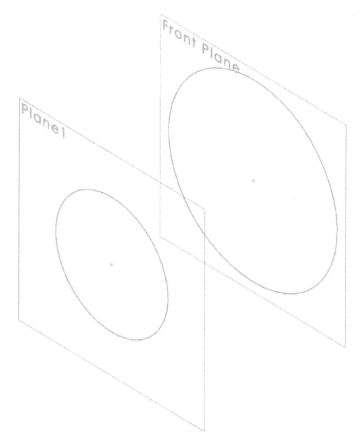

Figure 6.64 – The sketches of the two profiles

2. Using the right plane as a sketch plane, sketch the upper guide curve, as shown in the following diagram. Note that to fully define the sketch, the endpoints of the guide curve should have a **Pierce Relation** with the profile, as highlighted in the following diagram:

> **Note**
> The guide curves must intersect with the profiles.

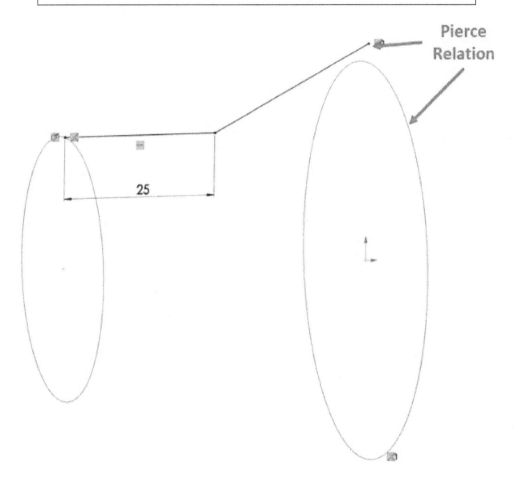

Figure 6.65 – The sketch of the guide curve

3. Mirror the first guide curve for the other side, as shown in the following diagram. We are doing this because the lower part of the curve is the same as the top one. Then, exit sketch mode.

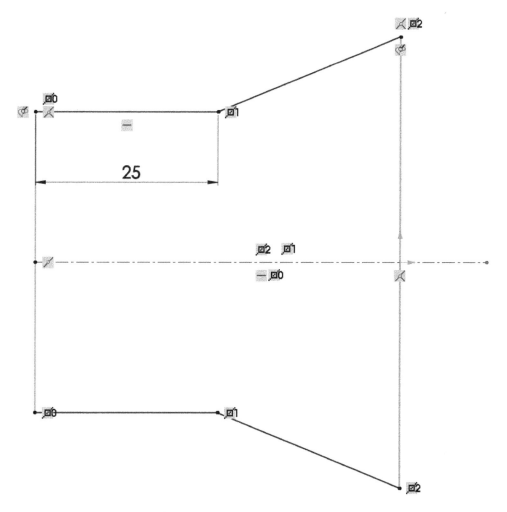

Figure 6.66 – The final two guide curves required for the 3D model

4. Now, let's apply the lofted boss feature. We will use all of the profiles and the guide sketches to apply it. Select the **Lofted Boss** command. Then, select the two circular profiles.

5. In the **Guide Curves** section, select the two guide curves we created. Note that we created the two guide curves under one sketch. Hence, when selecting one curve from the canvas, we will see the window that's shown in the following screenshot. As we can see, we can select the **Open Loop** option and then click on the green checkmark to apply it.

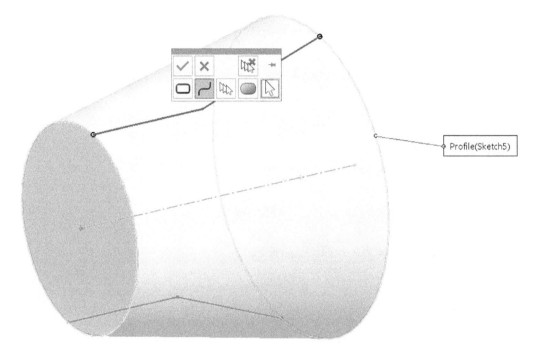

Figure 6.67 – Preview and selection procedure of one guide curve

> **Note**
> Our loft will shift toward the guide curve.

6. Following the same procedure that we completed in *step 5*, select the other guide curve. After that, our preview will look as follows:

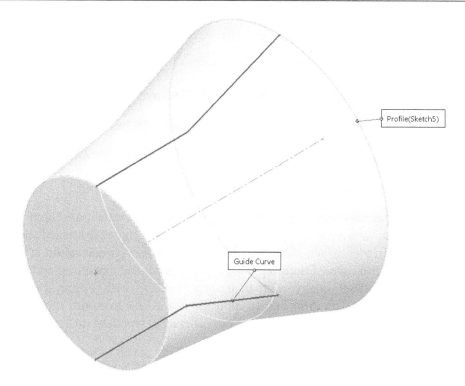

Figure 6.68 – The final preview after both guide curves are applied

7. Click on the green checkmark to apply the loft. The resulting model will be as follows:

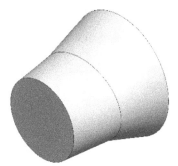

Figure 6.69 – The final resulting shape

Note that we have only applied two guide curves to govern the loft from the upper and lower sides. Hence, from the unguided side, we'll notice that the resulting shape is more elliptical. If we need more guidance when it comes to shape, we can increase the number of guide curves as we see fit. There is no limit to the number of profiles and guide curves we can have.

This concludes this section on lofted guide curves. We covered the following topics in this section:

- How to apply guide curves to control our loft
- How to select a part of a sketch and make it a guide curve

In this section, we covered the lofted boss and lofted cut features, how to apply them, and how to modify them. We also learned about guide curves.

Summary

In this chapter, we learned about a set of features that allow us to create more complex 3D models than what we were able to create in the previous chapters. We learned about plane reference geometries, which allow us to add new reference planes in addition to the default ones. We also covered the swept boss and swept cut and lofted boss and lofted cut features. Each feature set requires more than one sketch if we wish to apply them. For each, we learned their definition, how to apply them, and how to modify them. The features that we covered in this chapter allow us to generate 3D models, such as flexible tubing and irregularly shaped casings.

In the next chapter, we will learn about mass properties, which allow us to assign materials and calculate different properties, such as the mass of our 3D models.

Questions

Answer the following questions to test your knowledge of this chapter:

1. What are the eight methods of defining new planes?
2. Why do we need to define new planes?
3. What are swept boss and swept cut?
4. What are lofted boss and lofted cut?
5. Create the following model:

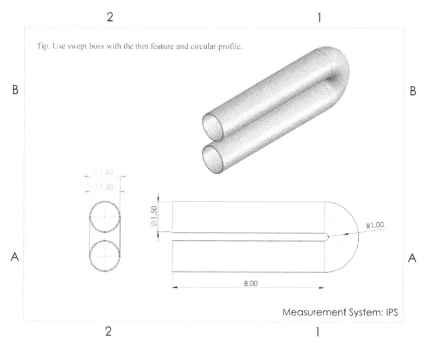

Figure 6.70 – The 3D model for question 5

6.   Create the following model:

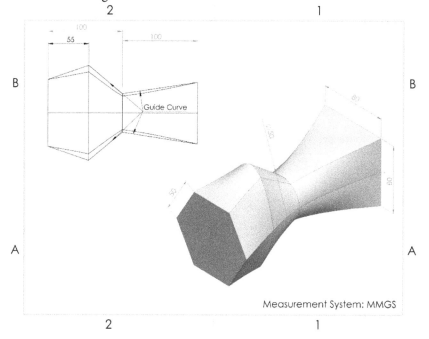

Figure 6.71 – The 3D model for question 6

7. Create the following model:

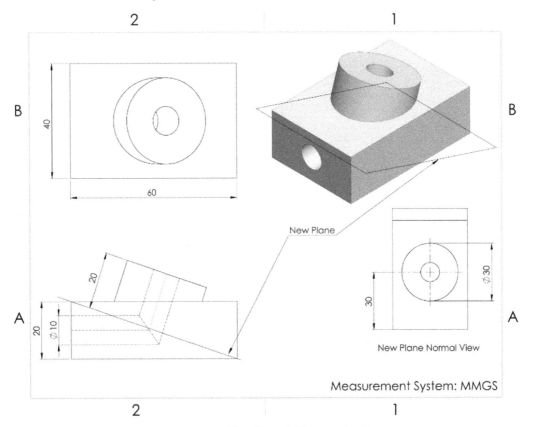

Figure 6.72 – The 3D model for question 7

Important Note
The answers to the preceding questions can be found at the end of this book.

Section 4 – Basic Evaluations and Assemblies – Associate Level

After building 3D models of parts, we might be required to assign them materials, evaluate their mass properties, and put them together to form an assembly. This section will introduce all that and cover everything that is expected of the associate level. These include assigning materials, evaluating mass properties, and applying standard mates to link different parts in an assembly.

This section comprises the following chapters:

- *Chapter 7, Materials and Mass Properties*
- *Chapter 8, Standard Assembly Mates*

7
Materials and Mass Properties

Whenever we design or model an object, we have to consider the structural materials to work with. In other words, should the object be made of steel, iron, plastic, wood, or any other materials? SOLIDWORKS provides a library with a variety of materials that we can choose from. It also provides tools that we can use to find mass properties of the object, such as the volume, mass, or center of mass of the modeled object.

The following topics will be covered in this chapter:

- Reference geometries – defining a new coordinate system
- Assigning materials and evaluating and overriding mass properties

By the end of this chapter, you will be able to assign materials to your parts and evaluate different associated mass properties. This will include finding the mass and the volume of a 3D model. We will also cover how to introduce a new coordinate system to your model. These skills will help your teams decide on what materials to choose for the products you are designing. It will also help you estimate different costs associated with materials and their related aspects, such as transportation and storage.

Technical requirements

This chapter will require that you have access to the SOLIDWORKS software.

Check out the following video to see the code in action: `https://bit.ly/3dSLQNG`

Reference geometries – defining a new coordinate system

By default, SOLIDWORKS provides us with one coordinate system. This system is centered on the origin. The origin is also the starting point of the three axes: X, Y, and Z. In certain cases, we require another base of a coordinate system. In this section, we will explore why we need new coordinate systems and how to define them. This will help us calculate different properties, such as the center of mass from a different perspective, which is a key skill that's required for collaborative work and a SOLIDWORKS professional.

What is a reference coordinate system and why are new ones needed?

In SOLIDWORKS, we can understand **reference geometries** as ones we can use as a base or as a reference for something else. For example, a plane is considered a reference geometry that we use as a base for sketches. A coordinate system is a reference geometry since we can use it as a base to determine specific locations within the canvas. In this section, we will learn what makes a coordinate system and how to define a new one in SOLIDWORKS.

In our sketch creation process, we always link our sketch to the origin. The origin is the base point for our coordinate system, which extends through the three axes: **X, Y, and Z**. In the lower-left corner of the SOLIDWORKS canvas, we can see the direction of the three axes:

Figure 7.1 – The indication of the axis that was found in the canvas

Using the coordinate system, we can locate any point in the canvas according to its X, Y, and Z coordinates. When in sketch mode, the lower-right corner of the canvas will show us the location of the cursor according to the **X**, **Y**, and **Z** locations. The following screenshot highlights where we can find these coordinates:

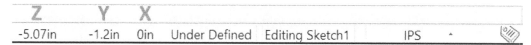

Figure 7.2 – The location of the cursor is indicated according to the axes below the canvas

> **Note**
>
> The order of the X, Y, and Z coordinates, as shown in *Figure 7.2*, will change according to the sketch plane.

Since SOLIDWORKS already provides us with a default coordinate system, why do we need additional ones? Here are two reasons why we may need a new coordinate system:

- **To determine the coordinates according to a different coordinate system with different orientations**: For example, if we measure the center of mass of an object, the measurement will be relative in the X, Y, and Z locations for the default coordinate system. With additional coordinates systems, we can determine the center of mass according to a different coordinates system of our choosing. We will explore this later in this chapter.

- **To switch our views of directions**: In certain applications, we may need to redefine our axes to change our sense of direction as we are creating a model. Additional coordinate systems can help us accomplish that. Also, when working with different people in a specific design, pointing toward different coordinate systems can ease communication.

We have just covered reference coordinate systems and why we may need additional ones for certain applications. Now, we will start learning how to define a new coordinate system within SOLIDWORKS.

How to create a new coordinate system

In this section, we will learn how to introduce a new **coordinate system**. To highlight this, we will create a simple triangular prism and introduce a **New Coordinate System**, as shown in the following diagram:

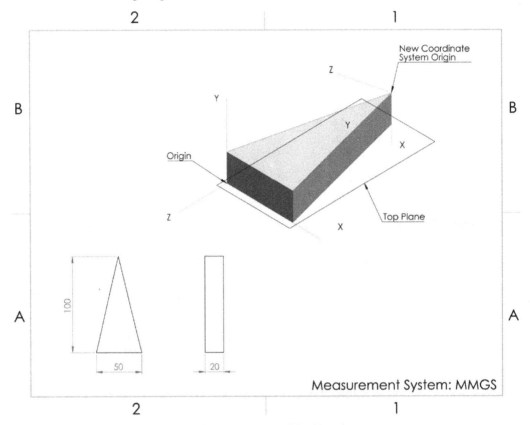

Figure 7.3 – The drawing we will build in this exercise

We will start creating this model by planning, sketching, and applying features. However, note that since we are familiar with the steps and with creating various shapes already, we will only explain this process in brief. Our plan for this model will be to start by creating the overall simple shape first and then introduce the new coordinate system. Follow these steps:

1. Sketch the isosceles triangular base using the top plane and apply extruded boss with the dimensions indicated in the preceding diagram. We should have the shape shown in the following screenshot:

Figure 7.4 – The first step is to build the 3D model

2. Under the **Features** tab, select **Coordinate System** under **Reference Geometries**, as shown in the following screenshot:

Figure 7.5 – The location of the new Coordinate System command

3. **PropertyManager** will appear on the left, showing the options that we need to define our new **Coordinate System**. To define a new coordinate system, we have to specify the origin and two axes. Set the origin by selecting the indicated vertex, as shown in the following screenshot.

4. We can do the same thing with **X axis** and **Y axis** by selecting edges. Note the arrows next to each axis selection of **PropertyManager**; they allow us to switch the direction of the axis. We can flip the direction of the axis until we get the right orientation, as shown in the following screenshot:

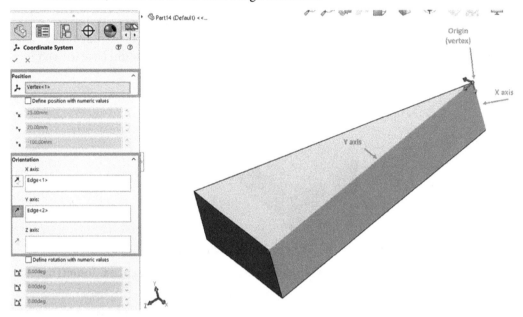

Figure 7.6 – The preview of the new coordinate system and its PropertyManager

5. Click on the green checkmark to approve the new coordinate system. Once we apply the new system, it will be shown on the model as **Coordinate System 1**, as follows:

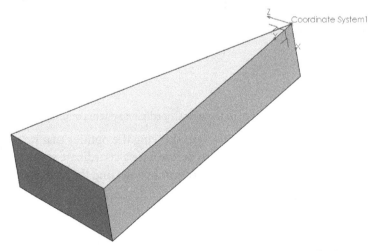

Figure 7.7 – The new coordinate system will be shown in the canvas after being applied

This concludes this exercise of creating the new coordinate system we needed. We will use this model again in the next part of this chapter to study mass properties. However, before moving on, let's explain the **Define position with numeric values** and **Define rotation with numeric values** options shown in **PropertyManager** in *Figure 7.6*:

- **Define position with numeric values**: Checking this option allows us to position the origin of our new coordinate system using numerical distance values in the X, Y, and Z axes as they relate to the absolute origin and axes. In contrast, the example we followed earlier positioned the new origin using an existing vertex.

- **Define rotation with numeric values**: Checking this option allows us to rotate the axes of the new coordinate system by inputting angular values concerning the absolute axes.

Both of these options can be helpful when we're looking to define a new coordinate system that is independent of the existing geometries of the part. With that, we can conclude our discussion on defining new coordinate systems. Next, we will start assigning a structural material and evaluate the mass property of our model.

Assigning materials and evaluating and overriding mass properties

Whenever we design physical objects, we have to think about what materials we will choose to build that object. These objects can be made out of plastic, steel, iron, and so on. SOLIDWORKS provides us with an array of different materials to choose from. It also provides us with the properties of each of those materials, such as their density, strength, thermal conductivity, and other properties related to the specific material. Also, once we assign a material to our model, we can evaluate the different **mass properties** of our objects, such as the mass and the center of mass.

In this section, we will learn how to assign materials and how to evaluate the mass properties of our models. We will also learn how to override the evaluated mass properties.

Assigning materials to parts

In this section, we will discuss how to assign specific materials to our parts. To do that, we will assign the **Steel: AISI 304** material to the model shown in the following diagram. Note that we created this model earlier in this chapter, as shown in *Figure 7.4*.

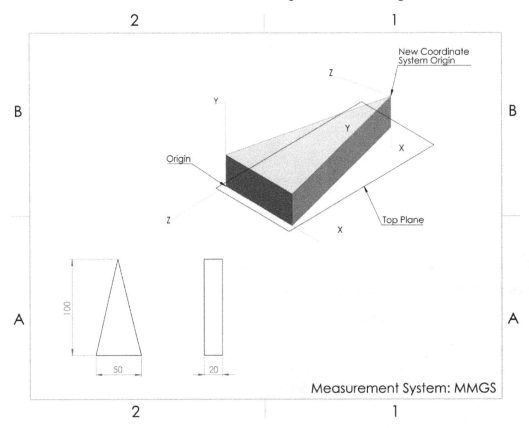

Figure 7.8 – The 3D model we will use for material assignment

To assign materials to our model, follow these steps:

1. On the design tree, right-click on the **Material <not specified>** entry. Then, click on the **Edit Material** option, as follows:

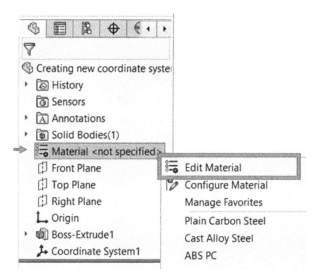

Figure 7.9 – The location of the Edit Material command

2. We will see a new window containing different options, as shown in the following screenshot. On the left, we have different categories of materials to choose from. Since our material is **Steel**, expand that option. Note that the **Steel** menu may be expanded by default when you get to the following window:

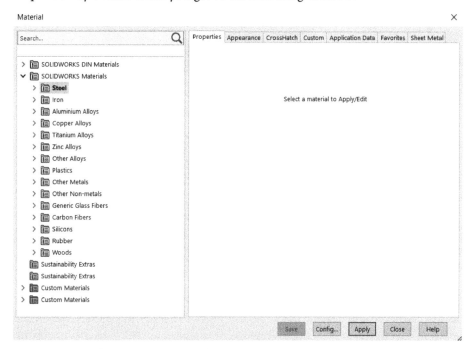

Figure 7.10 – The material library with a selection of different materials to choose from

> **Tip**
>
> At the top of the materials list, there is a search box that can help you find the required material faster.

3. Click on **AISI 304**. Our window will look as follows. Note the highlighted options, which show **Unit of Measurement** and **Material Properties**:

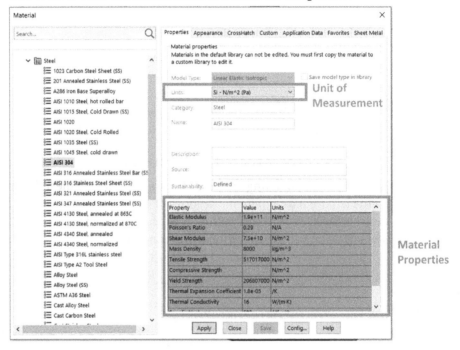

Figure 7.11 – We can choose the unit of measurement to view the material properties

> **Note**
>
> We can change the display unit to other units such as **Imperial** and **Metric**. This will change the unit that's displayed for the rest of the properties. The listed material properties include **Mass Density**, **Tensile Strength**, and **Poisson's Ratio**. We can click on other materials to find out what their properties are.

4. Click on **Apply** to apply the material to our part.

After applying the material, we will notice that the color of the part in the canvas will change to match the assigned material. For example, when assigning **AISI 304** stainless steel, the visual part in the canvas will change color to light gray. We can choose not to apply the material appearance from the **Appearance** tab before clicking **Apply**.

> **Tip**
> We can add our frequently used materials to the favorite tab by right-clicking on the material and adding it to our favorites.

This concludes this section on assigning materials to our parts. We learned about the following topics:

- How to assign different materials to an existing part
- How to find different material properties for each material

At this point, we have a material assigned to our part. Next, we will learn how to view the mass of the other mass properties of our part.

Viewing the mass properties of parts

In this section, we will learn how to view the different mass properties of our existing model. To demonstrate this, we will complete the tasks shown in the following diagram. Note that this is the model from the previous section:

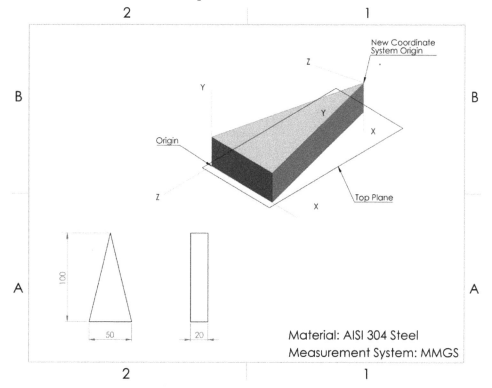

Figure 7.12 – The 3D model we will use to view the mass properties

In this exercise, we will complete the following four tasks:

- Find the mass of the model in grams

- Find the center of mass concerning the origin in millimeters

- Find the center of mass concerning the new coordinate system indicated in the diagram in millimeters

- Find the mass of the model in pounds

Note that all these tasks involve finding different mass properties, such as the mass and the center of mass. Hence, before we can accomplish these tasks, we need to discuss how to view the mass properties.

Viewing mass properties

To view mass properties, we can click on **Mass Properties** from the **Evaluate** tab, as highlighted in the following screenshot:

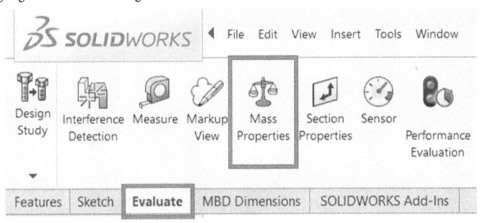

Figure 7.13 – The location of the Mass Properties command

This will show us the following window, which contains different **Mass Properties** related to our object. These include **Density**, **Mass**, **Center of mass**, **Volume**, **Moments of inertia**, and other properties. If you cannot find the **Mass Properties** option that's shown in the preceding screenshot, you can click on the **Tools** menu, and then go to **Evaluate**, where you will also find the **Mass Properties** option.

> **Note**
>
> The unit of measurement for the mass properties is the same as the units of measurements for the document. We will learn how to change that when we tackle the fourth task.

We will be able to find most of the information we need to resolve these tasks in the **Mass Properties** window. We will cover these tasks one by one:

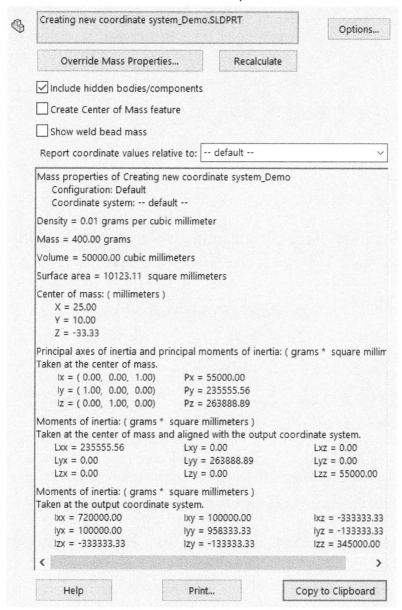

Figure 7.14 – The Mass Properties window will display different properties for the 3D model

Now, we can start accomplishing our four tasks by finding the mass of the model in grams.

Finding the mass of the model in grams

We can find the mass of the model in grams directly from the list of mass properties. Since the unit of measurement for the document is MMGS, the displayed mass is in grams. As shown in the following screenshot of the **Mass Properties** window, **Mass** is listed as **400 grams**:

Density = 0.01 grams per cubic millimeter

Mass = 400.00 grams

Volume = 50000.00 cubic millimeters

Figure 7.15 – Density, mass, and volume are among the displayed mass properties

Now, let's find the center of mass concerning the origin in millimeters.

Finding the center of mass concerning the origin in millimeters

From the **Mass Properties** window, we can find **Center of mass: (millimeters)**, as shown in the following screenshot:

Center of mass: (millimeters)
X = 25.00
Y = 10.00
Z = -33.33

Figure 7.16 – Center of mass, as shown in the Mass Properties window

Note that **Center of mass** is a relational value. In other words, the X, Y, and Z coordinates are concerned with the origin and the default coordinate system. By default, SOLIDWORKS calculates all of the relational values concerning the default coordinate system. We can adjust that to another coordinate system, which we'll do in the next task.

Next, we will find the center of mass concerning the new coordinate system.

Finding the center of mass concerning the new coordinate system in millimeters

In this task, we need to find the same information that we found in the second task, but concerning a different coordinate system. We created the new coordinate system for the model earlier in this chapter. To change the calculations so that they relate to the new coordinate system, you can change the field next to **Report coordinate values relative to** to the other coordinate system, as shown in the following screenshot:

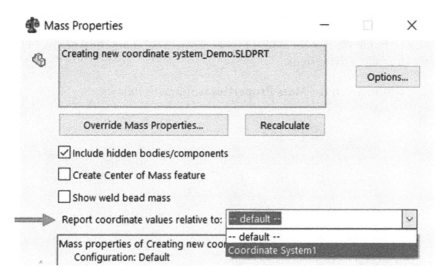

Figure 7.17 – We can evaluate the mass properties according to different coordinate systems

After selecting the new coordinate system, we will notice that the relational values, including **Center of mass**, will change, as shown in the following screenshot. The absolute values such as **Mass** and **Volume** are the same. Note that the new **Center of mass (millimeters)** is **X = 10.00**, **Y = 64.68**, and **Z = 16.17**, as shown in the following screenshot:

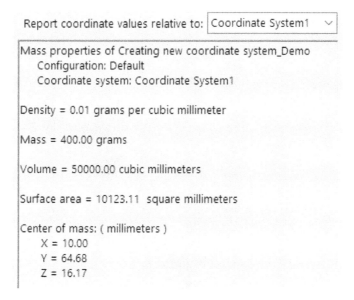

Figure 7.18 – Location-based values can change according to the coordinate system being used

Lastly, we will find the mass of the model in pounds.

Finding the mass of the model in pounds

For this task, we need to change the units of measurement for mass from grams to pounds. To do this, we can follow these steps:

1. Click on **Options...** in the **Mass Properties** window, as follows:

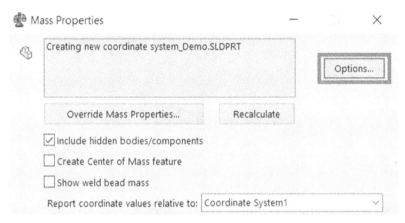

Figure 7.19 – The location of the Options… button

2. We will be taken to the **Mass/Section Property Options** menu, which will allow us to customize the measurement units for our mass properties. To change the mass unit, click on **Use custom settings**. Then, under **Mass**, select **pounds**, as shown in the following screenshot:

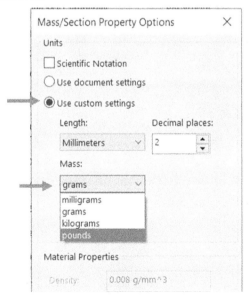

Figure 7.20 – We can change the unit of measurement that's used for the mass properties

3. Click **OK** to confirm our unit selection. This will apply the new mass unit to the **Mass Properties** window, as shown in the following screenshot:

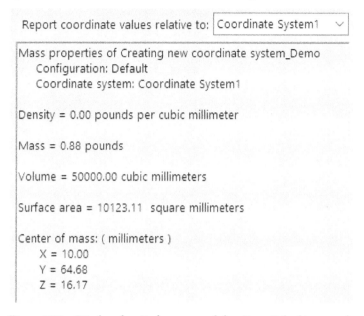

Report coordinate values relative to: | Coordinate System1 ✓ |

Mass properties of Creating new coordinate system_Demo
    Configuration: Default
    Coordinate system: Coordinate System1

Density = 0.00 pounds per cubic millimeter

Mass = 0.88 pounds

Volume = 50000.00 cubic millimeters

Surface area = 10123.11 square millimeters

Center of mass: (millimeters)
    X = 10.00
    Y = 64.68
    Z = 16.17

Figure 7.21 – Displayed units for mass and density switched to pounds

Note that the mass is now calculated in pounds with a value of **0.88 pounds**.

> **Note**
> You can change the displayed decimal places using the same **Options…** window.

This concludes this section on viewing mass properties for existing models. We learned about the following topics:

- How to view the mass properties for a model
- How to adjust mass properties calculations for a new coordinate system
- How to customize the unit of measurement for mass properties evaluation

So far, we have learned how to view the actual mass properties of our parts. These are based on calculations that SOLIDWORKS does based on the geometry and the material of our part. However, we also have the option to override those values with manual entries. We will learn about this in the next section.

Overriding mass properties

In this section, we will learn how to override mass property values. By default, SOLIDWORKS calculates the mass properties based on the material that's assigned, as well as the design itself. However, in some instances, we might want to override those calculated values. For example, we might have a part in an assembly where we know its final mass but not the exact design that will result in the mass. In this case, we can simply override the mass to our required value.

To demonstrate this, we will override the mass of our model from 400 grams to 500 grams. To accomplish this, follow these steps:

1. Open the **Mass Properties** window and click on **Override Mass Properties...**, as shown in the following screenshot:

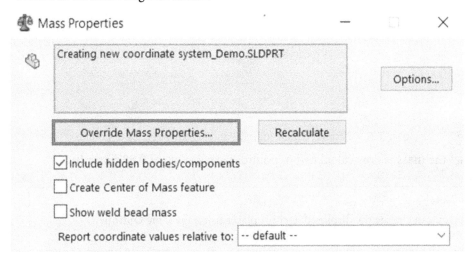

Figure 7.22 – The location of the Override Mass Properties... command

2. Check the **Override mass** box and input a value of 500, as shown in the following screenshot:

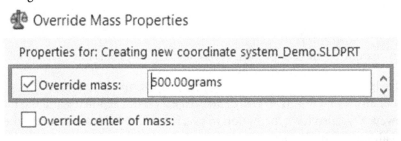

Figure 7.23 – We can change the mass properties values by overriding them

3. Click **OK** to implement the adjustment.

This will redefine the mass as **500 grams**. When viewing the mass properties, SOLIDWORKS will note the mass with the words **user-overridden**, as shown in the following screenshot:

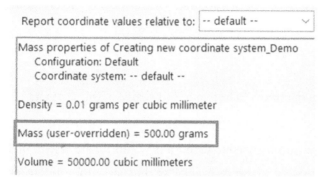

Figure 7.24 – The overridden mass will be indicated as such in the Mass Properties window

To change the mass back, we can go back to the **Override Mass Properties…** command and uncheck the box for **Override mass**. If we do this, our mass will go back to 400 grams as a calculated value. In addition to overriding the mass, we can also override the values for the center of mass and moments of inertia.

In practice, overriding the mass properties is more common when we're working with assemblies rather than parts. We do this if we want to keep a certain effect of a part in its interaction with other parts in an assembly. This is especially the case if our parts are not fully refined yet.

This concludes this section on assigning materials and evaluating mass properties. Being able to assign materials and evaluate mass properties for our model is essential in deciding what material we should use. Also, it plays an important role in calculating the cost of production, materials, transportation, and other applications that can relate to the evaluated mass properties.

Summary

In this chapter, we discussed reference coordinate systems and mass properties. We started by defining what a reference coordinate system is, why we need new ones, and how to define new ones in SOLIDWORKS. Then, we learned about mass properties.

First, we learned how to access the materials library we have in SOLIDWORKS and how to assign a particular material to our parts. Then, we learned how to find properties such as mass and volume for our models. Being able to determine mass properties will help us decide on what structural materials we can pick for the products we design. It will also help us to determine the costs associated with production, transportation, and so on.

In the next chapter, we will start working with assemblies. Assemblies are more than one part that's joined together in one file. Most of the products we use in our everyday lives, such as laptops, cars, and pens, consist of multiple parts that have been put together to form the final product or assembly. This makes these tools key to designing products that are used in our everyday lives.

Questions

Answer the following questions to test your knowledge of this chapter:

1. What are coordinate systems?
2. What parameters do we need to define a new coordinate system in SOLIDWORKS?
3. What information do we get by evaluating the mass properties of a part?
4. Create the following model and define the indicated **Coordinate System 1**:

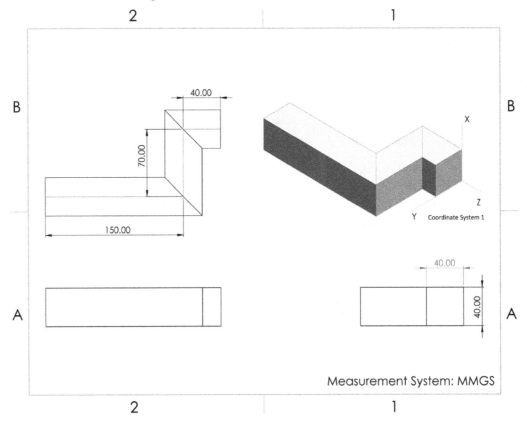

Figure 7.25 – The drawing for Question 4

5. Assign the **Aluminium Alloy: 1060 Alloy** material to the model we created in *Question 4*. What is the mass in grams? What is the center of mass (in millimeters) concerning **Coordinate System 1**, as indicated in the preceding diagram?

6. Create the following model and define the indicated coordinate system:

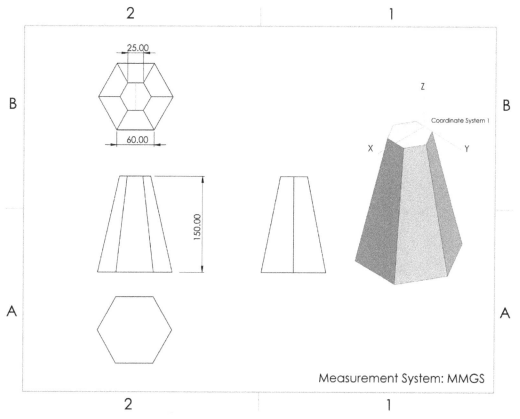

Figure 7.26 – The drawing for Question 6

7. Assign the **Plain Carbon Steel** material to the model we created in *Question 4*. What is the mass in pounds? What is the center of mass (in inches) concerning **Coordinate System 1**, as indicated in the preceding diagram?

> **Important Note**
> The answers to the preceding questions can be found at the end of this book.

8
Standard Assembly Mates

Most of the products that we interact with in our daily lives, such as laptops, phones, and cameras, are made up of many different components that have been put together; that is, they have been assembled. One of the major elements of mastering SOLIDWORKS is being able to use SOLIDWORKS assemblies, which allow us to put multiple parts together to create a single artifact. In this chapter, we will cover basic SOLIDWORKS assemblies and, in particular, standard mates.

In this chapter, we will cover the following topics:

- Opening assemblies and adding parts

- Understanding and applying non-value-oriented standard mates

- Understanding and applying value-driven standard mates

- Utilizing materials and mass properties for assemblies

By the end of this chapter, we will be able to put different parts together to form an assembly. The objective of this chapter is to get us to generate complex artifacts by linking different parts together and creating an assembly using standard mates.

Technical requirements

In this chapter, you will need to have access to the SOLIDWORKS software. The files for this chapter can be found at the following GitHub repository: `https://github.com/PacktPublishing/Learn-SOLIDWORKS-Second-Edition/tree/main/Chapter08`.

Check out the following video to see the code in action: `https://bit.ly/31ZCRra`

Opening assemblies and adding parts

In this section, we will take our first steps toward working with SOLIDWORKS assemblies. We will cover what SOLIDWORKS assemblies are, how to start an assembly file, and how can we add a variety of components to our assembly file. Opening an assembly file and adding different parts to it is the first step we need to take when we start any assembly.

Defining SOLIDWORKS assemblies

There are three main sections of SOLIDWORKS: *parts*, *assemblies*, and *drawings*. For each type, SOLIDWORKS creates a different file type with different file extensions. For assemblies, the file extension is `.SLDASM`, while a part has a file extension of `.SLDPRT`. In *Chapter 10, Basic SOLIDWORKS Drawing Layout and Annotations*, we will cover drawings, which have the file extension `.SLDDRW`.

With an assembly file, we can link more than one part file together to form one product. The following figures highlight two examples of assembly files. The following figure shows a simple assembly that consists of only three parts. We can see each and every part since they are highlighted by solid lines and borders that have been filled in with a variety of colors:

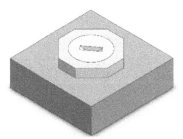

Figure 8.1 – A simple assembly of three parts

The following assembly is more complex than the first one. It is a mechanical assembly that consists of over 50 different parts:

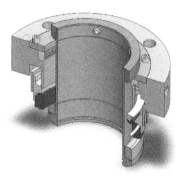

Figure 8.2 – An assembly of 50 parts

The different parts interact with each other via mates. Mates are very similar to the relations we used in sketching, for example, coincident, perpendicular, and tangent. Now that we know what assemblies are, we can move on and create our first assembly file.

Starting a SOLIDWORKS assembly file and adding parts to it

To demonstrate how to start an **assembly file** and add parts to it, we will start an assembly file and add the following parts to it. Make sure that you download the files that accompany this chapter. The parts you will download are as follows:

- A small triangular part:

Figure 8.3 – The rectangular prism is included in this chapter's downloads

- A larger base part, which will house the smaller triangular part:

Figure 8.4 – The part file for this shape is included in the chapter's downloads

Now that you've downloaded these parts onto your computer, we can start opening our assembly file.

Starting an assembly file

To start an assembly file, follow these steps:

1. Select **New** from the top of the SOLIDWORKS interface, as shown in the following screenshot:

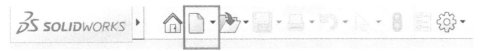

Figure 8.5 – Where to start a new file

2. Select **Assembly** and click **OK**:

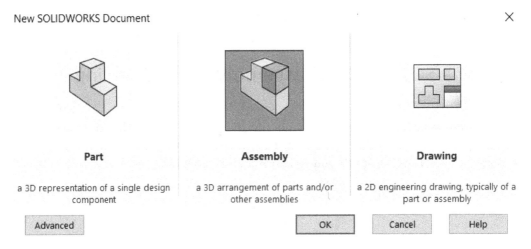

Figure 8.6 – A window showing the different new file types you can start with

Now that we have opened our assembly file, we can add the two parts we downloaded to it. We will do that next.

Adding parts to the assembly file

To add our two parts to the assembly file, follow these steps:

1. As soon as we open a new assembly file, SOLIDWORKS will prompt us to select a part to be added to the assembly. From here, you can navigate to the `Base.` `SLDPRT` file, which can be found in the SOLIDWORKS parts attached to this chapter. Click **Open** after selecting the file. Alternatively, we can double-click on the file to open it:

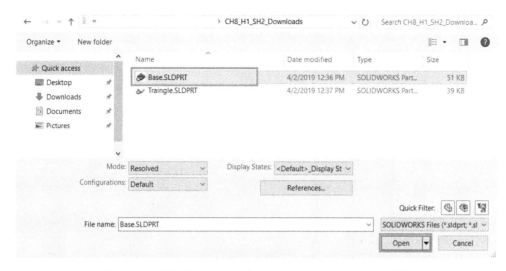

Figure 8.7 – The browser window to open a part in an assembly

Important Note
If SOLIDWORKS does not prompt you to add a part automatically, you can use the **Insert Components** option, as shown in *Figure 8.9*.

2. The part will appear in the assembly's canvas. At the bottom of the page, we will see some options that can help us orient the part if needed. Once we are satisfied with the part's orientation, we can left-click on the canvas using the mouse to place the part. Then, we need to click on the green check mark, which can be found at the top right or the top left of the page, as shown in the following screenshot:

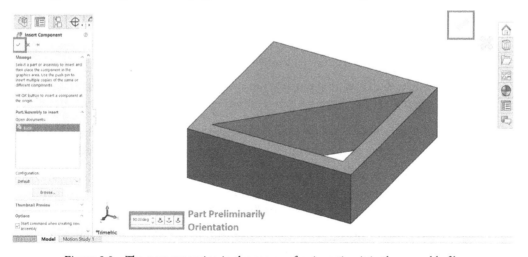

Figure 8.8 – The part appearing in the canvas after inserting it in the assembly file

3. To add the second part, we can click on **Insert Components**, as highlighted in the following screenshot. Now, navigate to the other part, that is, `Triangle.SLDPRT`. Select it, and click **Open**:

Figure 8.9 – The location of the Insert Components command

4. The triangle will appear on the canvas as well. We can left-click to place the part in the assembly. We will end up with the two parts on the canvas, as shown in the following figure. You may have a different placement for these two parts than what's shown here, though. This is not an issue at this point:

Figure 8.10 – The two parts will be in the assembly canvas

This concludes this exercise of adding parts to our assembly file. Before we move on, however, let's mention three key points when it comes to adding parts to our assembly:

- **Fixed parts**: The first part you insert into the assembly file is a fixed part by default. Fixed parts don't move in the assembly environment and are fully defined.

- **Floated parts**: The second part that's inserted is a floating part by default. Floating parts can be moved around the assembly environment since they are not defined. We will use mates to define floating parts later in this chapter.

- **Dragging parts**: You can click and hold the second part (triangle) and move it around the canvas. The first part is fixed by default, and so it cannot be moved.

We can change any part's status from fixed to floating and vice versa by right-clicking on the part and selecting the **Float** or **Fix** command. The **Float** command is highlighted in the following screenshot:

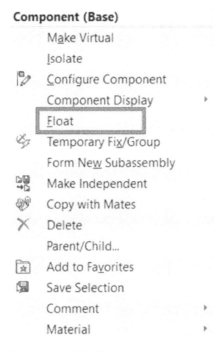

Figure 8.11 – The Float command location

Important Note

If the part is fixed, we will see the **Float** command, as shown in *Figure 8.11*. However, if the part is floating, then we will see the **Fix** command instead to make the part fixed.

At this point, we have inserted our two parts into an assembly file. However, the parts hold no linkage to each other. Next, we will look at mates, which will cause our parts to interlink.

Understanding mates

Mates are similar to sketch relations, but they act on assemblies. They govern how different parts interact with each other or move in relation to each other. As an example, examine the keys on a computer keyboard. Each key is stationed in a specific location and restrained by specific movements, such as up and down. We can think of this positioning and movement as being governed by an assembly's mates.

There are three categories of mates in SOLIDWORKS: **standard mates**, **advanced mates**, and **mechanical mates**. We will only cover standard mates in this chapter. Standard mates provide the following options:

- **Coincident**

- **Parallel**

- **Perpendicular**

- **Tangent**

- **Concentric**

- **Lock**

- **Distance angle**

Some standard mates require us to input a numerical value, such as the mate's distance and angle. We can refer to these as value-oriented mates. The rest of the mates do not require a numerical value. We can refer to these as non-value-oriented mates. We will learn more about these next.

Understanding and applying non-value-oriented standard mates

In this section, we will start linking different parts together in an assembly using the non-value-oriented standard mates. We will learn about the coincident, parallel, perpendicular, tangent, concentric, and lock mates. These mates don't need a numerical value input to be applied to them; instead, they are constructed based on their relationship with different geometrical elements. In this section, we will learn what those mates are and how to apply them. We will also cover the different levels of defining an assembly. We will start by defining each of those non-value-oriented standard mates.

Defining the non-value-oriented standard mates

The non-value-oriented standard mates are coincident, parallel, perpendicular, tangent, concentric, and lock. These are special in that they don't require any numerical input to be applied to them or defined for them. They are very similar to the sketching relations we applied while sketching. Here is a brief explanation of each standard mate:

- **Coincident**: This allows a coincident relation between two surfaces, a line and a point, and two lines.

- **Parallel**: This allows us to set two surfaces, two edges, or a surface and an edge so that they're parallel to each other.

- **Perpendicular**: This allows us to set two surfaces, two edges, or a surface and an edge so that they're perpendicular to each other.

- **Tangent**: This allows two curved surfaces to have a tangent between each other. This can also happen between a curved surface and an edge, as well as a straight surface.

- **Concentric**: This allows two curves to have the same center.

- **Lock**: This locks two parts together. When two parts are locked, they will copy each other's movements.

Now that we know what these standard mates do, we will start applying them in order to link different parts of our assembly. We will start with the coincident and perpendicular mates.

Applying the coincident and perpendicular mates

To explore how to make the mates coincident and perpendicular, we will make the assembly that's shown in the following figure. The assembly is made out of two parts – a **Base** part and a **Triangle** part. We opened these parts in an assembly file earlier in this chapter. We will continue from there:

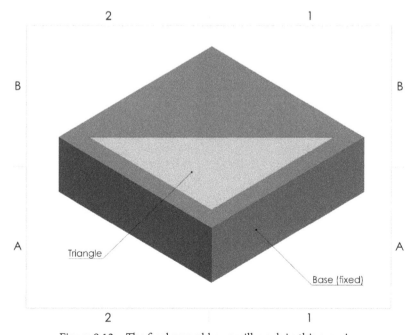

Figure 8.12 – The final assembly we will reach in this exercise

If you are starting over, you can download the two parts from the files for this chapter and open them in an assembly file. Our starting point will be having these two parts arranged on an assembly canvas, as shown here:

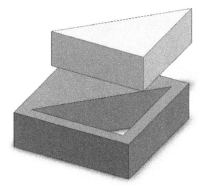

Figure 8.13 – The two parts placed in an assembly without mates

Now that we have the two parts in our assembly, we will apply the mates. We will start with the coincident mate, followed by the perpendicular mate.

Applying the coincident mate

To apply the coincident mate to our assembly, follow these steps:

1. To apply mates, select the **Mate** command, which can be found under the **Assembly** command, as shown in the following screenshot:

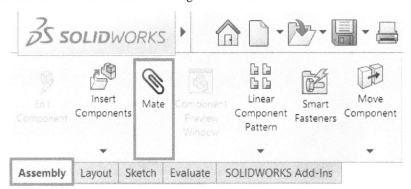

Figure 8.14 – The location of the Mate command

2. After selecting **Mate**, we will see a PropertyManager on the left-hand side of the screen. At this point, we will be asked to select which elements we want to mate. This can include surfaces, edges, and points. For this selection, we'll select the top surfaces of the base and the triangle, as shown in the following figure. These will fill in the **Mate Selections** space, which is highlighted in red in the following figure:

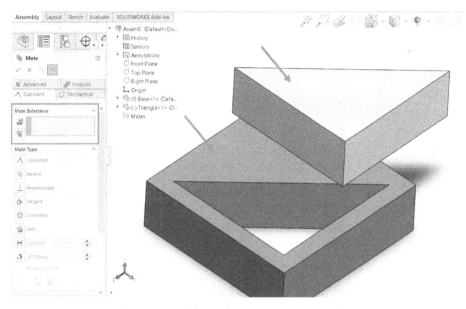

Figure 8.15 – The surfaces we can mate together

After the selection, we will see that the two parts move in relation to each other. Also, one mate will be selected automatically. In this case, the mate will be **Coincident**. We can change this mate if we want to use another one. However, in this case, **Coincident** will work for us. Our canvas will look as follows. To apply the mate, click on the green check mark on top of the mate's PropertyManager:

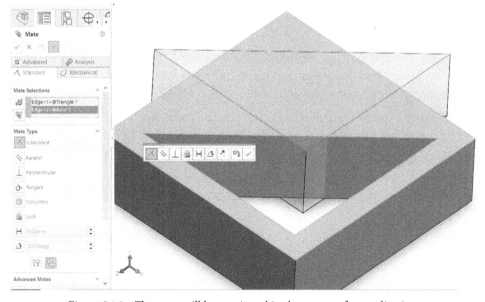

Figure 8.16 – The mate will be previewed in the canvas after application

> **Tip**
>
> To check what effect the applied mate has, click and hold the triangle and move it around. We will see that the movement of the part has been restricted due to the applied mate.

When we try moving the triangle around the canvas, we will notice that it can only move sideways to the base. However, it won't move up and down relative to the base. Now, let's apply another coincident mate to restrain the triangle more. Note that the **Mate** command is still active, which means we can apply more mates.

3. Select the edges shown in the following figure. These are the outer edges of the triangle and the inner edge of the base. Again, SOLIDWORKS will interpret that we want the coincident mate and automatically apply it:

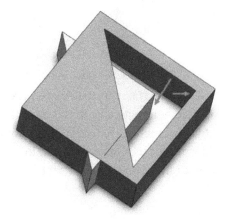

Figure 8.17 – The selection of the coincident mate

After selecting these two edges, we will see the following screenshot (*Figure 8.18*). Before applying the mate, please take note of the following:

- In the canvas, the two parts will move to preview the mate. Make sure that the preview matches our needs, as indicated with **A** in *Figure 8.18*.

- In the **Mate Selections** list in the PropertyManager, we can double-check whether we have selected two edges. If not, we can delete the undesired selections there. This is indicated with **B** in the following screenshot.

- In the **Standard Mate** selection in the PropertyManager, we can double-check that the selected mate is what we want to apply. If we want to use another mate, we can select it from there. This check is indicated with the letter **C** in the following screenshot:

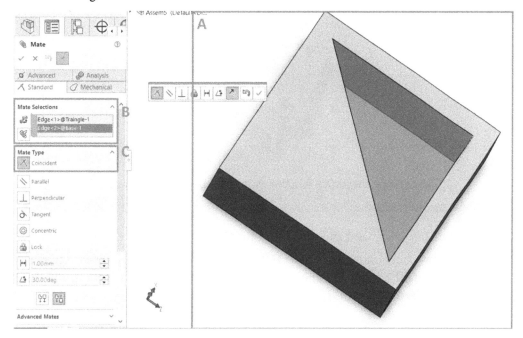

Figure 8.18 – Different checks to ensure we applied the correct mate

4. After conducting all the checks, we can click on the green check mark to apply the mates.

> **Tip**
> The *pin icon* next to the red cross allows you to keep the mate PropertyManager visible after applying the mate, making it faster to apply multiple mates one after the other.

This concludes our application of the coincident mate. Next, we will apply the perpendicular mate to our assembly.

Applying the perpendicular mate

To find out what restraints are missing from our assembly, we can click and hold the triangle and drag it. We will see that the triangle is hinged at the corner we just mated. To restrain this movement, we can apply the perpendicular mate between the faces, as shown in the following figure:

Figure 8.19 – The surface selection to apply the perpendicular mate

To apply the mate, follow these steps:

1. Select the **Mate** command and then select the **Perpendicular** mate.

2. Select the two faces under **Mate Selections**. Our view will look as follows. Again, before applying the mate, note the position of the parts, the mate selection, and the selected standard mates:

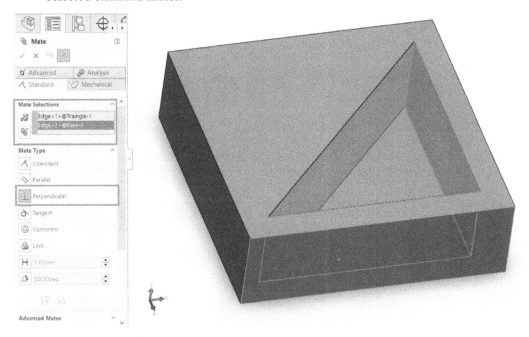

Figure 8.20 – The mate PropertyManager and a preview of the perpendicular mate

3. Click on the green check mark to apply the mate.

4. Since we don't need to apply any more mates, we can close the mate's PropertyManager by pressing the *Esc* key on the keyboard or clicking the red cross at the top of the PropertyManager.

At this point, our assembly will look as follows. Note that if we try to drag the triangle in any direction, it won't move. This indicates that our assembly is now fully defined:

Figure 8.21 – The final look of our assembly

This concludes the application of the coincident and perpendicular mates to our assembly. We will follow the same procedure to apply all the other mates. While applying the coincident and perpendicular mates, we learned about the following:

- How to access the mate command

- How to select different elements and apply the mates to restrain them

- How to check for unrestrained movements by holding and dragging parts

We have just finished fully defining our assembly by using the coincident and perpendicular mates. Next, we will work on another assembly, which will involve the parallel, tangent, concentric, and lock mates.

Applying the parallel, tangent, concentric, and lock mates

In this section, we will explore how to apply the parallel, tangent, concentric, and lock mates. To do that, we will apply the mates that are shown in the following figure. We will refer to this drawing as we apply the different mates:

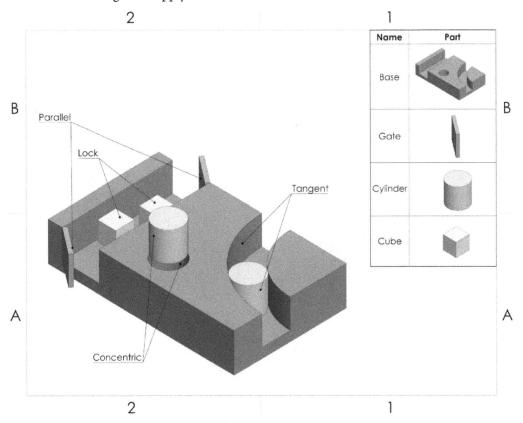

Figure 8.22 – The assembly we will work on in this exercise

To start, download the parts and the assembly file attached to this chapter. Our starting point, which is where we will apply all the mates, will be from the attached assembly file, which looks as follows:

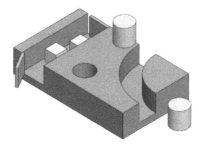

Figure 8.23 – The starting point of this exercise from the downloadable files

The attached assembly file already has a few coincident mates applied to it. The procedure of applying all the standard mates is similar, so we won't go into too much detail regarding the next four mates. Next, we will apply the **Parallel**, **Tangent**, **Concentric**, and **Lock** mates.

Applying the parallel mate

In this section, we will apply the parallel mate to the two gates that were shown in the initial drawing. These are also highlighted in the following figure. To do this, follow these steps:

1. Go to the **Mate** command and select the **Parallel** mate.

2. Under **Mate Selections**, select the two faces shown in the following figure:

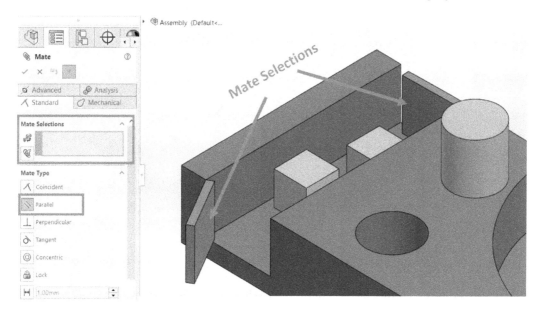

Figure 8.24 – The mate selection for the parallel mate

> **Important Note**
> The positions of the gates will shift as the mate takes effect.

3.   Apply the mate by clicking on the green check mark.

After applying the mate, it's good practice to drag the gates to see what effect it has. You will notice that, as we move one gate, the other gate will also move to keep the two faces parallel to each other. In this example, we applied the parallel mate to two faces, but we can also apply the mate in the same way to two straight edges or an edge with a face.

This concludes this exercise on applying the parallel mate. Next, we will start applying the tangent mate.

Applying the tangent mate

In this section, we will apply the tangent mate to the cylinder and base that were highlighted in the initial drawing. To do this, follow these steps:

1.   Go to the **Mate** command and select the **Tangent** mate.

2.   Under **Mate Selections**, select the two faces, as shown in the following figure. Note that the position of the cylinder will shift as the mate takes effect:

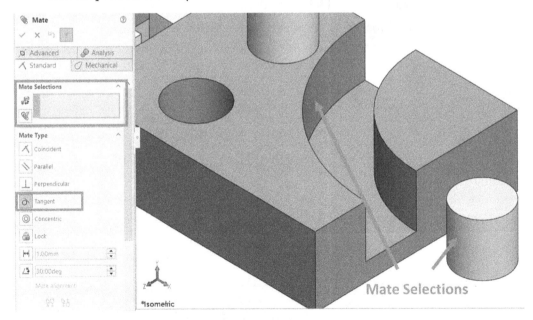

Figure 8.25 – The mate selection for the tangent mate

3.   Apply the mate by clicking on the green check mark.

After applying the mate, it's good practice is to drag the cylinder to see what effect it has. You will notice that the cylinder will move while keeping a tangent relation with the side we selected in the base part. In this example, we applied the tangent mate to two faces, but we can also apply the mate in the same way to two edges or an edge with a face.

This concludes this exercise on applying the tangent mate. Next, we will start applying the concentric mate.

Applying the concentric mate

In this section, we will apply the concentric mate to the cylinder and base that are highlighted in the initial drawing. To do this, follow these steps:

1. Go to the **Mate** command and select the **Concentric** mate.

2. Under **Mate Selections**, select the two faces shown in the following figure. Note the position of the cylinder will shift as the mate takes effect:

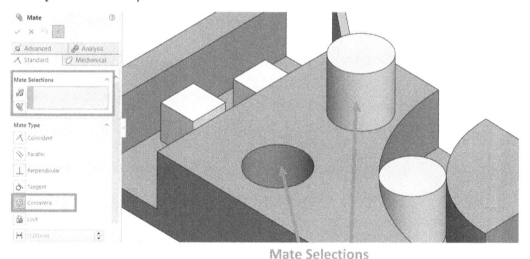

Mate Selections

Figure 8.26 – The mate selection for the concentric mate

3. Apply the mate by clicking on the green check mark.

After applying the mate, it's good practice to drag the cylinder to see what effect it has. You will notice that the cylinder will only move vertically, that is, up and down, so that the two rounded faces share the same center. In this example, we applied the concentric mate to two faces, but we can also apply the mate in the same way to two arc edges or an arc edge with an arc face.

This concludes this exercise on applying the concentric mate. Next, we will start examining the lock mate.

Applying the lock mate

In this section, we will apply the lock mate to the cubes that were highlighted in the initial drawing. To do this, follow these steps:

1. Go to the **Mate** command and select the **Lock** mate.

2. Under **Mate Selections**, select the two cubes shown in the following figure. Note that the position of the cubes will not change after we apply the lock mate:

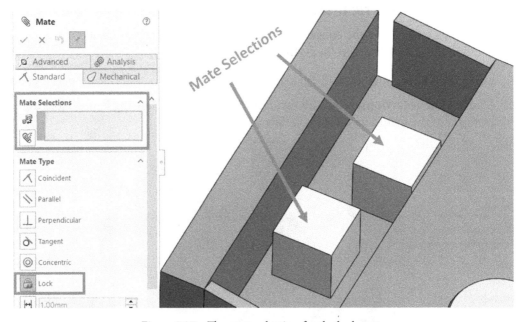

Figure 8.27 – The mate selection for the lock mate

3. Apply the mate by clicking on the green check mark.

After applying the mate, it's good practice to drag the cubes to see what effect it has. You will notice that as we move one cube, the other cube will move in the same way. This includes linear movements, as well as rotational movements. The lock mate can only be applied to whole parts.

This concludes applying the lock mate. At this point, we have covered how to apply all the non-value-oriented standard mates. Next, we will learn what fully defined means in the context of assemblies. We will also learn about other types of assembly definitions.

Under defining, fully defining, and over defining an assembly

When we finished the first assembly exercise, which is where we used the coincident and perpendicular mates, we noticed that all the parts are restrained from moving in any direction. Thus, the triangle will not move in any direction when it's dragged. This indicates that the assembly is now **Fully Defined** since both parts were fully restrained. This status is shown in the lower right-hand corner of the canvas, as shown in the following screenshot:

Figure 8.28 – The state of assembly is defined at the bottom of the canvas

However, in the second assembly exercise, where we looked at the parallel, tangent, concentric, and lock mates, we noticed that we could still drag the parts around, even after keeping certain movement constraints. When this happens, the status of the assembly will be **Under Defined**.

Similar to sketching, there are three different statuses and terms when it comes to defining an assembly. Those are under defined, fully defined, and over defined. However, the way we interpret them is slightly different in the context of assemblies. Here is a brief description of each status:

- **Under Defined**: There are one or more parts that are not fully constrained in terms of movement. In other words, if we click and hold that part and drag it, it will move.

- **Fully Defined**: In a fully defined assembly, all the parts are fully restrained. In other words, if we click and hold any of the parts and drag them, they will not move. Note that, in a fully defined assembly, we can apply more mates that serve the same purpose without over-defining the assembly. As such, even if the assembly is already fully defined, we can still apply more mates, that is, as long as they do not contradict each other. This aspect is different from defining sketches. In sketching, any relation that's added after fully defining a sketch will make the sketch over defined.

- **Over Defined**: In an over defined assembly, we have mates that contradict each other. Thus, we will need to delete or redefine some of the existing mates.

The indication at the bottom of the interface refers to the definition status of the whole assembly. Let's learn how we can find out the status of each individual part.

Finding the definition statuses of the parts

To find the definition status of each part in the assembly, we can look at the assembly design tree. At the beginning of each part's listing, SOLIDWORKS indicates what the statuses of the different parts are with symbols such as **(f)**, **(-)**, and **(+)**. The meaning of each symbol is as follows:

- **(f)**: Fixed

- (-): Under defined

- **No symbol**: Fully defined

- (+): Over defined

The following figure shows each of the symbols in the design tree:

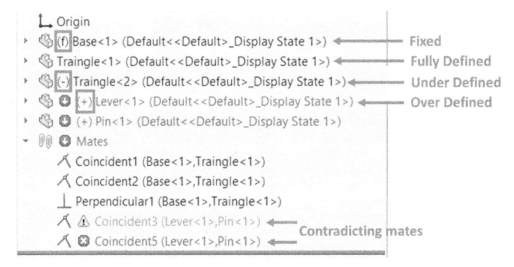

Figure 8.29 – The parts' statuses as shown in the assembly design tree

Finding out the status of each part will help us when we need to define our assemblies. With this, we can decide which part needs more mates or which mate we should reconsider. However, in the context of assemblies, which definition status is better? We will discuss that question next.

Which assembly definition status is better?

When defining assemblies, we should avoid having an over defined assembly. However, there are certain advantages of having our assembly under defined or fully defined. Here are some scenarios for both cases:

- If the assembly has a moving part, a common practice will be to have the assembly under defined so that the desired movement is visible if we were to drag and move the part around. For example, if we assemble a windmill, we can choose to have the blades under defined to show how they move and how all the parts interact with each other during that movement.

- If all the parts in the assembly are fixed, a common practice will be to have the assembly fully defined. A common example of fully fixed assemblies is tables, which don't have any moving parts.

As we can see, keeping our assemblies fully defined or under defined has certain advantages. As designers or draftsmen, we will have to weigh up the advantages of each and adapt our own approach. Next, we will learn how to view and adjust active mates.

Viewing and adjusting active mates

In the assembly design tree, we will see a list of all the parts in the assembly, as well as the mates that have been applied to those parts. The lowest part of the assembly design tree shows the mates. We can expand this list to view all the mates that were involved in making the assembly. The following figure shows the mates we used to make the assembly we constructed earlier, that is, two **Coincident** mates and one **Perpendicular** mate:

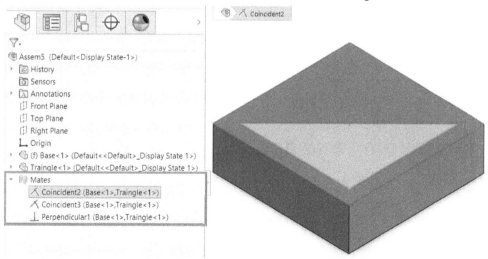

Figure 8.30 – Existing mates are listed in the assembly design tree

> **Tip**
>
> To see which elements of the parts are involved in the mates, we can click on the mate in the design tree. Then, the involved elements will be highlighted in the canvas, as shown in the preceding figure.

Now, we know how to view the active mates that we have in our assembly. Next, we will learn how to modify them.

Modifying existing mates

To modify a particular mate, we can *right or left-click* on the mate from the design tree. We will see the following menu. Here, we can choose to **Edit**, **Delete**, or **Suppress** the selected mate. Modifying mates follows the same procedure as modifying features:

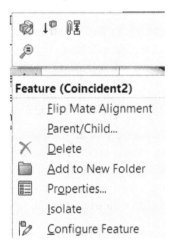

Figure 8.31 – A menu appears after right-clicking a mate, giving us different options

So far, we have learned how to use all the non-value-driven standard mates. We have also learned about the different statuses of assemblies, in addition to how to view and modify existing mates. Now, we can start learning about value-driven standard mates.

Understanding and applying value-driven standard mates

This section covers the standard mates that are defined by numerical values, that is, the distance and angle mates. We will learn about what do they do and how to apply them to an assembly. By the end of this section, we will be familiar with applying all the standard mates, which is our first step when it comes to working with SOLIDWORKS assembly tools.

Defining value-driven standard mates

Value-driven standard mates are those that depend on numerical values so that they can be set. They include two standard mates – distance and angle. Here is a brief definition of these two mates:

- **Distance**: This sets a certain fixed distance between two entities, such as edges and straight surfaces.

- **Angle**: This sets a certain fixed angle between two straight surfaces or edges.

Whenever we define one of these mates, we need to input a number that indicates the desired distance or angle. Now that we know what the distance and angle mates are, we can start applying them.

Applying the distance and angle mates

Here, we will apply the distance and angle mates to create the following assembly. You can download all the indicated parts from the download files that are linked to this chapter:

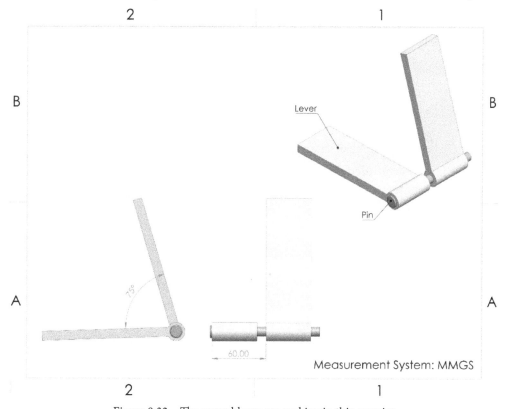

Figure 8.32 – The assembly we are making in this exercise

Note that, in the preceding drawing, the two levers are separated by a set distance of **60** mm. Also, the two levers have an angle of **75** degrees between them. This indicates that we can utilize the distance and angle relations to complete the assembly. To complete this assembly, we need to download the files that are attached to this chapter and open the `Lever-Pin Assembly.SLDASM` assembly file. The assembly will look as follows. Note that the assembly already has mates that are restraining them. However, we still need to add the distance and angle mates in order to achieve the assembly shown in the preceding figure. We can move the parts around in the assembly to find out how are they restrained:

Figure 8.33 – The initial status of the downloadable assembly for this exercise

Now that we have downloaded our parts, we can start applying the mates. We will start with the distance mate.

Applying the distance mate

To apply the distance mate, follow these steps:

1. Select the **Mate** command. In **Mate Selections**, select the two slides.

2. Select the **Distance** mate and select **60.00mm** in the distance space.

3. Apply the mate by clicking on the green check mark.

4. Note the **Flip dimension** check box below the distance value. Checking this box will switch the distance from being toward the left to being toward the right and vice versa:

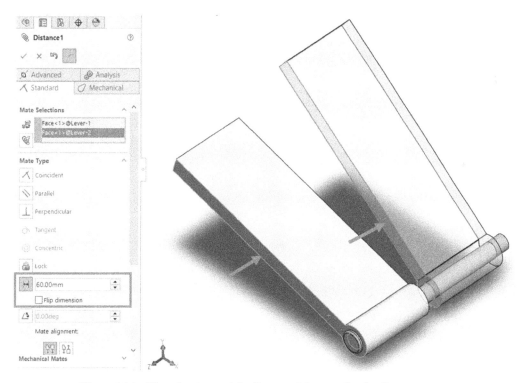

Figure 8.34 – The selection and the PropertyManager for the distance mate

After applying the mate, it's good practice to test its effect. We can do that by dragging the different parts to find out how the new mate is taking effect. In this case, we will notice that the levers can still rotate; however, they cannot move away from each other, that is, along the pin. Now, we will apply our next mate so that we can set the angle between the levers.

Applying the angle mate

To apply the angle mate, follow these steps:

1. Select the **Mate** command. In **Mate Selections**, select the two slides.

2. Select the standard mate **Angle** and input 75 degrees for the angle, as stated in the initial drawing.

3. Apply the mate by clicking on the green check mark.

Similar to the distance mate, we will get a **Flip dimension** checkbox. This will flip the dimension that the angle is measured in. Try checking the box to see what effect this has on the assembly. Once checked, the output will be reflected in the preview on the canvas.

Note that, when applying mates, we may get a shortcut menu showing the various mates. We can use this menu in the same way we use the PropertyManager:

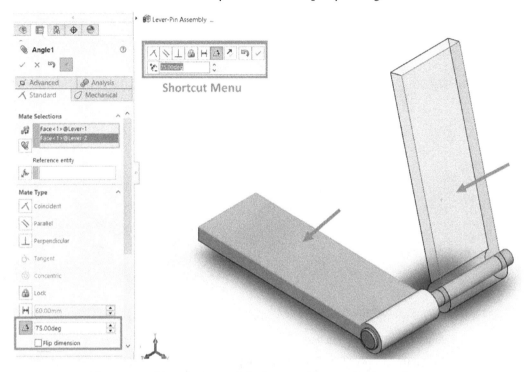

Figure 8.35 – The selection and the PropertyManager for the angle mate

Once we have applied the angle mate, the final shape will look as follows. Note that, if we drag any of the levers, the other lever will rotate with it while keeping the angle between them equal to 75 degrees. Also, note that the assembly is still under defined; however, we will keep it that way to show a simple simulation of how the different parts in the assembly move together:

Figure 8.36 – The final status of the assembly

This concludes this exercise on the distance and angle mates. We have learned about the following topics:

- What the distance and angle mates are and what they do
- How to apply the distance and angle mates

At this point, we have covered all the standard mates that can be used in a SOLIDWORKS assembly. These allow us to model products that consist of different parts that interact with each other. Next, we will start looking at materials and mass properties within the context of an assembly.

Utilizing materials and mass properties for assemblies

When creating assemblies, we may need to determine the mass, volume, center of mass, and other related mass properties. This information is necessary, as it helps us understand our product from a physical perspective. As a result, they will help us develop or modify our product in case we ever want to adjust the mass, volume, and so on to meet a specific requirement. Similar to when working with parts, we can evaluate mass properties within the context of assemblies. In this section, we will learn about setting new coordinate systems, editing materials for the parts within the assembly, and how to evaluate the different mass properties for our assembly.

Setting a new coordinate system for an assembly

In many cases, we may need to reorient models ourselves directionally and find the center of mass from different locations. These are more common practices when we're working with assemblies compared to when we're working with parts. This is due to it being less intuitive to build upon the default coordinate system within the assemblies' environment. The previous chapter examined this topic in more detail.

To define a new coordinate system, we can follow the same procedure that we followed when we defined coordinate systems for parts. To access the command, we can go to the **Assembly** commands category and select **Coordinate System** under **Reference Geometry**, as highlighted in the following screenshot:

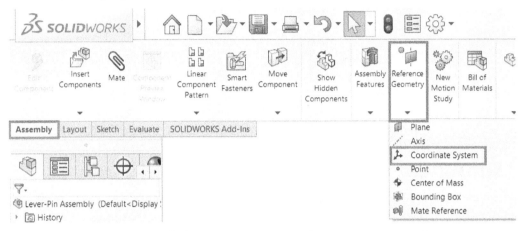

Figure 8.37 – The location of the new Coordinate System command

To define a new coordinate system, we have to define the origin, as well as the direction of the axes, similar to defining a coordinate system in parts. Introducing a new coordinate system in assemblies is a common practice when we're measuring coordinate-orientated mass properties such as the center of mass. Next, we will address how to edit materials within assemblies. Refer to *Chapter 7, Materials and Mass Properties*, for more information about coordinate systems. The procedure of setting and dealing with new coordinate systems is the same for both parts and assemblies.

Material edits in assemblies

There is no material assignment for the assembly. Instead, each part will carry its own material assignment. If the part was assigned a material when it was created, then this assignment will simply be transferred to the assembly. Within the assembly environment, we can still edit and assign materials to individual parts. We will learn how to do this here.

Assigning materials to parts in the assembly environment

We can assign a material to individual parts from the assembly environment. Follow these steps to do so:

1. Decide which part you would like to assign/adjust to/for the material.

2. Expand the part from the design tree. Then, select **Edit Material**, as shown in the following screenshot:

Figure 8.38 – We can edit parts' materials directly from the assembly

3. Assign or adjust the assigned material as needed.

Important Note

Once we assign the material to the parts, it will be updated in the original part file since they are now linked.

---

4. We can assign the same material to more than one part in one go by highlighting the parts, right-clicking, and then selecting the desired material, as shown in the following screenshot:

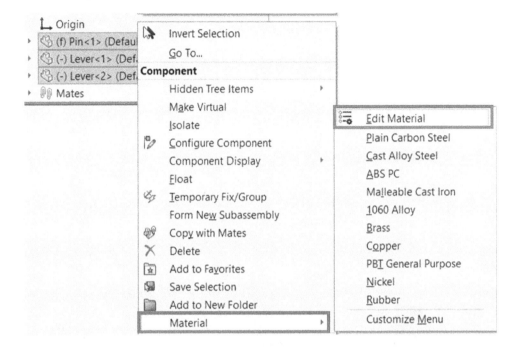

Figure 8.39 – We can edit the material for multiple parts at once

This concludes editing a part's material assignments within an assembly file. When editing materials in an assembly, take note of the following points:

- If we change the material within the assembly and save it, the part's material assignment will be updated as well. This is because the parts and the assembly file are connected.

- If we edit the material assignment for a repeated part (that is, we have more than one copy in the assembly), all the part's materials will be updated to match the new edit.

Now that we have materials assigned to our parts, we can start evaluating our mass properties.

Evaluating mass properties for assemblies

To evaluate the mass properties for an assembly, we can click on **Mass Properties** under the **Evaluate** commands category, as shown in the following screenshot:

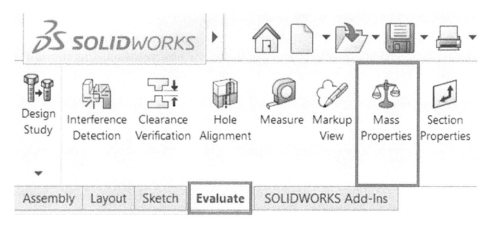

Figure 8.40 – The location of the Mass Properties command

This will show us the same information we received when we evaluated mass properties for parts. Refer to *Chapter 7, Material and Mass Properties*, for more information. The only difference is that the mass properties here will be a reflection on the whole assembly rather than on individual parts.

Note that the center of mass is calculated based on the position of the different parts that make up the assembly. If the assembly is under defined and we move the parts, the center of mass will change. This is in addition to all the other properties that are calculated based on the coordinate system's position, such as the moment of inertia. When moving the assembly, we may need to click on **Recalculate** in order to recalculate the mass properties, as shown in the following screenshot:

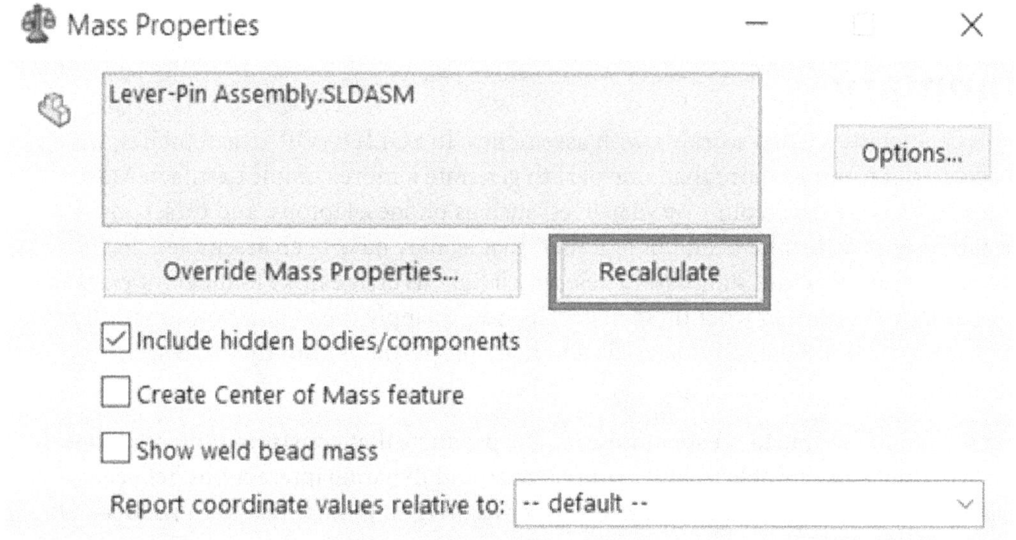

Figure 8.41 – The Recalculate command updates the values in the PropertyManager

Within assemblies, we can also calculate the mass properties of a specific part in relation to the assembly's coordinate system. To do this, we can select that part in the **Mass Properties** selection window, as highlighted in the following screenshot. This allows us to show the mass properties of a specific part within the assemblies' environment:

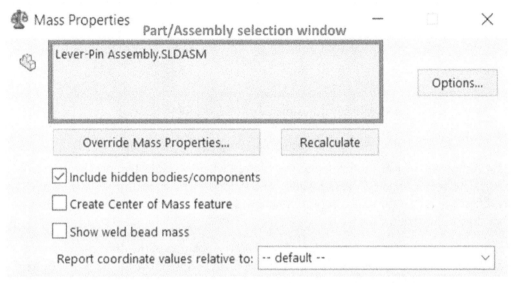

Figure 8.42 – We can find the mass properties of specific parts within the assembly environment

In this section, we have learned about mass properties in the context of assemblies. We have learned about setting new coordinate systems, adjusting the material assignments for our parts, and evaluating the mass properties for our assemblies.

Summary

In this chapter, we started working with assemblies. In SOLIDWORKS assemblies, we are able to put together more than one part to generate a more complex artifact. Most of the products we use in our everyday lives, such as phones, laptops, and cars, consist of multiple parts that have been put together; that is, they have been assembled. In this chapter, we learned about standard mates, which help us create links to different parts of the assembly. We learned what these mates do, how to apply them, and how to modify them. Then, we learned about materials and mass properties within the context of assemblies.

Now, we should be able to create more complex products that consist of more than one part. We should also be able to build simple static and dynamic interactions between those different parts. All of this brings us closer to designing more realistic products with SOLIDWORKS.

In the next chapter, we will start introducing 2D engineering drawings, which we will use to share our 3D models with individuals and organizations outside our circle or with those who don't have access to SOLIDWORKS. We will cover engineering drawings, why we need them, and how to interpret them.

Questions

Answer the following questions to test your knowledge of this chapter:

1. What are the SOLIDWORKS assemblies?
2. What are mates? What are the three different types of mates?
3. What are standard mates?
4. Download the parts linked to this question and assemble them to form the following drawing in *Figure 8.43*. The assembly should be fully defined:

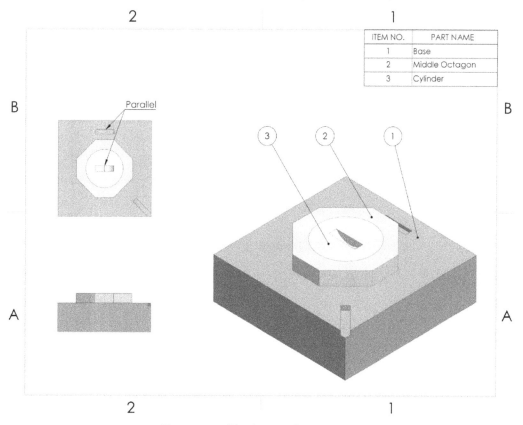

Figure 8.43 – The drawing for question 4

5. Using the assembly from the previous question, adjust the material for each part and define the coordinate system shown in the following drawing in *Figure 8.44*. Determine the mass in grams and the center of mass in mm according to the newly defined coordinate system:

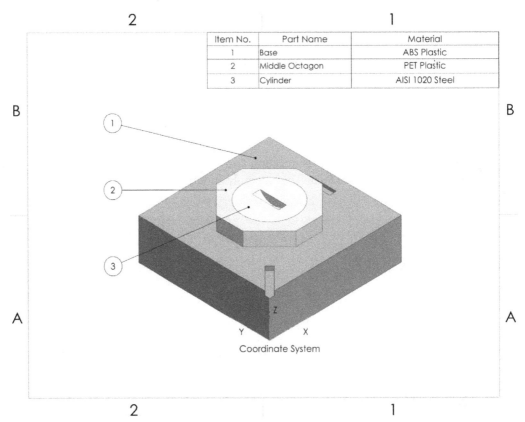

Item No.	Part Name	Material
1	Base	ABS Plastic
2	Middle Octagon	PET Plastic
3	Cylinder	AISI 1020 Steel

Figure 8.44 – The drawing for question 5

6. Download the parts linked to this question and assemble them to form the
    following drawing in *Figure 8.45*. The assembly should be fully defined:

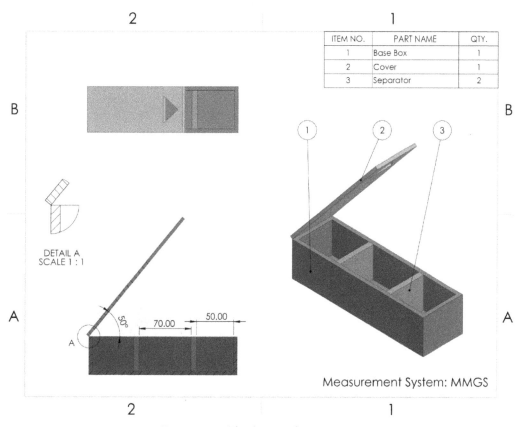

ITEM NO.	PART NAME	QTY.
1	Base Box	1
2	Cover	1
3	Separator	2

Measurement System: MMGS

DETAIL A
SCALE 1 : 1

50°   70.00   50.00

A

Figure 8.45 – The drawing for question 6

7. Using the assembly from the previous question, adjust the material for each
    part and define the coordinate system for the following drawing in *Figure 8.46*.
    Determine the mass in pounds and the center of mass in inches, according to the
    newly defined coordinate system:

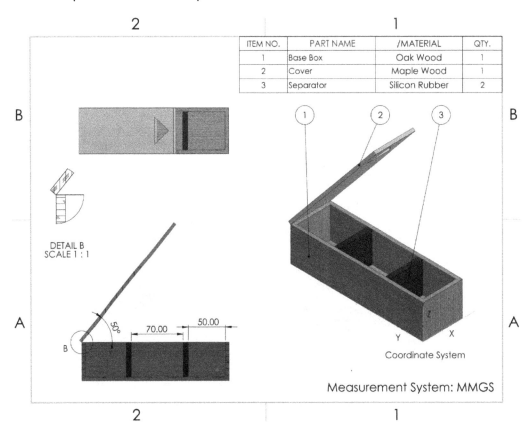

ITEM NO.	PART NAME	/MATERIAL	QTY.
1	Base Box	Oak Wood	1
2	Cover	Maple Wood	1
3	Separator	Silicon Rubber	2

DETAIL B
SCALE 1 : 1

50.00

70.00

50°

Coordinate System

Measurement System: MMGS

Figure 8.46 – The drawing for question 7

Important Note

The answers to the preceding questions can be found at the end of this book.

Section 5 – 2D Engineering Drawings Foundation

2D engineering drawings are key to the design workflow. They enable you to document and communicate your work. This section will cover the foundations of engineering drawings' interpretations, how to construct simple engineering drawings, and how to generate and adjust bills of materials in the SOLIDWORKS drawing environment. It also includes a project requiring you to interpret drawings and 3D model a pair of glasses.

This section comprises the following chapters:

- *Chapter 9, Introduction to Engineering Drawing*
- *Project 1, 3D - Modeling a Pair of Glasses*
- *Chapter 10, Basic SOLIDWORKS Drawing Layout and Annotations*
- *Chapter 11, Bills of Materials*

9
Introduction to Engineering Drawings

Whenever we want to communicate a specific design to others, for example, makers, manufacturers, or evaluators, a 2D engineering drawing is often required. Such drawings usually communicate the shape, materials, and dimensions of any product. This chapter will cover basic knowledge of engineering drawings and how to interpret different layouts found in them. Interpreting drawings is an essential skill for us to be able to generate our own drawings and to collaborate with other people by interpreting their drawings.

The following topics will be covered in this chapter:

- Understanding engineering drawings
- Interpreting engineering drawings

By the end of this chapter, you will have gained knowledge about different engineering drawing concepts. In addition, you will be able to interpret the different types of lines and views often found in engineering drawings.

Understanding engineering drawings

Engineering drawings are what we use to communicate designs to other entities. Whenever we produce a design for a specific product, we are often required to present an engineering drawing with it to communicate the design. Within an engineering drawing, we can communicate the shape of the design, the materials, the suppliers, and any other information we want to communicate.

Also, when engineers and technicians maintain a certain plant or a facility, they interact with engineering drawings in their day-to-day jobs. This is to identify what the machine comprises, how to maintain it, and the materials required for that. A couple of examples of engineering drawings are as follows. The following figure is of a simple part, communicating only the shape and overall dimensions of the part:

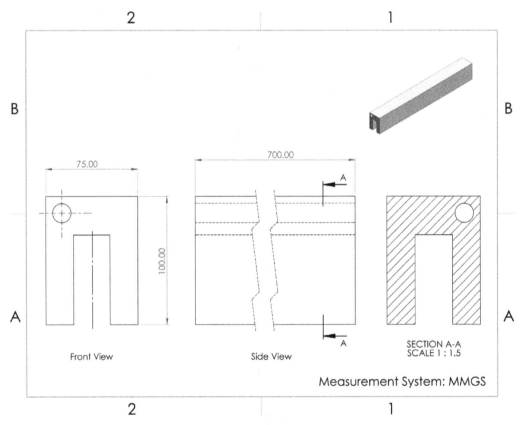

Figure 9.1 – A drawing communicating a simple part

The following figure is of a more complex assembly. Note that this drawing does not communicate dimensions; rather, it communicates the different parts included in the assembly. In other words, it communicates the bill of materials:

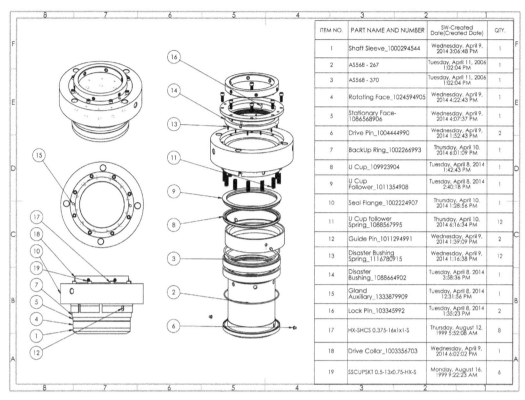

ITEM NO.	PART NAME AND NUMBER	SW-Created Date(Created Date)	QTY.
1	Shaft Sleeve_1000294544	Wednesday, April 9, 2014 3:06:48 PM	1
2	AS568 - 267	Tuesday, April 11, 2006 1:02:04 PM	1
3	AS568 - 370	Tuesday, April 11, 2006 1:02:04 PM	1
4	Rotating Face_1024594905	Wednesday, April 9, 2014 4:22:43 PM	1
5	Stationary Face-1086568906	Wednesday, April 9, 2014 4:07:37 PM	1
6	Drive Pin_1004444990	Wednesday, April 9, 2014 1:52:43 PM	2
7	BackUp Ring_1002266993	Thursday, April 10, 2014 6:01:09 PM	1
8	U Cup_109923904	Tuesday, April 8, 2014 1:42:43 PM	1
9	U Cup Follower_1011354908	Tuesday, April 8, 2014 2:40:18 PM	1
10	Seal Flange_1002224907	Thursday, April 10, 2014 1:28:56 PM	1
11	U Cup follower Spring_1088567995	Thursday, April 10, 2014 6:16:34 PM	12
12	Guide Pin_1011294991	Wednesday, April 9, 2014 1:39:09 PM	2
13	Disaster Bushing Spring_1116780915	Wednesday, April 9, 2014 1:16:38 PM	12
14	Disaster Bushing_1088664902	Tuesday, April 8, 2014 3:58:36 PM	1
15	Gland Auxiliary_1333879909	Tuesday, April 8, 2014 12:31:56 PM	1
16	Lock Pin_103345992	Tuesday, April 8, 2014 1:35:23 PM	2
17	HX-SHCS 0.375-16x1x1-S	Thursday, August 12, 1999 5:52:08 AM	8
18	Drive Collar_1003356703	Wednesday, April 9, 2014 6:02:02 PM	1
19	SSCUPSKT 0.5-13x0.75-HX-S	Monday, August 16, 1999 9:22:23 AM	6

Figure 9.2 – A drawing showing an assembly and its bill of materials

Engineering drawings vary in complexity according to what they communicate. Also, the information displayed on a drawing sheet can vary from one organization to another. However, all drawings follow the same standards in terms of communicating different aspects of the drawing. Engineering drawings became an essential tool for communication due to their flexible distribution. They can be printed on paper or sent as images or PDFs for viewing with common software, such as an image viewer or a PDF reader.

One major element of SOLIDWORKS is drawings. These enable us to create engineering drawings for our parts and assemblies. To be able to create drawings in SOLIDWORKS, it is important for us to have some understanding of basic drawing standards and communication practices.

In this section, we have learned about engineering drawings and their purpose. Now, we can start learning about some key standards used when creating drawings to help us interpret them.

Interpreting engineering drawings

Being able to interpret engineering drawings is an essential part of creating them. In this section, we will cover essential drawing competencies, including how to interpret different types of lines and different types of drawing views. We will start by understanding lines, then views, and then projections. Interpreting drawings is a skill that grows with time as we are exposed to more drawings. We will learn about some key standards that are followed when generating drawings. Those common standards will help us interpret drawings regardless of their source.

So first, we will start by interpreting the most essential drawing element – lines.

Interpreting lines

In simple terms, we can look at drawings as different lines connected. However, the shape of a line gives it a different meaning. The following figure shows the most common types of lines found in engineering drawings:

Type and shape of the line	Line indication
Visible object lines	Visible object lines show the visible outline of the object.
Hidden line	This shows the hidden outline of the object from the drawing's viewpoint. This includes any details that are at the back of the object.
Centerline	This indicates the center of any two entities. For example, the center of two edges and the center of a circle.
Dimension	This line is not part of the object; rather, it indicates the dimension of a drawing entity. Note that dimension lines are much lighter in comparison to visible object lines.
Break line	This indicates a break in the object in the drawing. This is often used to fit relatively long objects within a drawing, such as long construction beams. Note that there are many different types of break lines. The three types shown here are the most common ones. They are jagged cut, zig-zag cut, and small zig-zag cut, as seen from top to bottom.
Section / hatch lines	Section lines are inclined lines indicating a cut in a section. We will see them in section views.
Section cutting line	Section cutting lines highlight the viewing location and angle of a section cut.

Figure 9.3 – Common different lines found in engineering drawings

The following figure highlights a model and its 2D drawing. All the lines in the table are highlighted in the following drawing for easy reference:

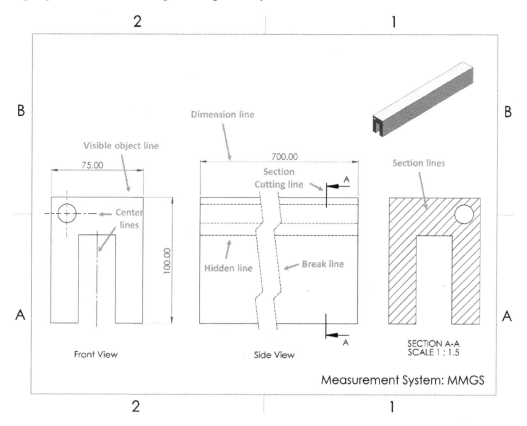

Figure 9.4 – An engineering drawing utilizing different types of lines

Note that when we create drawings with SOLIDWORKS, the software will generate all those lines according to the international standard. However, it is important for us to be able to identify the different types of lines when we see them. Now that we know how to interpret lines, we can start learning how to interpret views.

Interpreting views

In a general sense, a drawing consists of different views of a specific object. Each of the views can give us a deep insight into the shape of the object. As views are also indicated with lines, there is a lot of common knowledge between lines and views. We will look at the most common views at this level and how we can interpret them. We will briefly discuss auxiliary views, section views, detail views, broken-out section views, and crop views.

To investigate all the views, we will examine them using the following model:

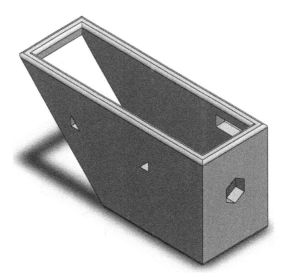

Figure 9.5 – We will use this part to demonstrate the different drawing views

Using the model in *Figure 9.5*, we will explore the following views:

- Orthogonal views
- Auxiliary views
- Section views
- Detail views
- Broken-out section views
- Crop views

For each view, we will define its purpose and highlight how it looks in relation to the preceding model.

Orthogonal views

Orthogonal views are the most common views we will come across in engineering drawings. They are basically a combination of the front, side, top, bottom, and back views. There are two common standards in constructing orthogonal views. They are **First Angle projections** and **Third Angle projections**. Both standards are different, based on the interpretation of the top, right, left, and bottom views surrounding the base front view. The following figure shows the orthogonal view for our model on the third angle projection:

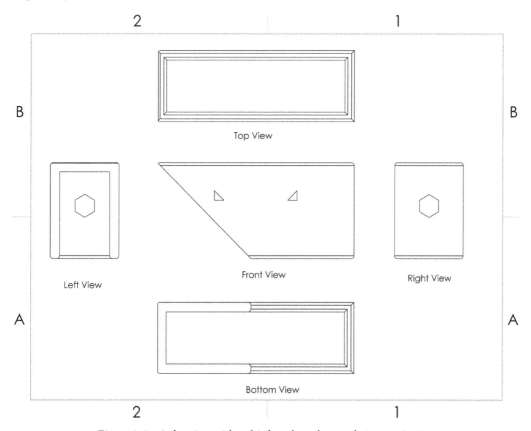

Figure 9.6 – A drawing with a third angle orthogonal view projection

The following figure shows the same orthogonal views following the first angle projection standard:

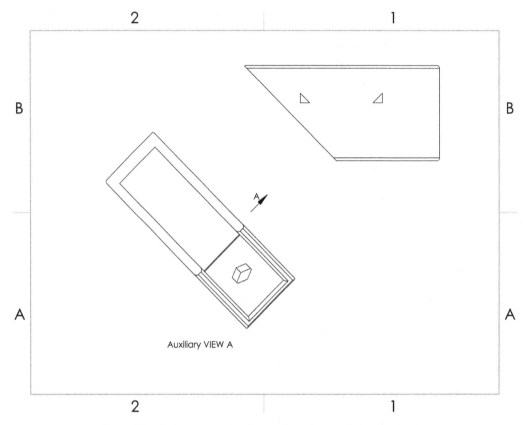

Figure 9.7 – A drawing with a first angle orthogonal view projection

The two standards are more prominent in different countries. For example, first angle projections are more common in Europe and India, while the third angle projections are more prominent in the United States of America and Japan. You can follow the link in the *Further reading* section at the end of the chapter for more information about those standards.

Auxiliary view

The **auxiliary view** shows the view of the model if we look at it from a selected surface or edge, that is, a perpendicular projection of a surface. The following figure highlights the front view of the model as well as the auxiliary view from a tilted angle. Note that the indicated arrow shows the view angle of the auxiliary view. Auxiliary views are often used to show the true size of a specific angle or to communicate specific details that are not clear enough from basic orthographic views:

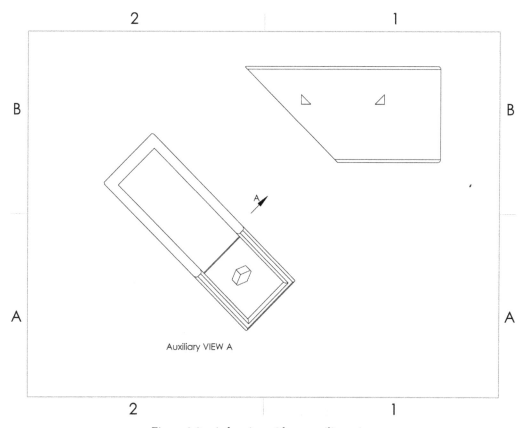

Figure 9.8 – A drawing with an auxiliary view

Section views

Section views allow us to see cross sections of our models on a selected plane. This type of view allows us to see details that otherwise would be hidden from normal orthogonal views. The following figure shows the front view as well as a section view of our model.

The section line in the front view indicates where the cut was made, while the arrow indicates which side we are looking at:

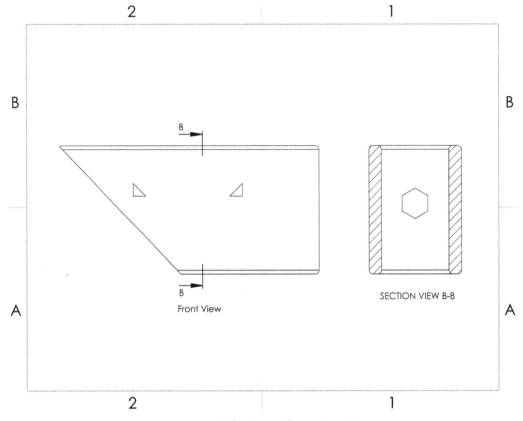

Figure 9.9 – A drawing with a section view

Detail views

Detail views allow us to see and note small details that are otherwise hard to notice. *Figure 9.10* shows the front view as well as a detailed view of the small triangle:

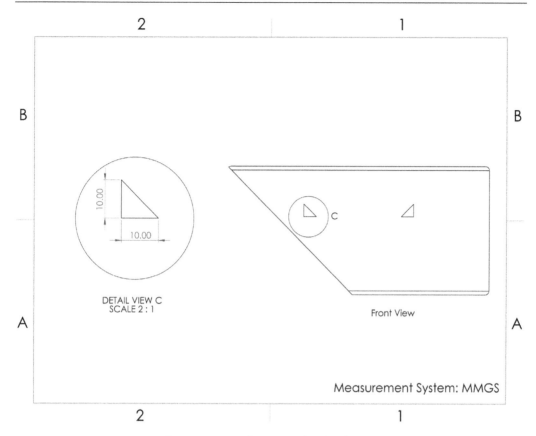

Figure 9.10 – A drawing with a detail view

Note that the front view has a circle indicated by **C** to show the area/zone from where the view is taken. The same letter is used to name the view; thus, our detail view is also named **C**. Also, note that the detail view has its own scale to show how big it is. Generally, the detail view would have a larger scale compared to the original view.

Broken-out section views

Broken-out section views are local section views that do not require a section line. They allow us to see what is behind a specific surface. The following figure highlights the front view with and without a broken-out section.

Broken-out section views allow us to see details that are hidden from view:

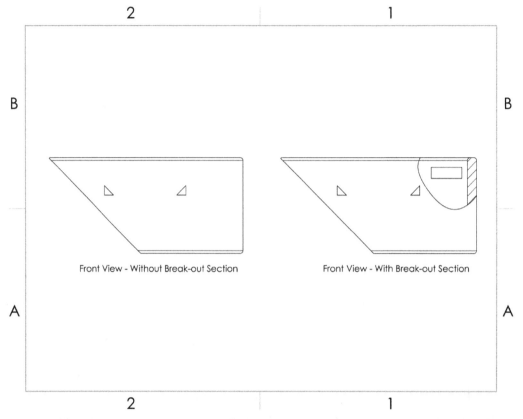

Figure 9.11 – A drawing with a broken-out section view

Crop views

In a **cropped view**, we can crop a specific part of a drawing and show it as a standalone view. The following figure highlights the full right view of our model as well as a cropped view of it:

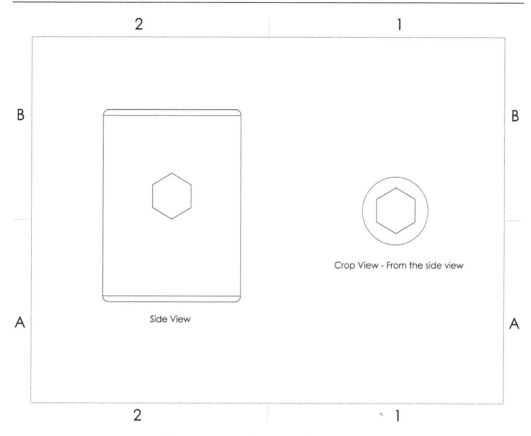

Figure 9.12 – A drawing with a crop view

Now that we have learned the major types of views that we will get exposed to when working with SOLIDWORKS, or when interpreting drawings provided to us, next, we will learn about axonometric projections.

Axonometric projections

Simply put, we can understand axonometric projections as 3D views of the model we are creating the drawing for – in other words, the object tilted to a certain degree in comparison to the plane of projection, which is our drawing sheet. There are three common types of axonometric projections: **isometric**, **dimetric**, and **trimetric**. To understand the difference between the three projections, we will examine them in relation to a simple cube. The drawings illustrate the same cube, showing the three types of projection:

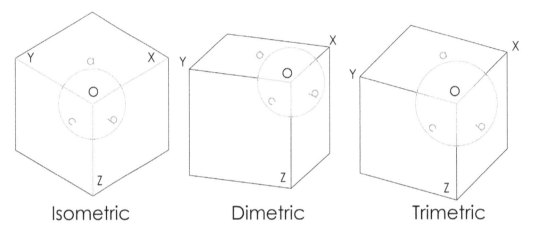

Figure 9.13 – A cube shown in isometric, dimetric, and trimetric views

Here are short descriptions illustrating the difference between the three types of axonometric projection:

- **Isometric**: Here, angles *a*, *b*, and *c* are equal. Also, lines O–X, O–Y, and O–Z are equal. This projection is the most common projection we will see and use.

- **Dimetric**: Here, angles *a* and *b* are equal. Also, lines O–Y and O–Z are equal.

- **Trimetric**: All the indicated angles and lines are unequal.

This concludes our brief introduction to axonometric projections. At the level of this book, we will not learn how to generate all of the highlighted views and projections we have covered. However, it is important for us to be able to interpret them.

Summary

Engineering drawings are essential to communicate our designs to manufacturers, maintenance teams, or any other entity. Engineering drawings not only communicate dimensions; they can also communicate materials, part specifications, tolerances, and whatever an organization/individual takes as a practice or a standard to follow. In this chapter, we learned what engineering drawings are and their purpose. We also learned how to interpret different standards related to lines and drawing views. Being able to interpret drawings is a fundamental skill for working with SOLIDWORKS. However, as with other skills, it takes time and practice to master.

Next, we will work on a 3D-modeling project to create a pair of glasses. The project will provide you with comprehensive practice of all the topics covered from the beginning of the book to this chapter.

Questions

Answer the following questions to test your knowledge of this chapter:

1. What are 2D engineering drawings?

2. What does the following line indicate?

Figure 9.14 – The line referred to in question 2

3. What is the difference between visible object lines and dimension lines?

4. What are hidden lines, what do they indicate, and what do they look like?

5. What are detail views and when do we use them?

6. What is a cropped view?

7. In the following figure, what is the name of the view shown on the right? What is its purpose?

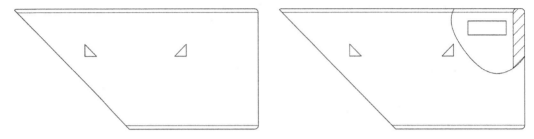

Figure 9.15 – The drawing view referred to in question 7

Important Note
The answers to the preceding questions can be found at the end of this book.

Further Reading

More information about the first and third angle projections in orthogonal views can be found here: `https://en.wikipedia.org/wiki/Multiview_orthographic_projection`.

Project 1: 3D-Modeling a Pair of Glasses

SOLIDWORKS is a 3D-design tool. As with all tools, the more you use it, the better you become at it. In this project chapter, you will be provided with work that you can complete to hone your skills. In this project, you will be 3D-modeling and assembling a pair of glasses from a set of engineering drawings.

This project chapter will cover the following topics:

- Understanding the project
- 3D-modeling the individual parts
- Creating the assembly

By the end of this chapter, you will have more confidence in using the different SOLIDWORKS tools for practical projects.

Technical requirements

You will need to have access to SOLIDWORKS to complete the project.

Understanding the project

Understanding what the project entails is essential before starting the work. This will allow you to draw a plan and manage your work expectations when completing the project. For this exercise project, you will be 3D-modeling a pair of glasses, as shown in the following figure:

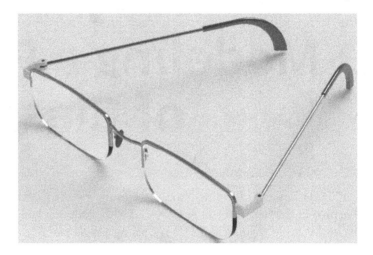

Figure P1.1 – The pair of glasses you will 3D-model in this project

The pair consists of 15 parts, 8 of which are unique. The following figure highlights the bill of materials, showing the names of the parts, their quantity, and position in the assembly:

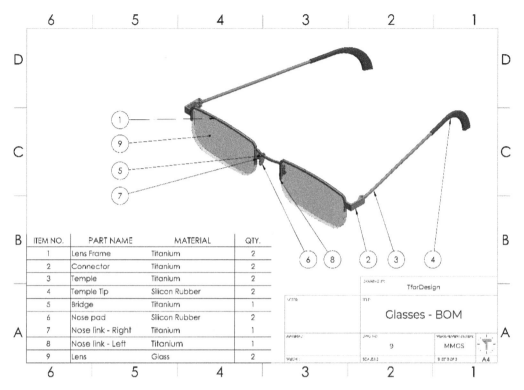

ITEM NO.	PART NAME	MATERIAL	QTY.
1	Lens Frame	Titanium	2
2	Connector	Titanium	2
3	Temple	Titanium	2
4	Temple Tip	Silicon Rubber	2
5	Bridge	Titanium	1
6	Nose pad	Silicon Rubber	2
7	Nose link - Right	Titanium	1
8	Nose link - Left	Titanium	1
9	Lens	Glass	2

Figure P1.2 – The pair and the bill of material

At this point, you already have an idea of the project's outcome and the complexity of the required parts and assembly. In this next section, we will provide you with the engineering drawings needed to replicate all the parts and assembly. Now that we have an idea of the project's final output, we can discuss how you can tackle it in the context of this writing.

> **Important Note**
>
> The drawings and 3D models presented in this project are for practice purposes rather than manufacturing purposes.

There are two ways in which you can tackle this project, depending on your 3D-modeling level. They are as follows:

- **Moderate level**: Take a look at the provided drawings and hints to complete the project.

- **Advanced level**: Only take a look at the drawings without using the hints.

Other than the two suggested approaches, you can also design your own way to 3D-model the glasses utilizing the provided drawings and selected hints.

> **Tip**
> You can treat the project as your own and customize the provided glasses to end up with your own unique design.

In this write-up, we will first explore the individual parts and then move on to the assembly. So, let's get started with the parts. We will also provide you with hints that can assist you with your work.

3D-modeling the individual parts

In this section, we will explore the different part drawings to make up the pair of glasses. The provided drawings have enough information for you to replicate all the parts to end up with an identical result to the one shown in *Figure P1.1*.

Thus, one option you have for going about the project is to create an exact replica of the given drawings. However, you can also choose to customize and adjust different elements of the design to make it your own. Keep in mind that this is your project, so feel free to treat it as such.

Creating the individual parts

The provided pair of glasses consists of 15 parts. However, 8 parts are unique, which you will need to 3D-model, as highlighted in *Figure P1.2*. The parts you will need to 3D-model are as follows:

1. Lens frame
2. Lens
3. Bridge
4. Nose link
5. Nose pad
6. Connector
7. Temple
8. Temple tip

> **Important Note**
>
> The names of the parts presented in the bill of materials might be different to the names practiced in different parts of the world.

Your task is to use the presented drawings to 3D-model the individual parts. As you are 3D-modeling the different parts of the glasses, keep in mind that there is no one correct way of 3D-modeling any of the parts. However, we will provide you with some hints that can push you forward if you find yourself getting stuck. You can also feel free to customize your own design using the given drawings as a base of inspiration.

> **Important Note**
>
> The order in which the following drawings are listed is arbitrary.

Let's start exploring the drawings one after the other. The first drawing is for the *lens frame*:

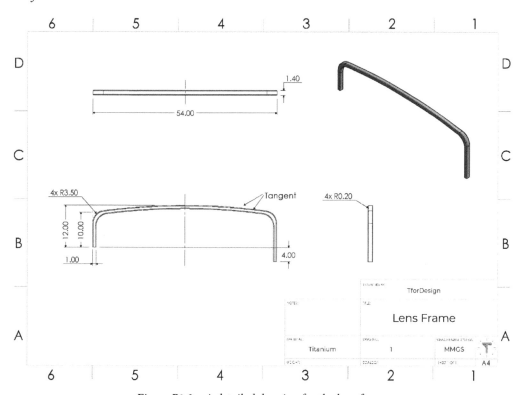

Figure P1.3 – A detailed drawing for the lens frame

Here is a sample procedure for 3D-modeling the lens frame:

1. You can start with an extruded boss to create the main shape of the frame, as shown in the following figure:

Figure P1.4 – An extruded cut can be used to start the lens frame

2. Apply fillets to get the rounded edges.

Next, we can look at the *lens* itself, which will be held by the *lens frame*:

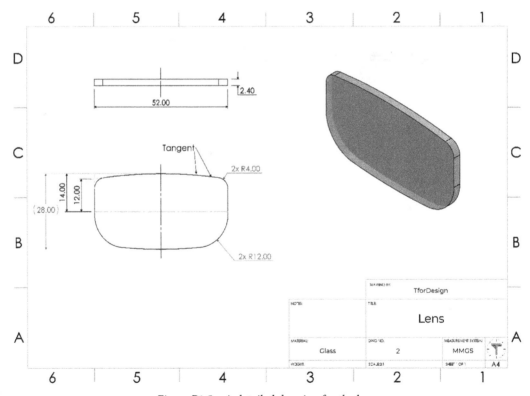

Figure P1.5 – A detailed drawing for the lens

Here is a sample procedure for 3D-modeling the *lens*:

1. Use an extruded boss to build the following shape:

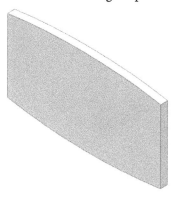

Figure P1.6 – A possible first step in creating the lens using an extruded boss

2. Use fillets to round the corners.

After the lens, we can explore the *bridge*, which will be at the center of the glasses.

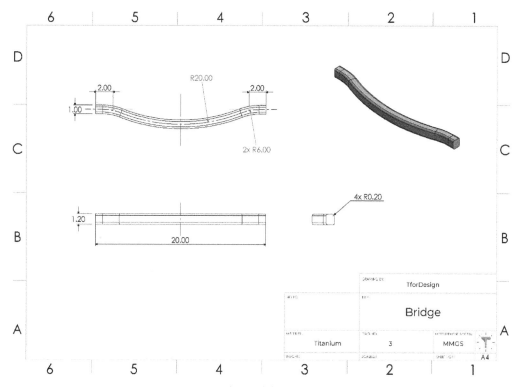

Figure P1.7 – A detailed drawing for the bridge

Here is a sample procedure for 3D-modeling the *Bridge*:

1. Use a swept boss with a square profile to create the basic structure, as shown in the following figure:

Figure P1.8 – A swept boss can be used to create the bridge

2. Use the fillet feature to round out the rounded corners, as shown in *Figure P1.7*.

After the *bridge*, we can take a look at the *nose pad*, which will be connected to the *nose link* and resting on the nose of the user.

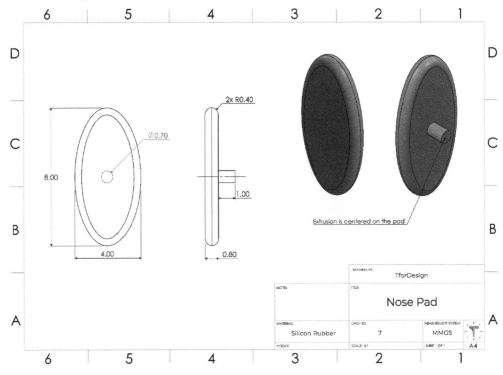

Figure P1.9 – A detailed drawing for the nose pad

Here is a sample procedure for 3D-modeling the *nose pad*:

1. Use an extruded boss with an elliptical sketch to create the following shape:

Figure P1.10 – An extruded boss can create the bulk of the nose pad

2. Apply another extruded boss to create the cylindrical extrusion, as shown in the following figure:

Figure P1.11 – A second extruded boss can be used to create the small step on the nose pad

3. Use the fillet feature to create the rounded R0.4 mm edges.

Next, we can have a closer look into the *connector*, which will be linked to the *lens frame* on one side and the *temple* rod on the other side.

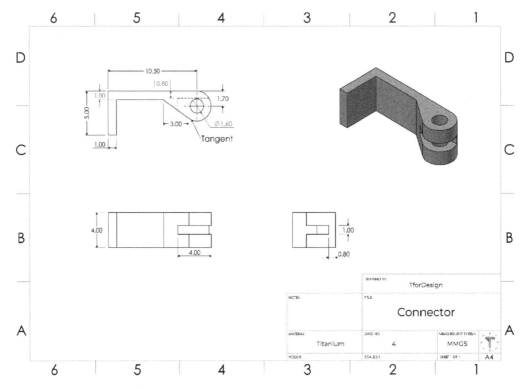

Figure P1.12 – A detailed drawing for the connector

Here is a sample procedure for 3D-modeling the *connector*:

1. Use an extruded cut to create the main shape of the connector, as shown in the following figure:

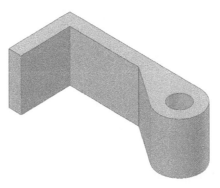

Figure P1.13 – Most of the connector can be done with one extruded boss

2. Use an extruded cut to create the middle cut to end up with the following shape:

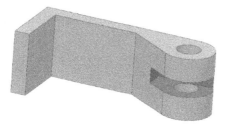

Figure P1.14 – The slot in the connector can be created with an extruded cut

After the *connector*, we start working on the *temple*, which is the long part that links to the connector and helps hold the glasses against the user's ears.

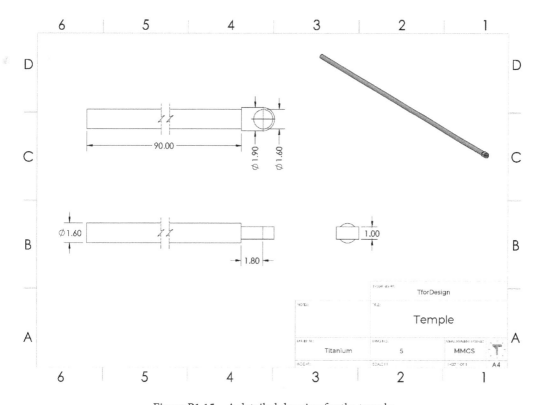

Figure P1.15 – A detailed drawing for the temple

Here is a sample procedure for 3D-modeling the *temple*:

1. Use the Extruded Boss feature to create the following shape:

Figure P1.16 – An extruded boss can generate the ring-end of the temple

2. Apply another extruded boss to create the long cylindrical rod ending with the final temple, as shown here:

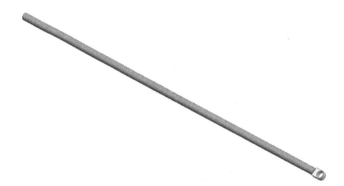

Figure P1.17 – The 90 mm-long cylindrical part can be created with an extruded boss

Note that the drawing in *Figure P1.15* utilizes break lines to shorten the length of the temple rod in the drawing sheet.

Next, we can start looking at the last part, the *temple tip*, which is the relatively softer part touching the user's ear.

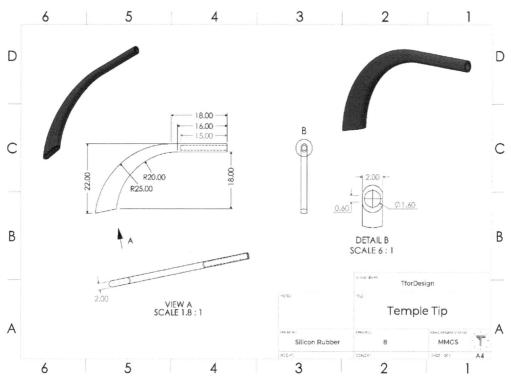

Figure P1.18 – A detailed drawing for the temple tip

Here is a sample procedure for 3D-modeling the temple tip:

1. We can use the Lofted Boss feature with guide curves to create the shape of the temple tip. We can use sketches that resemble the following:

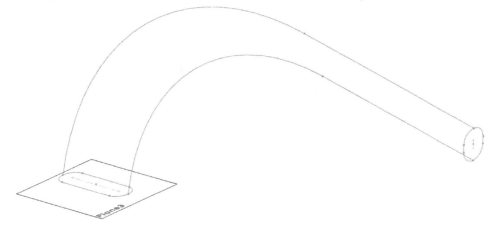

Figure P1.19 – The profiles and guide curves used to create the base lofted boss

2. Using the sketched profiles and guide curves, use the Lofted Boss feature to end up with the following shape:

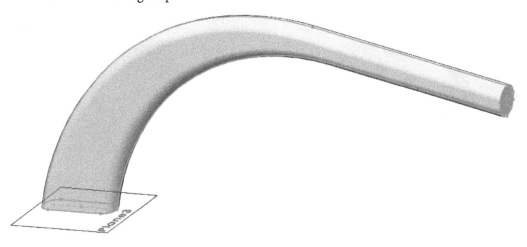

Figure P1.20 – The resulting shape after applying the lofted boss

3. Use an extruded cut to create the 1.6 mm diameter hole, as shown in the following figure:

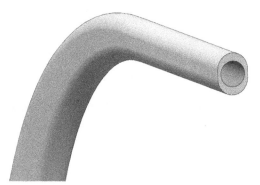

Figure P1.21 – The hole in the temple tip can be made with an extruded cut

Next, we can have a closer look at the *nose link*, which will link the frame of the glasses to the *nose pad*.

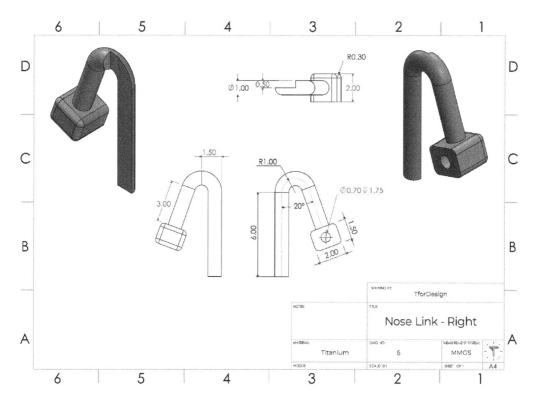

Figure P1.22 – A detailed drawing for the nose link

Here is a sample procedure for 3D-modeling the *nose link*:

1. Use a swept boss with a circular profile to create the tube-looking shape of the *nose link*, as shown in the following figure:

Figure P1.23 – A swept boss can be used to start creating the nose link

2. Use an extruded boss to create the cubical block, as shown in the following figure:

Figure P1.24 – An extruded boss built using a face resulting from the swept boss

3. Use the Extruded Cut feature to generate the 0.7 mm diameter circular hole on the block, as shown in the following figure:

Figure P1.25 – The circular hole in the nose link can be created with an extruded cut

4. Use an extruded cut to cut off a rectangular part of the *nose link*, as indicated in the following figure:

Figure P1.26 – A rectangular sketch can be used with an extruded boss to create the cut

5. Use the Fillet feature to create the R0.3 mm rounded edges.

By this stage, we are done 3D-modeling all the unique parts for our glasses. Feel free to treat the glasses as your own by adjusting the sizes and changing the design. There are no right or wrong answers to how to 3D-model. Next, we will create the mirrored part for our *nose link*.

Creating a mirrored part

A mirrored part has the same features as the original part, but is flipped around a specific plane. They are similar to our right and left arms. Both arms have the same features. However, they are mirrored off one another. Another common example around us is the right and left earphones, right and left shoes, right and left casings, and so on.

The *nose link* part has right and left configurations in the pair of glasses we are creating for this project. We have already created the right *nose link*. We can create the left part by following these steps:

1. Open up the *nose link* part we created earlier.

2. Select the mirroring plane, as shown in the following screenshot:

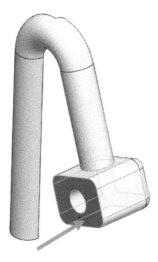

Figure P1.27 – The face to use as a mirror plane to mirror the part

> **Tip**
> Any plane, including new reference planes, can be used to create the mirrored part.

3.   With the mirror plane selected, go to **Insert**, and then select **Mirror Part…**:

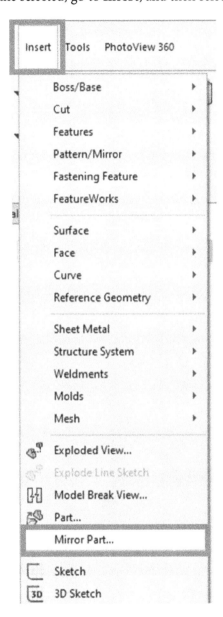

Figure P1.28 – The location of the Mirror Part command

4.   This will open up a new part with the menu shown in *Figure P1.29*. This will allow us to pick the elements we want to transfer to the new mirrored part. In our case, we need to transfer **Solid bodies**. Make sure that box is checked, as shown in the following screenshot:

Figure P1.29 – The initial transfer options when creating a mirrored part

5. Click on the *green checkmark* to create the new mirrored part:

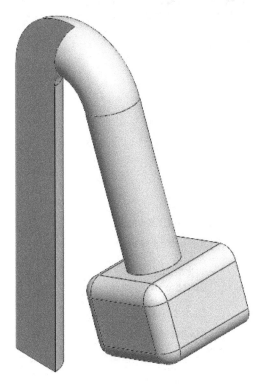

Figure P1.30 – The final mirrored nose link part

6. Save the new part with a new name.

By doing this, we have a new mirrored *nose link* to complement our original one. Keep in mind that the mirrored part takes all its features from the original part. Thus, any changes to the original part will be reflected on the mirrored one. However, new features added to the mirrored part will not be added to the original one.

Now that we have 3D-modeled all the parts, we can join them together in an assembly to form the full pair of glasses.

Creating the assembly

Now that we have all the parts 3D-modeled, we can start exploring the assembly and joining all the parts together. We will do that in this section. We will start by talking about the fixed part, and then explore the different parts and the accompanying mates. The following drawing highlights the fully assembled pair of glasses as well as the names of all the parts:

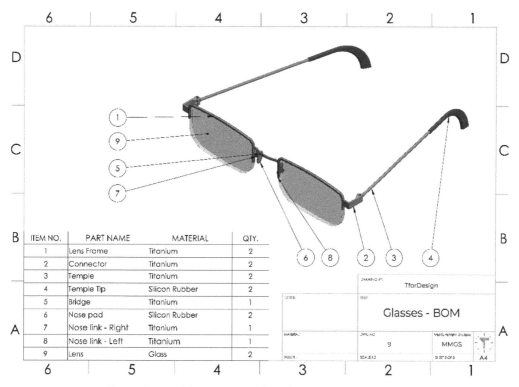

ITEM NO.	PART NAME	MATERIAL	QTY.
1	Lens Frame	Titanium	2
2	Connector	Titanium	2
3	Temple	Titanium	2
4	Temple Tip	Silicon Rubber	2
5	Bridge	Titanium	1
6	Nose pad	Silicon Rubber	2
7	Nose link - Right	Titanium	1
8	Nose link - Left	Titanium	1
9	Lens	Glass	2

Figure P1.31 – The pair assembly and its arrangement of parts

The first part we insert into an assembly becomes a fixed part by default. There is no one answer to which part should be used as a fixed part in the assembly. However, a good practice is to select an inherently non-moving part. A good candidate is the *bridge*.

The following figure highlights the major mates for the pair. You may add more mates as required to get the desired result.

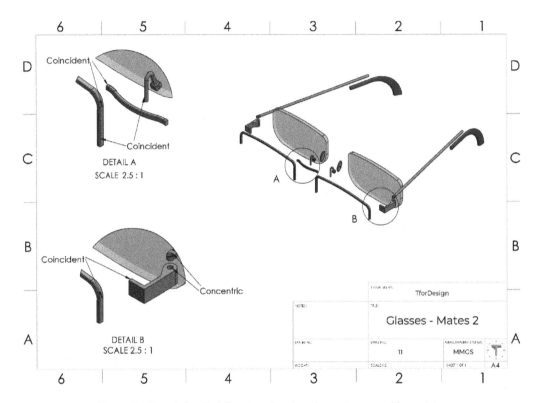

Figure P1.32 – A detailed drawing showing the major assembly metrics

The next figure provides some extra details regarding the assembly, such as the angle of the *nose pad* in relation to the *lens frame*:

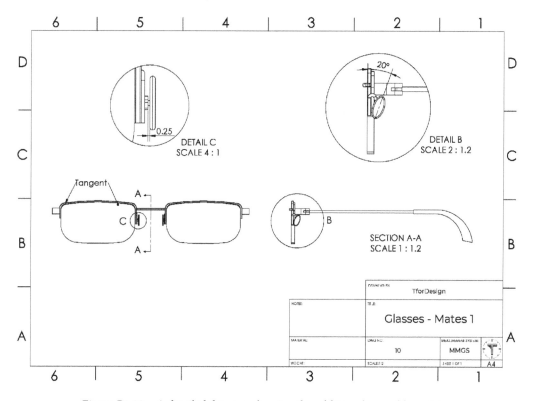

Figure P1.33 – A detailed drawing showing the additional assembly metrics

Putting an assembly together is more open-ended than 3D-modeling a part. So, you can formulate your own way and sequence for joining the different parts together. However, the following figure presents a possible sequence that you can follow if you are unsure where to start. Note that where the mate coincident is the only one mentioned, it was applied more than once.

	Part added	Mates	Preview
Step 1	Bridge (fixed)	NA	
Step 2	Lens Frame	Coincident	
Step 3	Lens	Tangent Coincident	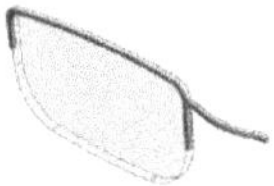
Step 4	Connector	Coincident	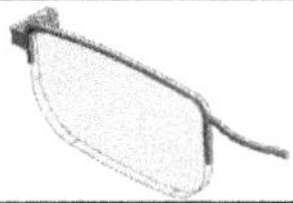
Step 5	Temple	Coincident Concentric	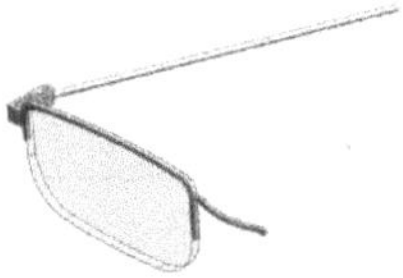
Step 6	Temple Tip	Coincident Concentric Parallel	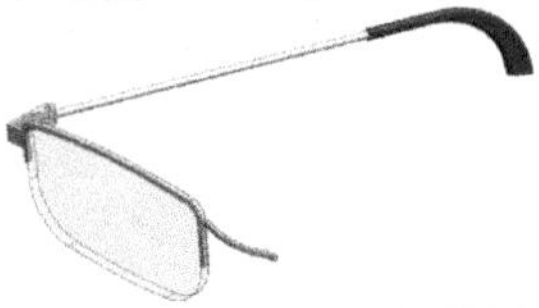
Step 7	Nose Link	Coincident	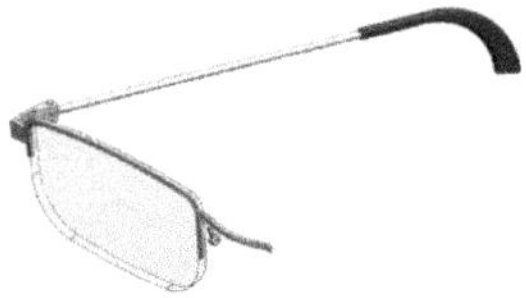
Step 8	Nose Pad	Concentric Distance Angle	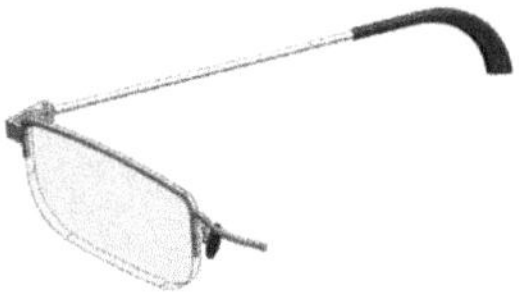

Figure P1.34 – A possible sequence in building the assembly

By the end of *step 8*, half of the glasses would be assembled. You can follow *steps 2–8* again to assemble the other half. Alternatively, you can use the **Mirror Components** command to duplicate applicable parts to the other side.

As you are building the assembly, note that the given mates will not fully define the assembly; instead, it will allow some movements that can easily demonstrate the functionality of the glasses. For example, the *temple* will rotate, keeping one end linked to the *connector* in a similar mechanism to a standard pair of glasses. You can choose to add additional mates to lock the assembly in a specific position.

By completing the assembly, you have completed the project work to 3D-model a pair of glasses.

Summary

In this project chapter, you worked toward 3D-modeling a set of glasses. To that end, you had to interpret engineering drawings, 3D-model different parts, and then join them together in an assembly. The skills used to complete the project include the essential 3D-modeling skills you will use for any project.

In the coming chapters, we will start addressing more advanced commands and features that will allow us to optimize our 3D-modeling approach and 3D-model more complex geometries faster.

10
Basic SOLIDWORKS Drawing Layout and Annotations

It is important for designers and engineers to be able to explain their 3D models to other teams or manufacturers. This could be for the purpose of reviewing the design at hand or manufacturing it. In this chapter, we will learn how to use SOLIDWORKS' drawing tools to do that. We will cover how to generate simple drawings with orthogonal views, how to communicate dimensions and drawing information, and how to export drawings as shareable images or PDFs.

In this chapter, we will cover the following topics:

- Opening a SOLIDWORKS drawing file
- Generating orthographic and isometric views
- Communicating dimensions and design
- Utilizing the drawing sheet's information block
- Exporting the drawing as a PDF or image

By the end of this chapter, you will be able to generate simple engineering drawings to explain your design to individuals or groups that are not SOLIDWORKS users. You will also be able to produce drawings that can be used for manufacturing, documentation, and archiving.

Technical requirements

In this chapter, you will need to have access to the SOLIDWORKS software.

The project files for this chapter can be found at the following GitHub repository: https://github.com/PacktPublishing/Learn-SOLIDWORKS-Second-Edition/tree/main/Chapter10.

Check out the following video to see the code in action: https://bit.ly/3m3UOvQ

Opening a SOLIDWORKS drawing file

In practice, whenever we want to create a drawing in SOLIDWORKS, the first thing we will do is open up a new SOLIDWORKS drawing file. This will have a different format than parts and assemblies. In this section, we will learn how to open a drawing file. This will be our first step when we start working with SOLIDWORKS drawings. To open a new SOLIDWORKS drawing file, follow these steps:

1. Click on **New** at the top of the interface, as shown in the following screenshot:

Figure 10.1 – The location of the button that allows you to open a new file

> **Tip**
> You can press *Ctrl* + *N* on the keyboard for a shortcut to open a new file.

2. Select **Drawing** and click **OK**, or double-click on **Drawing**.

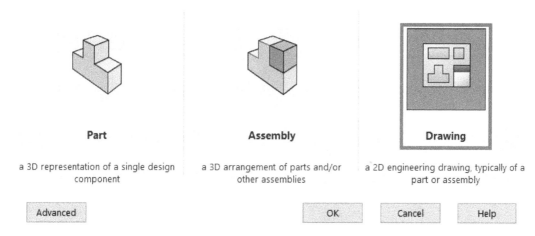

Figure 10.2 – We will work in a drawing document

3. This will open up a drawing file, as shown in the following screenshot. The first window will prompt us to select the size of the drawing sheet we want to use. The list contains all the major standard sheet sizes that are used in the industry. In this exercise, we will select the first one, **A (ANSI) Landscape**, and then click **OK**:

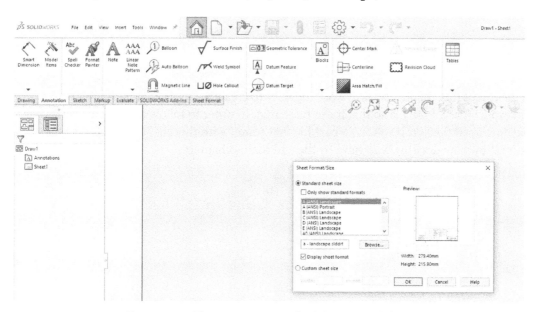

Figure 10.3 – There are many standard sheets to pick from

This will open the following sheet and interface. We will be working with this throughout this chapter:

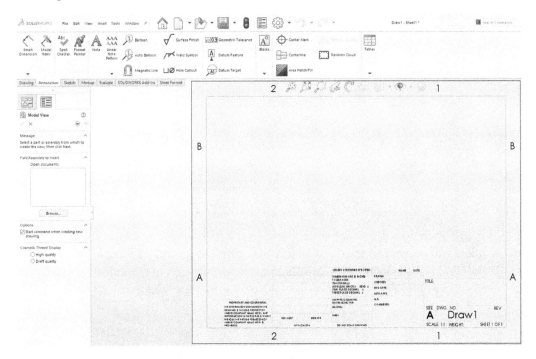

Figure 10.4 – The A (ANSI) Landscape sheet

In the rest of this chapter, we will work together to create a simple engineering drawing and export it as a PDF file so that it can be shared. In order to follow along, make sure that you download the SOLIDWORKS part file that comes with this chapter. The following figure shows the drawing we'll have by the end of this chapter:

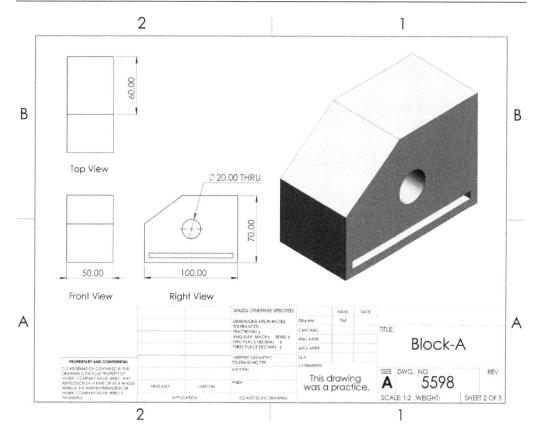

Figure 10.5 – The final drawing we will produce in this chapter

Now that we know how to open a SOLIDWORKS file, we need to generate the standard orthographic views and isometric projection for our model.

Exploring and Generating orthographic and isometric views

The drawing views that we will use the most for our drawings are orthographic and isometric. These views are the simplest to interpret. In this section, we will create a drawing file and input orthographic and isometric views into it. We will also cover the scales and display styles for our drawing. We will start by selecting our targeted model. Then, we will create and adjust the views so that we can coordinate our ideas.

Selecting a model to plot

SOLIDWORKS' drawing tools are based on parts or assemblies that have already been modeled. A drawing file will be linked to the parts or assemblies it communicates with. Thus, after opening a drawing file, our first step is to select the part or assembly file that we want to include in the drawing. Throughout this practical exercise, we will use the following model to create the drawing. To follow along, download the model that's linked to this chapter:

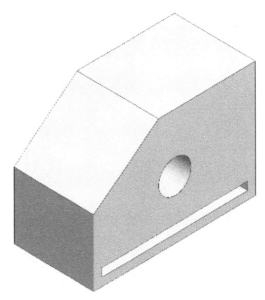

Figure 10.6 – The 3D model we will generate a drawing for

To start drawing, follow these steps:

1. Open a new drawing file. We covered this process in the previous section.

2. After opening a new drawing file, you will notice that the **Model View** selection will be shown on the left, as highlighted in *Figure 10.7*. If not, we can select it by clicking on **Model View** in the **Drawing** tab.

3. Click on **Browse...** to find the model you want to use.

Figure 10.7 – The Model View button to insert a 3D model

4. Browse for the model linked to this exercise, select it, and then click **Open**:

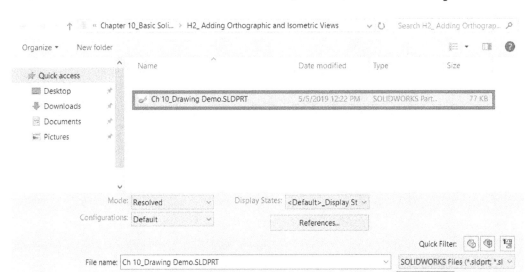

Figure 10.8 – The 2D drawing is generated from a 3D model

This will open a link between our part and our drawing file. Next, we will generate our orthographic and isometric views.

Generating orthographic and isometric views

Once we've selected the model from the model view, we can automatically input our orthographic and isometric views. Note that orthographic views are third-angle projections. After generating the third-angle orthographic projections, we will cover how to change them into first-angle projections. To generate the orthographic views, we can follow these steps:

1. Move the cursor onto the body of the sheet. A rectangular outline will appear. By default, this indicates the front view of the model. This is shown in the following screenshot:

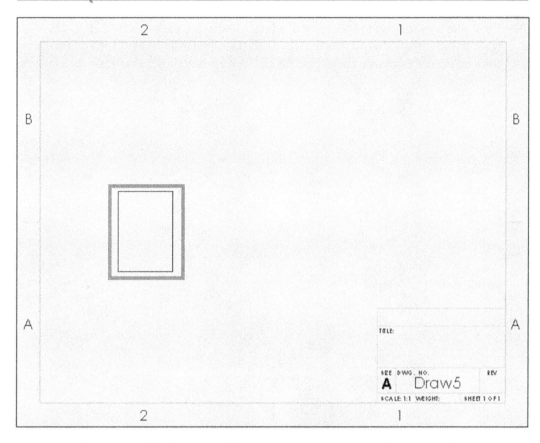

Figure 10.9 – The front view will appear when dragging a part into the drawing

2. Left-click the mouse to input the front view. After that, move the mouse to the right of the view. You will see the right view appear, as shown in *Figure 10.10*. Left-click the mouse again to input that view.

3. Then, move the mouse to the top of the front view to see the top view. Left-click again to input that. Then, move the mouse to the upper-right side of the front view. You will see the isometric view appear. Left-click again to input that view:

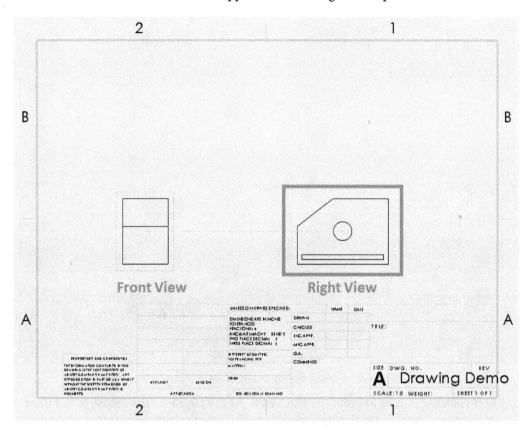

Figure 10.10 – The orthographic views are generated dynamically

> **Note**
>
> This is a dynamic way of inputting orthographic and isometric views. If we move the mouse in any direction, we will notice that the view changes. These additional views are referred to as projected views.

4. After inputting the front, right, top, and isometric views, we can simply press *Esc* on the keyboard or click on the green checkmark at the top right to confirm that the views are correct, as shown in the following screenshot:

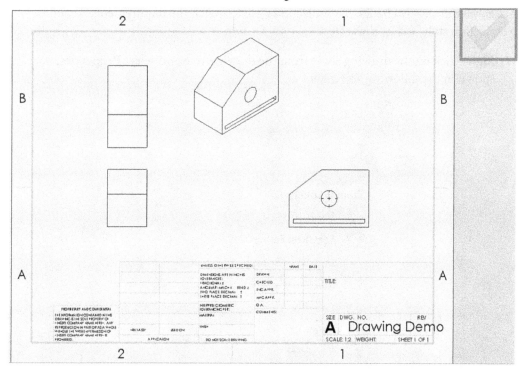

Figure 10.11 – After inserting the views, we can click the indicated checkmark to confirm them

At this point, we should have a few orthographic views, as well as an isometric view, in our drawing canvas. Before we make further adjustments to our drawing views, we need to address how to change our orthographic projections from third-angle to first-angle and vice versa. In addition, we will address some principles that are related to our initial views, that is, parent and child views, another way we can insert views, and how to delete views. We will start by learning about parent and child views.

Changing from third-angle to first-angle projections

The orthographic projections made earlier in this exercise were third-angle projections. However, you might be requested to produce drawings in first-angle projections. If that is the case, we can simply change the projection style to match the requirements. To adjust the projection style, follow these steps:

1. Right-click on the drawing sheet from the drawing tree and select **Properties…**, as shown in the following screenshot:

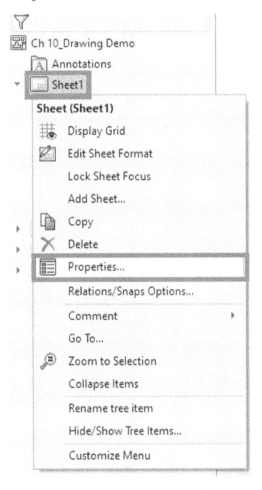

Figure 10.12 – The sheet properties can be found in the drawing design tree

2. Under the **Sheet Properties** tab, there is an option for the type of projection, as shown in the following screenshot. You can select the required type, then click **Apply Changes**:

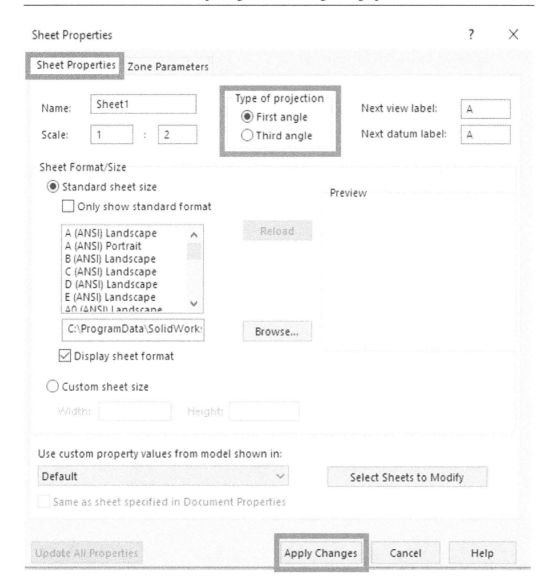

Figure 10.13 – The sheet properties allow you to adjust an active sheet

Once the changes are applied, the existing views will switch from being third-angle projections to first-angle projections. The same procedure can be followed to change the projection type from a first angle to a third angle. Now, we can move on to exploring other drawing principles, starting with the parent and child views.

Parent and child views

Note that in the preceding drawing, the right, top, and isometric views were created based on the front view. As such, we can understand the front view as the parent view of all the other views that it inspired in terms of their creation. The following figure highlights the parent view and child view of the drawings we just created. This allows us to propagate changes to more than one view at a time. When we move the parent view, we will see that all the child orthographic views move with it. Also, when changing aspects, such as the drawing scale in the parent view, they will be copied to all the child views:

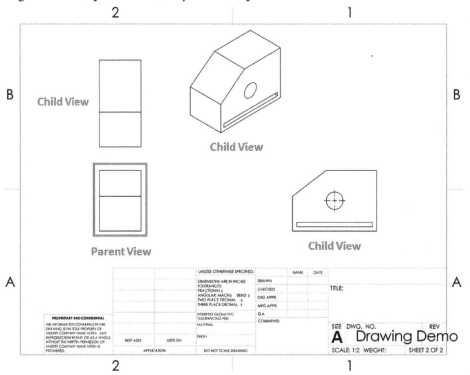

Figure 10.14 – Parent and child views

By default, child views will copy the features of the parent view. However, we can stop this from the drawing view's property manager. Next, we will explore another way of adding views: by using the View Palette.

Adding views via the View Palette

Other than adding views via the model view feature, we can add separate views more flexibly via the View Palette. We can access the View Palette via the Task Pane on the right-hand side of the screen. If the Task Pane is not visible, you can display it by clicking on **View | User Interface | Task Pane**, as shown in the following screenshot:

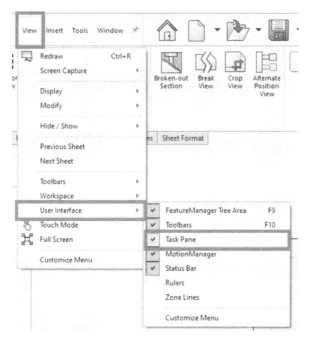

Figure 10.15 – The Task Pane has to be shown in the user interface to use it

Now that we have the Task Pane visible, we can use it to insert views via the View Palette. To do it, we can follow those steps:

1. Click on the View Palette from the shortcuts menu on the right-hand side of the screen, as shown in the following screenshot:

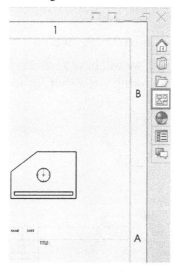

Figure 10.16 – The location of the View Palette

2. If the model is already selected, we will be able to see it in the drop-down menu at the top of the options, as highlighted in the following screenshot. If not, we can browse to select a part or assembly:

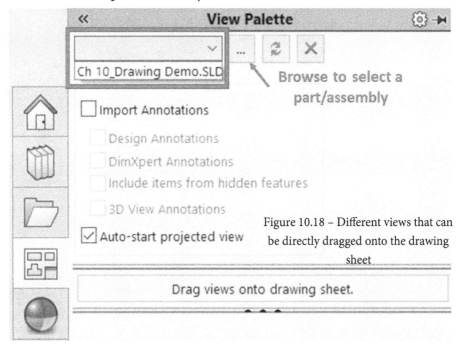

Figure 10.18 – Different views that can be directly dragged onto the drawing sheet

Figure 10.17 – A drop-down menu showing the active 3D model in the drawing

3. Once we've selected the model, we will have a number of viewing options to choose from, as highlighted in the following screenshot. To add a view, we can simply drag it onto the drawing sheet:

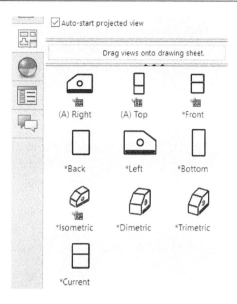

Figure 10.18 – Different views that can be directly dragged onto the drawing sheet

4. Drag the **Trimetric** view from the View Palette onto the drawing canvas. This will insert that view into the drawing, resulting in a drawing as in the following figure:

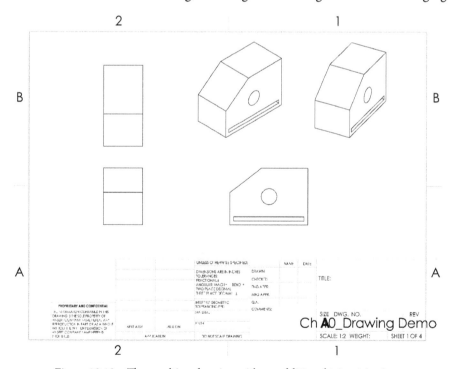

Figure 10.19 – The resulting drawing with an additional trimetric view

Using the View Palette to add views provides us with a quicker way to directly drag and drop specific drawing views from the side of the interface. The current view represents the last part of the orientation in the original .SLDPRT file. Next, we will learn how to delete views.

Deleting views

Often, we may input a drawing view and decide to delete it later. For example, after inserting the trimetric view, which we did earlier, we came to the conclusion that it adds no value; thus, we would like to delete it. There are two methods we can follow when it comes to deleting a drawing view:

- **Method 1**: Select the drawing view on the drawing sheet and then press the *Delete* key on the keyboard. We will get a **Confirm Delete** message, as shown in the following screenshot. To delete the view, click **Yes**:

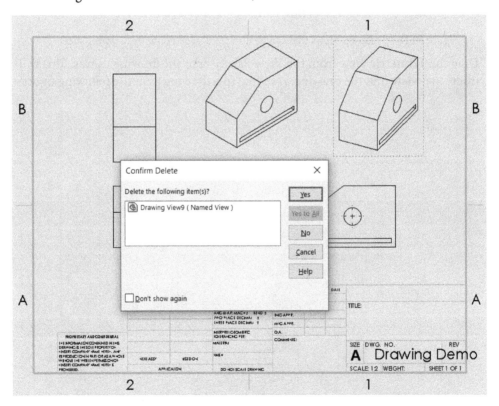

Figure 10.20 – A confirmation message appears when deleting a view

- **Method 2**: Right-click on the view and select **Delete**, as shown in the following screenshot:

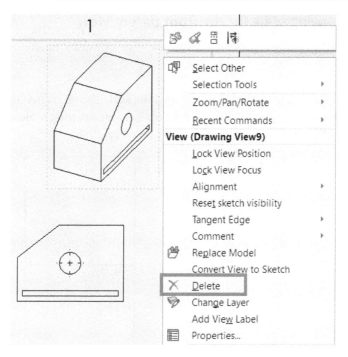

Figure 10.21 – We can delete a view using the Delete option

This concludes the two common methods we can follow to delete a certain view from our drawing's canvas.

In this section, we have added orthographic and isometric views to our drawing file. However, note that we still have lots of empty space in our drawing sheet, which means we can make our drawing views larger. We will address the drawing scale and display in the next section.

Adjusting the drawing scale and the display

With SOLIDWORKS drawings, we also have the option of adjusting the size of our displayed view and what it looks like. In this part, we will continue working on our drawing by adjusting both the size and the display style.

Adjusting the scale of our drawing

When we input the drawing views into the drawing sheet, SOLIDWORKS automatically sets a certain suitable scale for our drawing. The scale refers to the size of the drawing as it's displayed in the drawing sheet. Regardless of the drawing that's displayed, we can change it. To highlight how we can change the drawing scale, we will adjust the scale for the front view (a parent view) and the isometric view (a child view).

Changing the drawing scale for the front parent view

To change the drawing scale for the front parent view for our drawing, follow these steps:

1. Click on the front (parent) view.

2. In the PropertyManager, scroll down to find the scale listing. When clicking on the drop-down menu, we will see multiple options for drawing scales. We can also select **User Defined** if we wish to define our own scale:

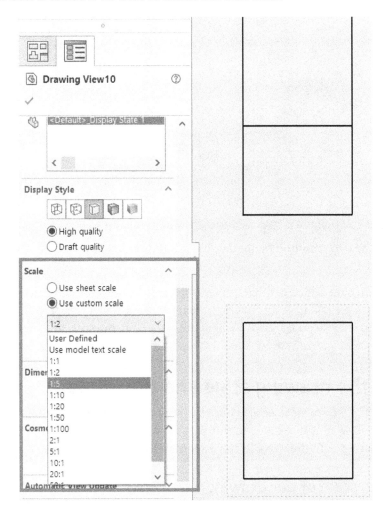

Figure 10.22 – Adjusting the scale using the PropertyManager

3. Set the scale to **1:5**. We will notice that all the drawing views become smaller, as shown in the following figure. Note that all the views changed in scale as well. This is because they are linked to the parent view by default:

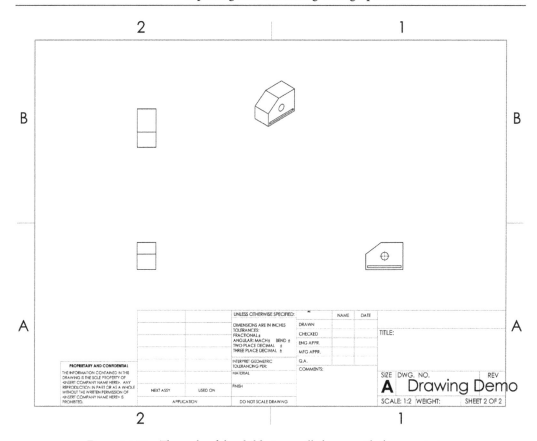

Figure 10.23 – The scale of the child views will change with the parent view

> **Note**
> After changing to the new scale, the views are very small in relation to the
> drawing sheet. It would be better if we were to enlarge the scale more.

4.   Select the front view again and change the scale to **1:2**.

Note that there is no right or wrong scale. The best way to find the best scale is to try
different ones until you are satisfied with the size and proportion of your drawing views.
Now that we've looked at the parent view, we will learn about changing the scale of a
child view.

Changing the drawing scale of the isometric child view

To change the drawing scale of the isometric child view, we can follow the same steps that we followed for the front view. However, there is one slight difference. Follow these steps to change the scale:

1. Select the isometric view and find the scale menu in the PropertyManager. Note that the scale is set to **Use parent scale**. This is why the scale of the isometric view changed as well when we changed the scale of the front view. To change the scale, select **Use custom scale**. Then, adjust the scale to **1:1**:

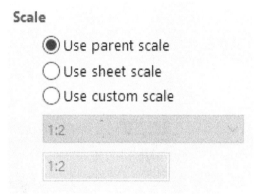

Figure 10.24 – We can set a different scale for a child view compared to the parent

2. As we can see, the front view became much bigger, while all the other views remained the same size. The resulting drawing will look as follows:

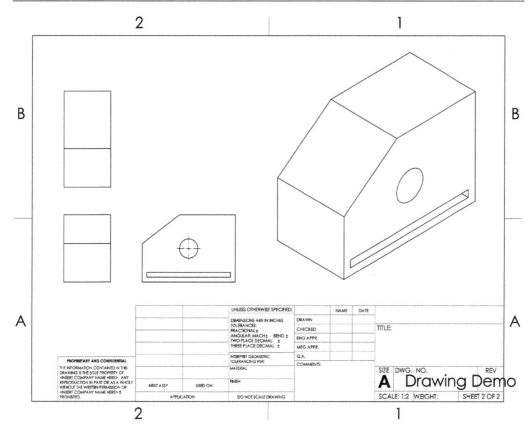

Figure 10.25 – The drawing after adjusting the scales

Tip

You can move the views around by clicking and holding the left mouse button. We usually do this to arrange our views.

Now, we have the final scale for our drawing set. However, before we move on to adjusting the display, let's talk about the scale ratios that are provided within SOLIDWORKS.

Understanding scale ratios

In the drawing we created earlier, we had two scales: 1:2 and 1:1. Let's take a look at what the first and second numbers refer to:

- **First number**: The first number refers to the actual object size from modeling. In other words, it refers to the dimensions we used when we made the part in the first place.

- **Second number**: The second number refers to the size on the drawing sheet. This relates to the final printing size of the drawing paper. Remember that when we first started our drawing file, we had to select a drawing sheet standard, which included the size of the drawing paper.

Now, let's put these two numbers together. If we were to manufacture the actual object and print the drawing sheet for our model, we would notice the following:

- The drawing sizes of the front, side, and top views are half the size of the actual object; thus, the scale is 1:2.

- The drawing size of the isometric view is exactly the same size as the actual object; thus, the scale is 1:1.

Now that we understand the scaling ratio and how to adjust our scales, we can start learning about the different displays and how to adjust them.

Different display types

Apart from the drawing view scale, we can also adjust how the drawing is displayed. There are five different displays we can use with our drawing views. These are highlighted in the following table and relate to the isometric view we have in our previous drawing:

Name of Display	Example View
Wireframe	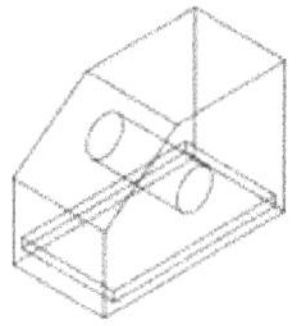
Hidden Lines Visible	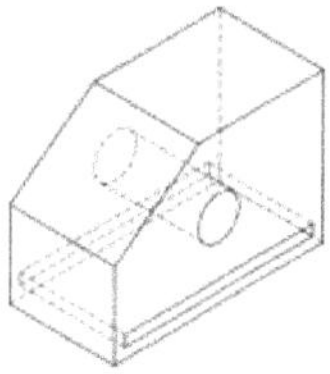
Hidden Lines Removed	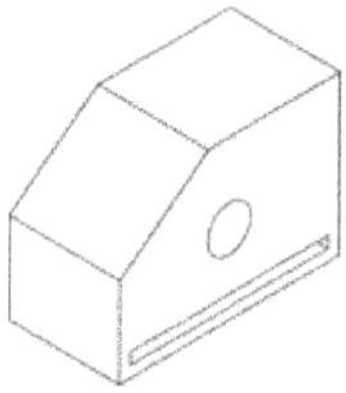
Shaded With Edges	
Shaded	

Figure 10.26 – The different display types we can use with the drawing views

To change the display of our isometric view to one of the ones shown in the preceding table, follow these steps:

1. Select the isometric view in the drawing sheet.

2. From the PropertyManager, find **Display Style**, as shown in the following screenshot. The small icons that are circled indicate the different display styles that were shown in the preceding table. Note the **Use parent style** option here, too—we can leave that checked if we want it to match the parent display style. Selecting any of the other view styles will automatically uncheck that box:

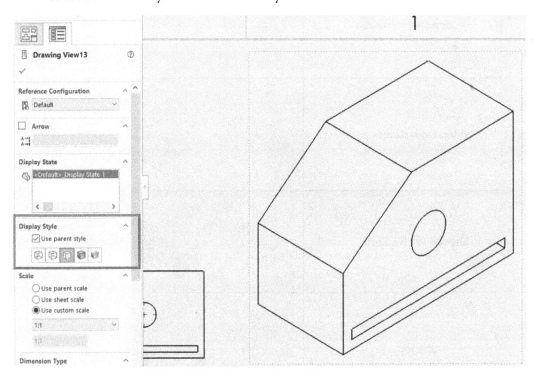

Figure 10.27 – The location of the Display Style option in the PropertyManager

3. Change the display to **Shaded With Edges**, which is the fourth icon from the left. After this, our drawing will look as follows:

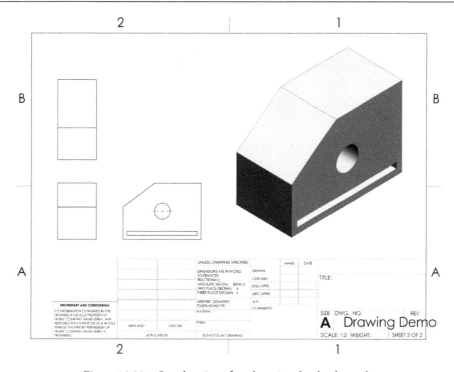

Figure 10.28 – Our drawing after changing the display style

Similar to drawing scales, there is no right or wrong view. It all depends on us as designers and draftsmen. We have to make the best decision when it comes to which view communicates our message the best.

So far, we have our drawing views, along with our desired display style, in our drawing canvas. These are used to communicate the shape of the model. Next, we will learn how to communicate dimensions in the drawing sheet.

Communicating dimensions and design

Now that we have different views in our drawing sheet, we can start adding information so that we can communicate the different elements of our drawing. In this section, we will cover how to add dimensions to our views and how to add different annotations, such as centerline and hole callout, to communicate our drawing in a clearer way. Having dimensions in our drawings is often necessary since we are often designing physical products. Dimensions help us communicate the size of our objects. Other annotations, such as centerline, notes, and hole callout, help us communicate the specifications of holes, centers of circles, and general notes we want to convey to whoever is viewing our drawing. We will start by learning how to display numerical dimensions using the Smart Dimension tool.

Using the Smart Dimension tool

The Smart Dimension tool allows us to easily display dimensions in our drawings. We will continue working with our previous drawing and add the selected dimensions to our drawing sheet. We will add the dimensions that are highlighted in the following drawing:

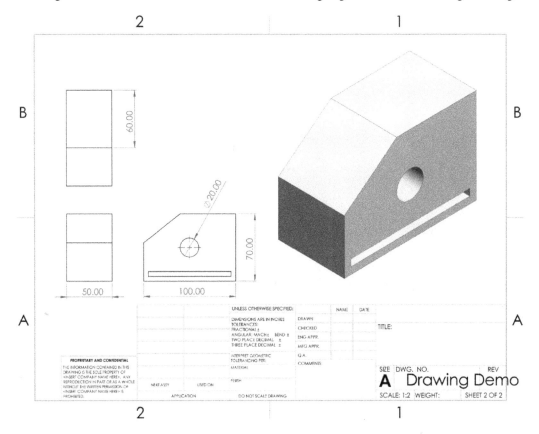

Figure 10.29 – Our end drawing after adding the dimensions

To add dimensions, follow these steps:

1. Under the **Annotation** tab, select the **Smart Dimension** option, as highlighted in the following screenshot:

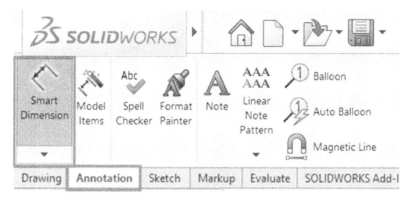

Figure 10.30 – The Smart Dimension option

2. Now, we can simply click on whatever part of the drawing we like to dimension it. This works very similarly to the Smart Dimension feature we used in sketching.

3. Click on the lines arrowed in the following figure to add dimensions to them. The first click will show the dimension, while the other click will confirm it:

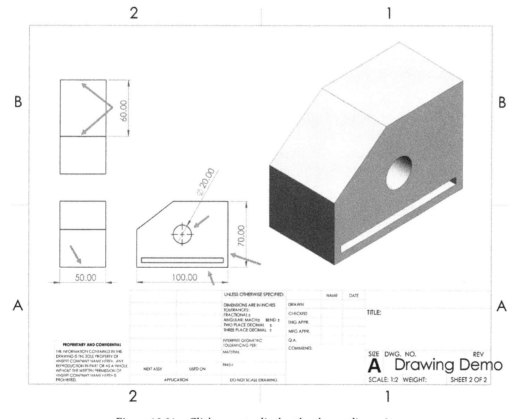

Figure 10.31 – Click areas to display the shown dimensions

This concludes how we can use the Smart Dimension tool in drawings. In addition to dimensioning line lengths and circle diameters, we can also dimension angles or any distance between two selected points.

Note

Using the Smart Dimension tool within drawings doesn't change the dimensions of the model. By default, it will display the dimension set in the 3D model itself.

If we mistakenly input a dimension we don't want, we can delete it. To delete a dimension, we can do one of two things:

- Right-click on the dimension and select **Delete**.
- Select the dimension and press the *Delete* key on the keyboard.

Now that we have our drawing views dimensioned, we can start inputting additional annotations to make it easier for others and ourselves to understand the drawing. In the next section, we will input centerlines, notes, and the hole callout.

Centerlines, center marks, notes, and hole callout annotations

In addition to dimensions, we can further clarify our drawings by adding additional annotations, such as centerlines and notes. SOLIDWORKS drawings provide an array of annotations we can use. However, we will only be covering the following in this section:

- **Centerlines**
- **Center marks**
- **Notes**
- **Hole callout**

Centerlines

As the name suggests, centerlines highlight the center of drawing entities. They can highlight the center between two lines. In the drawing we created earlier, we will add the following centerline, which is highlighted in the right-hand view:

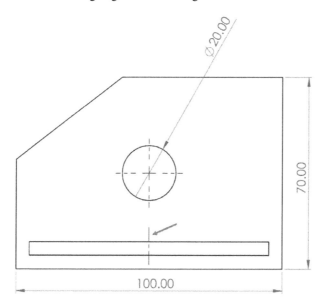

Figure 10.32 – The centerline we will add in the exercise

To add a centerline, follow these steps:

1.   Under the **Annotation** tab, select **Centerline**, as shown in the following screenshot:

Figure 10.33 – The location of the Centerline option

2. Now, we can click on the two entities we would like to create a centerline between. In our drawing, click on the two lines shown in the following screenshot. This will automatically put the centerline between them:

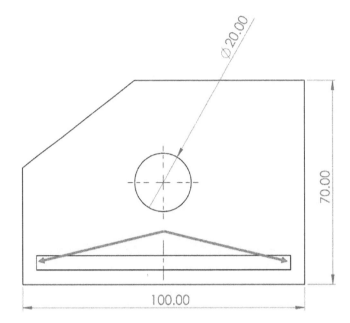

Figure 10.34 – Click areas to add the required centerline

This concludes how we can generate a centerline in SOLIDWORKS drawings. Centerlines can make it easier to interpret parts and design intents from drawings by indicting a central location between any two lines in the drawing.

> **Tip**
> We can extend the centerline as needed by dragging one end of it in a certain direction.

Next, we will address center marks.

Center marks

Center marks mark the center of circles, fillets, and slots to make them easier to identify when evaluating a drawing. The following figure indicates a center mark of a circle:

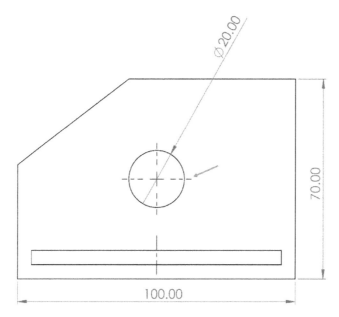

Figure 10.35 – A center mark indicating the center of a circle

To add a center mark, follow these steps:

1. Under the **Annotation** tab, select **Center Mark**, as shown in the following screenshot:

Figure 10.36 – The location of the Center Mark option

2. Now, we can directly click on circles, fillets, and slots to insert a center mark.

The **Center Mark** PropertyManager also has the option to auto-insert a center mark in our drawing views. This can save us time if we intend to add center marks to all circles, fillets, and slots. The **Auto Insert** option is highlighted in the **Center Mark** PropertyManager in the following screenshot:

Figure 10.37 – The Auto Insert option inserts center marks automatically to applicable entities

> **Tip**
> You can have SOLIDWORKS insert center marks automatically with holes, fillets, and slots as you are inserting the drawing views. To enable this, you can go to **Tool | Options | Document Properties | Detailing**. Then, under the **Auto insert on view creation** title, you can check the options for center marks. This can save you time and effort if you are inserting center marks all the time.

Next, we will learn how to add notes to our drawings.

Notes

Notes are text indications that we can add to our drawings to highlight any specific aspect of it. We can understand it as an open text box for us to write anything that we are trying to convey. To highlight how can we use notes, we will add the indicated notes to the following drawing.

To add a note, follow these steps:

1. Under the **Annotation** tab, select the **Note** option, as highlighted in the following screenshot:

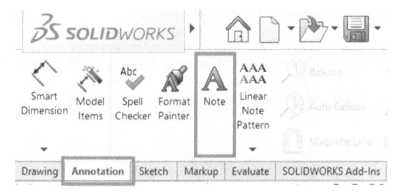

Figure 10.38 – The location of the Note option

2. Move the cursor to the drawing sheet and click under the front view. This will open a text box. Write Front View into it. Take note of the text format popup, as shown in the following screenshot. This allows us to easily adjust the format of the note, including its size, color, and font:

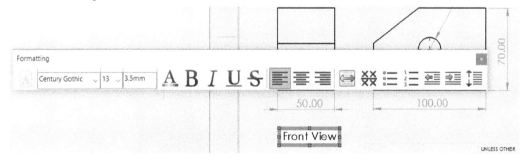

Figure 10.39 – The text format shortcut allows us to easily modify our note's format

3. Input the Right View and Top View notes in the same way.

> **Tip**
> While inputting a note, if we move the cursor toward lines in the drawing, SOLIDWORKS will automatically generate an arrow pointing toward that location for the note.

This concludes how can we add notes to our drawing. Next, we will learn how to use the hole callout command with holes.

Hole callout

The hole callout feature allows us to easily present information related to a particular hole in our model. This information includes the diameters of the hole, the length of the hole, and any standards related to the creation of the hole. To demonstrate this feature, we will use it on the hole shown in the right-hand view of our drawing. Note that the following figure also highlights the difference between the normal **Smart Dimension** we used earlier and the **Hole Callout** feature:

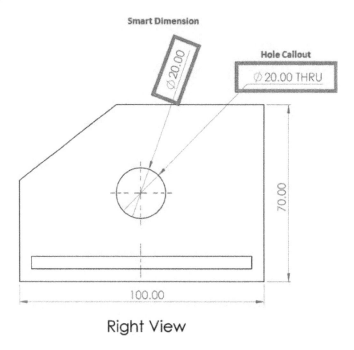

Figure 10.40 – The outcome from the Smart Dimension versus the Hole Callout commands

To add a hole callout, follow these steps:

1. Under the **Annotation** tab, select **Hole Callout**, as highlighted in the following screenshot:

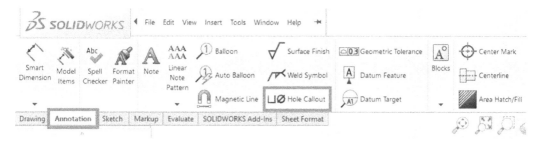

Figure 10.41 – The location of the Hole Callout option

2. Click on the circular hole we want to call out. In this case, it is the circle indicated in the right-hand view, as shown in the following figure. Once we click on the hole, the hole callout will appear, along with all the information linked to that particular hole:

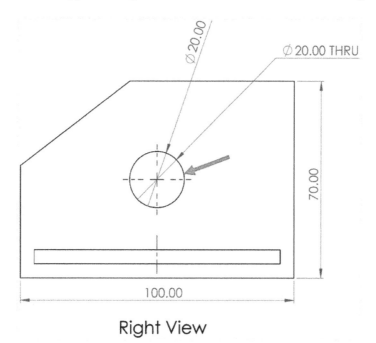

Right View

Figure 10.42 – The click area to get the hole callout

3. Delete the smart dimension input to make the drawing look cleaner.

Note that in this example, the information that's linked to the hole is its depth, marked as **THRU**, which indicates that the hole goes through all the models. If the hole has more information linked to it, such as a different hole type, it will be displayed within the callout as well, as shown in the following drawing:

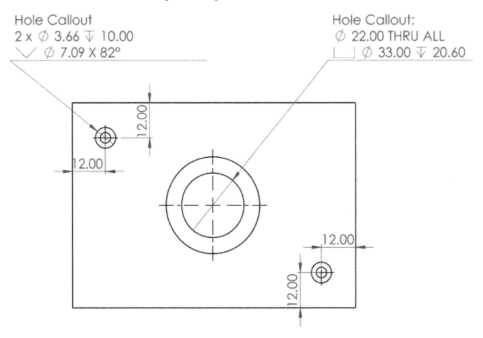

Figure 10.43 – Different types of hole callouts

This concludes how we can use a hole callout. At this point, our drawing has all the required views, display types, dimensions, and annotations. Next, we will learn how to adjust the information block, which contains information that's relevant to us, such as its name, number, company, and designer.

Utilizing the drawing sheet's information block

In this section, we will cover how to edit the information block that's located at the bottom of our drawing sheet. This information block displays information such as material, mass, drawer, reviewer, and drawing number. The information block in our current drawing is highlighted with a red box in the following screenshot. We will cover how to edit the existing information and how to add new information to the block:

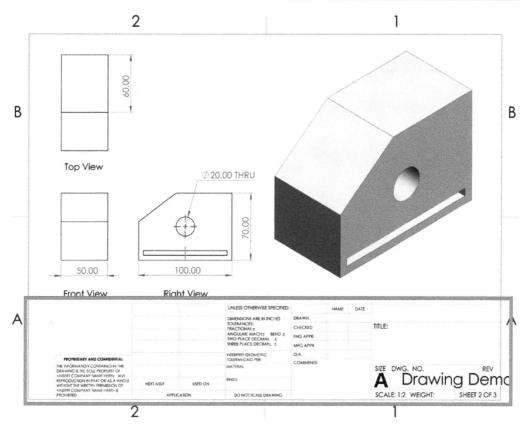

Figure 10.44 – The information block shows information related to the drawing

Now that we know what an information block is, we can start adjusting it.

Editing the information block

In this exercise, we will make the following edits:

- **TITLE:** Block-A
- **DWG. NO.:** 5598
- **DRAWN NAME:** TM

To edit the information in the sheet, follow these steps:

1. Right-click anywhere in the drawing sheet. Then, select **Edit Sheet Format**, as shown in the following screenshot. Make sure that you don't right-click the drawing view:

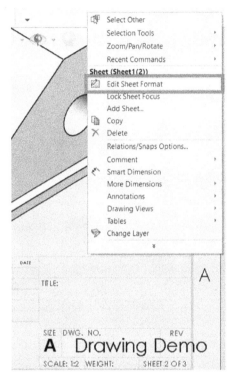

Figure 10.45 – The menu showing the Edit Sheet Format option

2. Now, the drawing lines surrounding the information block will turn blue, as shown in the following screenshot. This means we can edit the information block. The boxes for **TITLE, DWG. NO.**, and **DRAWN NAME** already have text boxes. All of them are empty except for **DWG. NO.**, which will take the name of the file by default. However, the empty fields already have text boxes that we can fill out. We can see these boxes by clicking in the middle of the empty box. The text box for **TITLE** is highlighted in *Figure 10.46* as well. To edit the **TITLE** box, we can double-click on it and input the text we want. Edit the text so that it displays the following information:

    *   **TITLE:** Block-A

    *   **DWG. NO.:** 5598

    *   **DRAWN NAME:** TM

The output is shown in the following screenshot:

Figure 10.46 – Double-clicking on the empty boxes enables you to modify them

3. Exit the information block's edit mode by clicking on the icon highlighted in the following screenshot. This is located on the top right-hand side of the drawing canvas:

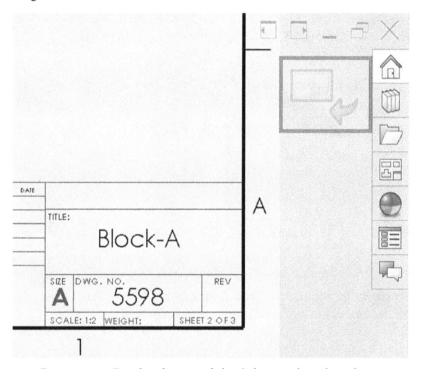

Figure 10.47 – Exit the editing mode by clicking on the indicated icon

Tip

When in editing mode, we can modify the location of a text box by dragging it.

This concludes how we can edit information in our drawing information block. However, we've only learned how to edit existing information. Next, we will learn how to add new information boxes to the information block.

Adding new information to the information block

In addition to editing information and filling in existing text boxes, we can also add new text boxes to our information block. In this exercise, we will add the following text under **COMMENTS**: This drawing was a practice.. To do this, follow these steps:

1. Right-click the drawing sheet and select **Edit Sheet Format** to start editing the information sheet.

2. Note that the note field doesn't have an existing text box. To add text, we need to add a normal note. We can do this by selecting the **Annotation** tab and then selecting **Note**, as highlighted in the following screenshot. Then, we can insert the note under the **COMMENTS** heading in the information sheet:

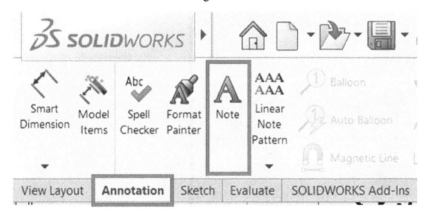

Figure 10.48 – The location of the Note option

3. After inserting the note, we can type in the text This drawing was a practice., as shown in the following screenshot:

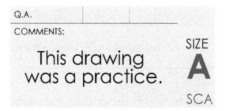

Figure 10.49 – The note typed in the comments section

4. Exit the sheet format editing mode. The final drawing will look as follows:

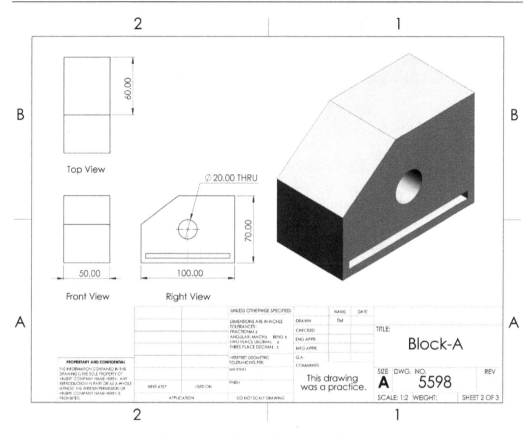

Figure 10.50 – The final look of our drawing

This concludes how to add additional text to our drawing information block. At this point, our drawing is complete for the purpose of this exercise. You can always add more edits and information to the information block to meet your requirements.

To share the drawing with other individuals, especially if they don't have access to SOLIDWORKS, we will need to export the drawing as an image or PDF so that they can view it. We will learn how to do this next.

Exporting the drawing as a PDF or image

Now that we've completed creating a drawing with SOLIDWORKS' drawing tools, we need to export it as a PDF or image file so that we can share it with individuals who don't have access to SOLIDWORKS. This is what we'll do in this section. First, we will export the drawing as a PDF file, and then as an image.

Exporting a drawing as a PDF file

To export a drawing as a PDF file, follow these steps:

1. Click on the **Save As** option, as shown in the following screenshot:

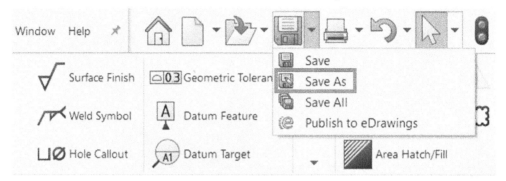

Figure 10.51 – The Save As option

2. Under **Save as type**, open the drop-down menu and select **Adobe Portable Document Format (*.pdf)**, as highlighted in the following screenshot:

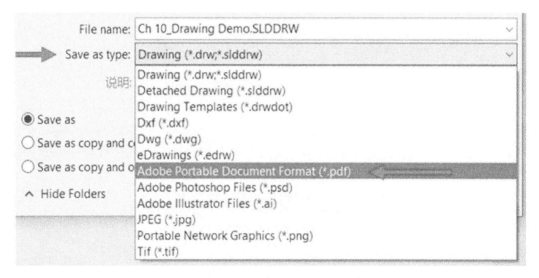

Figure 10.52 – Saving the drawing as a PDF document

3. Click on **Save**. This will save the file as a PDF in the designated folder.

This concludes how to save the drawing sheet as a PDF file. Next, we will learn how to export it as an image.

Exporting the drawing as an image

Exporting the drawing as an image is similar to exporting the drawing as a PDF. Follow these steps to do so:

1. Click on the **Save As** option.

2. Under **Save as type**, open the drop-down menu and select **Portable Network Graphics (*.png)**, as highlighted in the following screenshot. We can also select other image formats if needed:

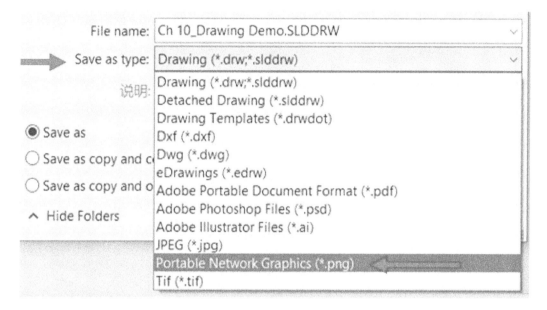

Figure 10.53 – Saving the drawing as a PNG image

3. Click on **Options...**, as shown in the following screenshot:

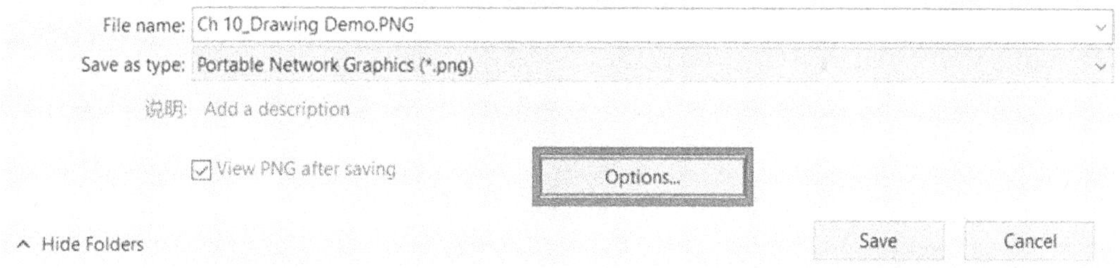

Figure 10.54 – The Options menu allows you to set the quality of the saved image

4. Double-click on the options that are highlighted in the following screenshot. Click **OK** after fixing these options. The marked options indicate the following:

- **Print capture**: This exports the image as a print. When exporting the image for sharing, it is best to use the **Print capture** option. The **Screen capture** option prints whatever is shown on your drawing canvas.

- **DPI**: This stands for **dots per inch**. The higher the number, the better the quality of our exported drawing print.

- **Paper size**: This allows you to determine the paper print size of the export:

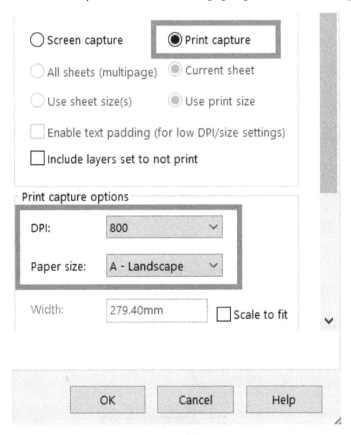

Figure 10.55 – The settings used in this exercise

5. Click on **OK** and then **Save** to save and export the image. This will save the file as a PNG in the designated folder.

This concludes how to save the drawing sheet as an image. We can now share our drawings with others as an image or a PDF. These formats can be accessed by larger groups of people without them needing access to special software such as SOLIDWORKS.

Note that, throughout this chapter, we have focused on generating a drawing to communicate a SOLIDWORKS part. However, the same principles apply when communicating an assembly.

Summary

Engineering drawings are what engineers and designers use to communicate their designs to other parties, such as manufacturers. SOLIDWORKS provides us with comprehensive tools that we can use to generate those drawings. In this chapter, we learned how to input the most basic drawing views, that is, orthographic and isometric views. Then, we learned how to adjust the drawing scale and display style for a particular view. After that, we learned how to add dimensions and different annotations, such as centerlines and hole callouts. Finally, we learned how to adjust the information block and export the drawing as an image or PDF file.

The skills we learned about in this chapter allow us to communicate our designs to external entities. Drawings present the link between us as SOLIDWORKS users and others who don't have access to or expertise in the software. This is what makes the topics in this chapter important.

In the next chapter, we will discuss how to add a **Bill of Materials** (**BOM**) to our drawings to highlight the different parts that are used in an assembly.

Questions

Answer the following questions to test your knowledge of this chapter:

1. How can we open a new drawing file?
2. What different display styles can we use with our drawing views?
3. What is the best scale to use for our drawing views?
4. What is the information block we can often find at the bottom of a standard drawing sheet?

5. Download the model linked to this chapter and duplicate the shown drawing (sheet specs: **A (ANSI) Landscape**, scale 1:3). The measurements system has been set to MMGS:

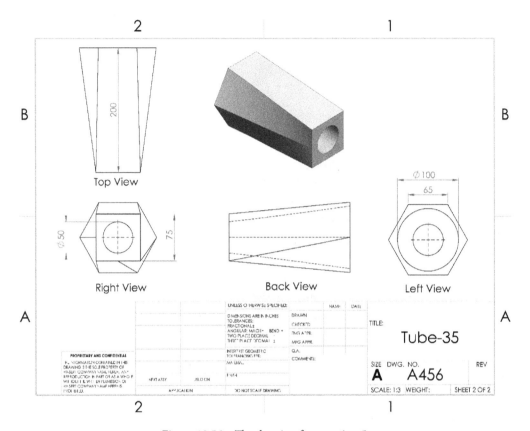

Figure 10.56 – The drawing for question 5

6. Download the model linked to this exercise and duplicate the shown drawing (sheet specs: **A4 (ANSI) Landscape**, scale 1:3). The measurements system has been set to MMGS:

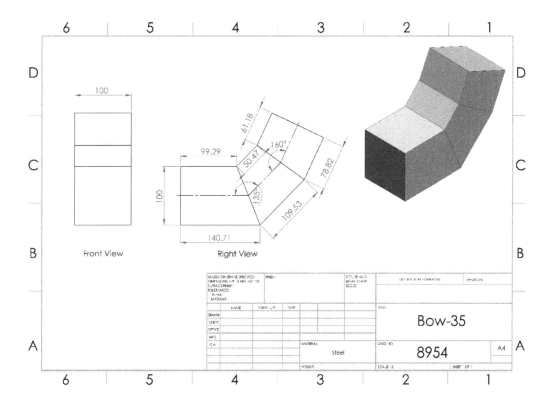

Figure 10.57 – The drawing for question 6

7. Create the shown model from scratch and then duplicate the shown drawing (sheet specs: **B (ANSI) Landscape**, scale: 1:1). The measurements system has been set to MMGS:

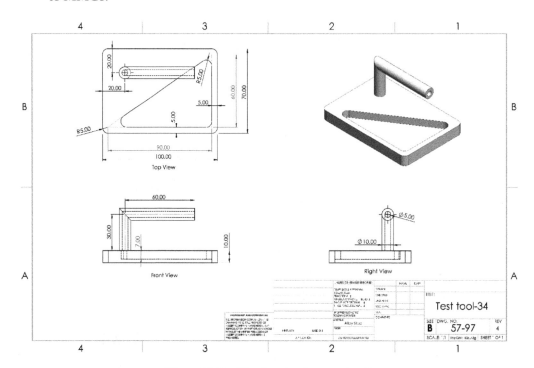

Figure 10.58 – The drawing for question 7

Important Note

The answers to the preceding questions can be found at the end of this book.

11
Bill of Materials

Most assemblies usually consist of multiple parts. For example, a simple coffee table would have four legs, a top, and perhaps some screws. Other more complex assemblies, such as engines, would contain hundreds of different parts. A **Bill of Materials (BOM)** can help us list and communicate those different assembly parts. Apart from the list of parts, it also helps communicate any other desired information, such as cost, materials, and part numbers. This chapter will enable us to create standard BOMs. It will also enable us to modify and utilize equations in our BOMs.

The following topics will be covered in this chapter:

- Understanding BOMs
- Generating a standard BOM
- Adjusting information in the BOMs
- Utilizing equations with BOMs
- Utilizing parts callouts

By the end of the chapter, we will be able to generate a standard communicative BOM that goes with our drawings. We will also be able to fine-tune the information displayed to fully match our needs.

Technical requirements

This chapter will require access to SOLIDWORKS software.

The project files for this chapter can be found in the following GitHub repository: `https://github.com/PacktPublishing/Learn-SOLIDWORKS-Second-Edition/tree/main/Chapter11`.

Check out the following video to see the code in action: `https://bit.ly/3s6sZXy`

Understanding BOMs

A BOM is an essential part of any engineering drawing representing an assembly. This is because it shows relevant information about the different parts that are present in our final product. Before we make BOMs, we need to understand what they are and their purpose. In this section, we will learn about BOMs and introduce the BOMs we will create in this chapter. Let's start.

Understanding a BOM

BOMs are often part of engineering drawings, specifically with drawing sheets of assemblies. They show more specific information about the product we are working on. For example, a typical list of materials might contain the following information:

- Names of the parts in the assembly
- Part numbers
- Quantity of each part
- Sequential listing of each entry

However, they are not limited to this information. BOMs are customizable, depending on the established practices and the application needs. Other information that can often be found in BOMs is cost, manufacturer, materials, store locations, reference numbers, and so on. The following are two examples of assembly drawings with BOMs. Note that each table is considered a BOM. The drawing shown in *Figure 11.1* is for a mechanical cap. The BOM there includes **PART NAME**, **PartNo**, **DESCRIPTION**, **QTY.**, **Cost** per part, and **Total Cost** per part. It also highlights the **Total Cost** sum for all the parts and the **Highest Cost** value for one part, and also includes a subassembly and its parts, noted under **Damper Assembly**:

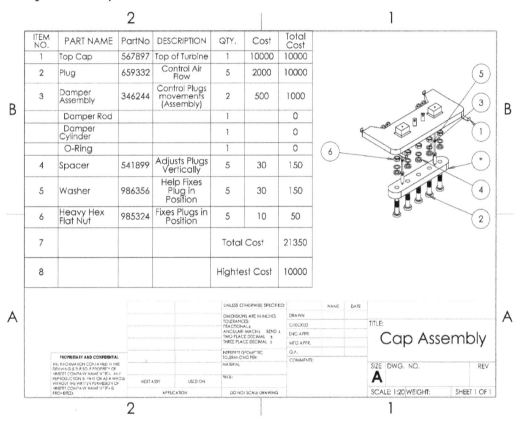

ITEM NO.	PART NAME	PartNo	DESCRIPTION	QTY.	Cost	Total Cost
1	Top Cap	567897	Top of Turbine	1	10000	10000
2	Plug	659332	Control Air Flow	5	2000	10000
3	Damper Assembly	346244	Control Plugs movements (Assembly)	2	500	1000
	Damper Rod			1		0
	Damper Cylinder			1		0
	O-Ring			1		0
4	Spacer	541899	Adjusts Plugs Vertically	5	30	150
5	Washer	986356	Help Fixes Plug in Position	5	30	150
6	Heavy Hex Flat Nut	985324	Fixes Plugs in Position	5	10	50
7			Total Cost			21350
8			Hightest Cost			10000

Figure 11.1 – A drawing with a BOM of a cap assembly

The following diagram and BOM are of a simple coffee table. The bill includes **PART NAME**, **COST PER PART (USD)**, **QTY.**, and **TOTAL COST PER PART (USD)**. Note that the information in this bill is different than the one shown previously:

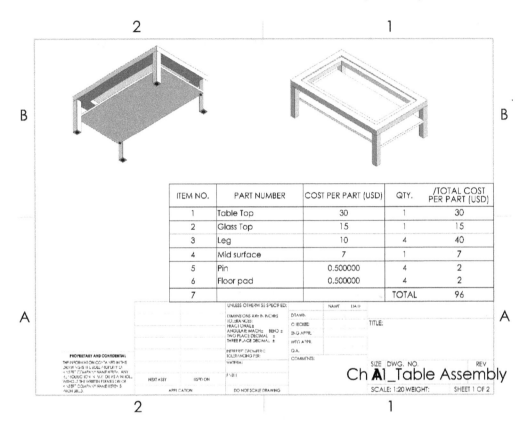

ITEM NO.	PART NUMBER	COST PER PART (USD)	QTY.	/TOTAL COST PER PART (USD)
1	Table Top	30	1	30
2	Glass Top	15	1	15
3	Leg	10	4	40
4	Mid surface	7	1	7
5	Pin	0.500000	4	2
6	Floor pad	0.500000	4	2
7			TOTAL	96

Figure 11.2 – An assembly drawing with a BOM for a coffee table

Throughout this chapter, we will be working to create the preceding drawing sheet and BOM from scratch. You can download all of the parts and assembly files for this chapter. Our first step will be to generate a standard BOM, which we will do next.

Generating a standard BOM

In this section, we will learn how to generate a standard BOM using the tools provided by SOLIDWORKS drawings. We will start by inserting views of our model into the drawing and then generate a BOM. A standard BOM is our starting point when generating those bills within SOLIDWORKS drawing tools. After generating the standard bill, we will be able to modify it further so that it matches our specifications.

Inserting an assembly into a drawing sheet

A BOM often accompanies assemblies. This is because the BOMs will indicate the parts that exist in the assembly. Hence, we will start by adding a drawing of the assembly to our sheet. Inserting an assembly works the same way as inserting a part. We can use the following steps:

1. Open a new drawing file and pick the **A (ANSI) landscape** sheet format.

2. Go to **View Palette**, browse for the assembly file we downloaded with this chapter, and select it:

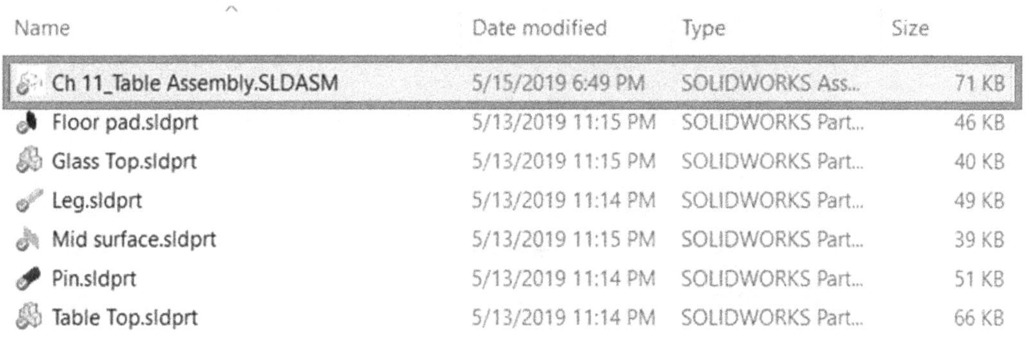

Name	Date modified	Type	Size
Ch 11_Table Assembly.SLDASM	5/15/2019 6:49 PM	SOLIDWORKS Ass...	71 KB
Floor pad.sldprt	5/13/2019 11:15 PM	SOLIDWORKS Part...	46 KB
Glass Top.sldprt	5/13/2019 11:15 PM	SOLIDWORKS Part...	40 KB
Leg.sldprt	5/13/2019 11:14 PM	SOLIDWORKS Part...	49 KB
Mid surface.sldprt	5/13/2019 11:15 PM	SOLIDWORKS Part...	39 KB
Pin.sldprt	5/13/2019 11:14 PM	SOLIDWORKS Part...	51 KB
Table Top.sldprt	5/13/2019 11:14 PM	SOLIDWORKS Part...	66 KB

Figure 11.3 – Inserting the assembly file into the drawing

3. In **View Palette**, select the two views, that is, the *Isometric view and **Bottom View**, and insert them into the sheet, as shown in the following screenshot:

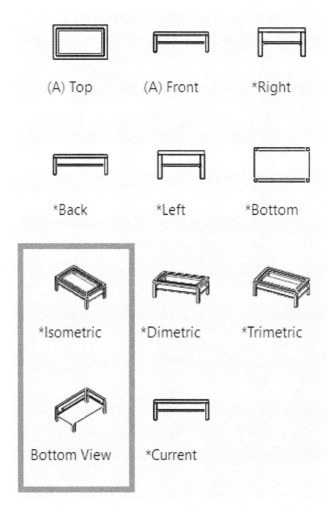

Figure 11.4 – Inserting the Isometric and Bottom View into the drawing

4. Change the scale to 1:15 and the display to **Shaded With Edges** for both views. Our drawing sheet should look as follows:

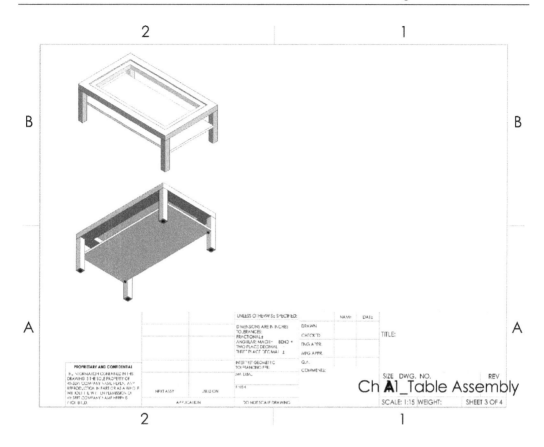

Figure 11.5 – The drawing sheet with the coffee table assembly

This concludes this section on adding an assembly to a drawing sheet. Note that it follows the same rules as inserting parts. Next, we will generate our standard BOMs for this assembly.

Creating a standard BOM

Now that we have inserted the assembly file into the drawing sheet, we can generate a standard BOM for the assembly. To generate that, we can follow these steps:

1. From the **Annotation** tab, select the **Tables** drop-down menu and select **Bill of Materials**, as shown in the following screenshot:

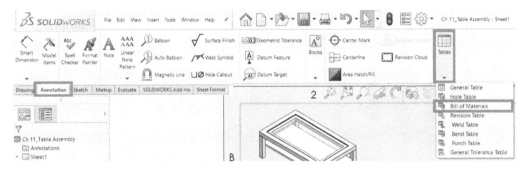

Figure 11.6 – The command to insert a BOM

> **Important Note**
>
> You can also insert a BOM by right-clicking on a drawing view, and then going to **Tables** and selecting **Bill of Materials**.

2. Then, select the view for the model that you want to create the BOM for. Since both the drawing views are the same, we can click on either of them.

3. This will show the **Bill of Materials** PropertyManager toward the left of the screen. As we are creating a standard default bill, we can simply leave all of the default options as they are, as highlighted in the following screenshot. Click on the *green check mark* to confirm the table:

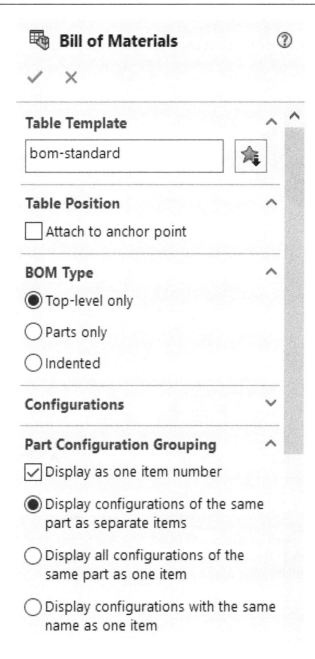

Figure 11.7 – Initial options for inserting a BOM

4. After that, a standard BOM will be generated and displayed in our drawing sheet. We can place the bill on the sheet to get the drawing sheet shown in the following figure. We can adjust the table location by dragging it:

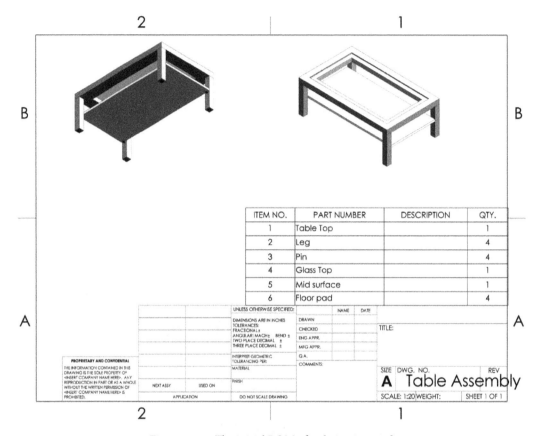

ITEM NO.	PART NUMBER	DESCRIPTION	QTY.
1	Table Top		1
2	Leg		4
3	Pin		4
4	Glass Top		1
5	Mid surface		1
6	Floor pad		4

Figure 11.8 – The initial BOM after being inserted

This concludes how to generate a standard BOM. This will be our usual first step whenever we generate a BOM. Next, we will be working on adjusting the information in our bill to match the bill that was shown earlier in this chapter.

Adjusting information in the BOMs

Often, the information in the standard BOM doesn't exactly match our requirements. Hence, we need to be able to adjust the information shown to match our needs and requirements. In this section, we will learn how to adjust the table by changing information and adding information. By the end of this section, our drawing will look as follows. Note that the headings and information are different from how they were previously:

Figure 11.9 – The drawing by the end of this section

We will start by adjusting our bill's titles and information category.

Adjusting listed information in the BOM

Here, we will learn how to change information that is already listed in the table by making the following adjustments:

- Changing a title in the BOM
- Changing a column category in the BOM

Changing a title in the BOM

Here, we will change the title of **PART NUMBER** to **PART NAME**, as highlighted in the following figure:

Figure 11.10 – Double-clicking on the cell title allows us to edit it

To do that, we can follow these steps:

1. Double-click on the title cell. This will open up the title for editing.

2. Change the title from PART NUMBER to PART NAME.

3. Click anywhere outside the table to confirm the adjusted information.

When it comes to editing information, we can think of the table in a similar way to tables in Microsoft Excel.

Changing a column category

When generating a BOM, usually, each column is linked to a series of information that is gathered from the model itself; for example, column quantity (**QTY.**) automatically links to the number of parts that are present in the assembly. This will then display the quantity without us manually inputting it. We can still add more columns that are not linked with more manual information if needed. We can also adjust the information category for a specific column. In this section, we will change the **DESCRIPTION** column to **COST PER PART (USD)**, as highlighted in the following figure. This new column will automatically be filled with linked cost information:

Figure 11.11 – We will adjust the DESCRIPTION column to COST PER PART (USD)

To adjust the column from **DESCRIPTION** to **COST PER PART (USD)**, we can follow the following steps:

1. Hover the mouse cursor over the table, and then select the letter **C** at the top of **DESCRIPTION** to select the whole column. Then, click on the **column** property, as highlighted in the following screenshot:

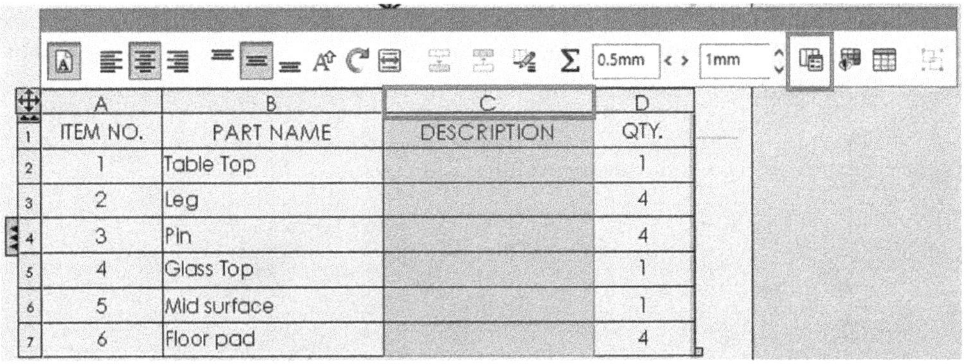

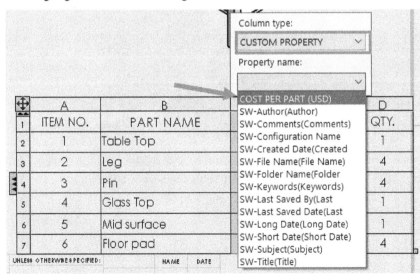

Figure 11.12 – The column property allows us to link the table to different information

2. From the shown options, make the following selections:

-   **Column type | CUSTOM PROPERTY**

-   **Property name | COST PER PART (USD)**

This is highlighted in the following screenshot:

Figure 11.13 – The COST PER PART (USD) as found under the custom property

3. Once we select that, the whole column will change, as well as all of the values in it. It will be filled with assigned values for the costs of each part. The BOM will look as follows:

ITEM NO.	PART NUMBER	COST PER PART (USD)	QTY.
1	Table Top	30	1
2	Leg	10	4
3	Pin	0.5	4
4	Glass Top	15	1
5	Mid surface	7	1
6	Floor pad	0.5	4

Figure 11.14 – Values are automatically filled for COST PER PART (USD)

> **Important Note**
>
> The values for the cost per part are not automatically generated by SOLIDWORKS. Rather, they are inputted manually into each part during the design process. The custom property function can then auto-call those values in the BOM.

This concludes how to adjust the column category for a specific column in our BOMs. Being able to adjust categories is a necessary skill that will allow us to extract information linked to our models and put it into our BOMs. Next, we will start changing the order of listed information in our bill by sorting it.

Sorting information in our BOMs

We can also re-sort the information in our BOMs to put it in a specific order. In our case, we will order the information based on **COST PER PART (USD)** in descending order. To do this, we can follow these steps:

1.   Right-click anywhere in the table and select **Sort**:

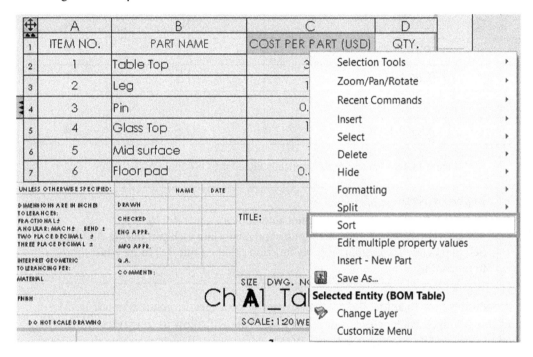

Figure 11.15 – The location of the Sort command

2. We will get the following window. We can adjust the first **Sort by** option to **COST PER PART (USD)**, and then select the **Descending** option next to it. **Method** should be selected as **Numeric.** Click **OK** after that:

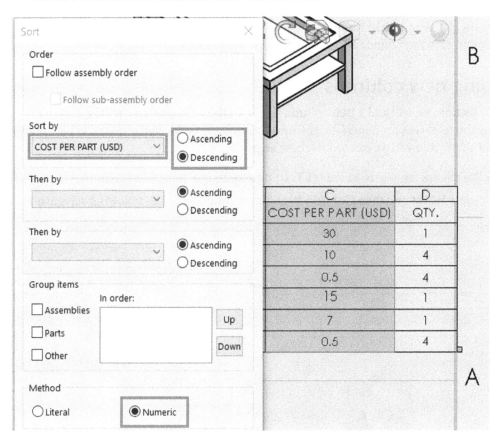

Figure 11.16 – The adapted setting for the Sort function

This will automatically re-sort the order of the whole table so that it's in descending order based on **COST PER PART (USD)**. Finally, our BOM will be as follows:

ITEM NO.	PART NAME	COST PER PART (USD)	QTY.	
1	Table Top	30	1	
2	Glass Top	15	1	
3	Leg	10	4	
4	Mid surface	7	1	
5	Pin	0.500000	4	
6	Floor pad	0.500000	4	

Figure 11.17 – The Sort function allows you to order the BOM from most to least expensive

This concludes how to sort BOMs according to a specific order. We will often end up sorting our bills for a variety of reasons that will make it easier for us to find information. For example, we would want to sort a bill by cost per part to make it easier to identify the most and least costly parts. In an alternative scenario, we might sort the part names by alphabetical order to make finding a part easier by name. Next, we will learn how to add new columns to our table to accommodate new information.

Adding new columns

In this section, we will add a new column to our BOMs. The new row will be used to display the total cost for materials. The new column will be located after the **QTY.** column. To add a new column, we can follow these steps:

1. Right-click anywhere in the **QTY.** column or in any cell in that column.

2. Select **Insert** and then **Column Right**, as highlighted in the following screenshot:

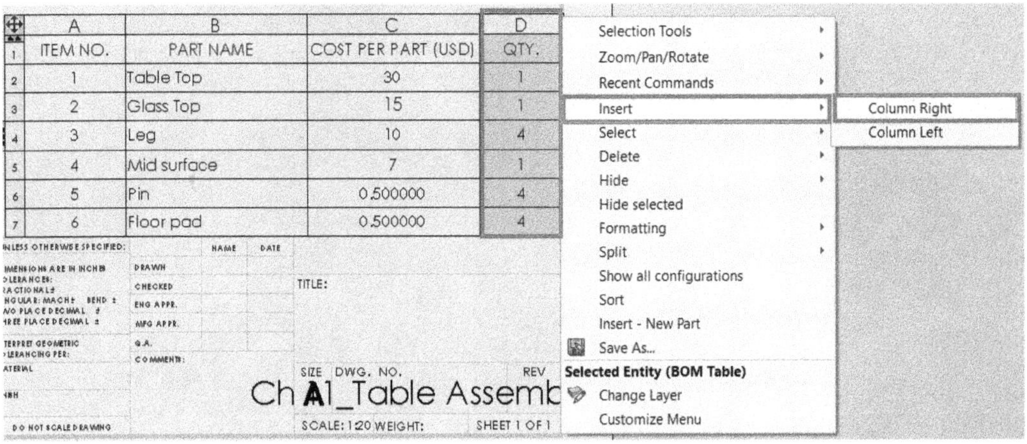

Figure 11.18 – Inserting a new column

This will insert a new empty column to the far right so that we get the following table:

Figure 11.19 – We can insert new columns into our BOM for more information

This concludes how to add a new column to our BOMs table. Note that to add a new row, we can follow the same procedure for adding a column. When we insert a new BOM, the number of columns and rows is limited. Hence, it will be important for us to add rows and columns to include more information. In the next section, we will fill out the new column we added using equation functions.

Utilizing equations with BOMs

Equations in SOLIDWORKS drawings allow us to create simple calculations within our BOMs. This will allow us to perform operations such as addition, subtraction, division, and multiplication. In this section, we will learn more about what can we do with drawing equations. We will also use this feature for additional calculations in our BOMs.

What are equations in SOLIDWORKS drawings?

Equations allow us to perform several mathematical equations in our BOMs without leaving the SOLIDWORKS drawing interface. With equations, we can think of our BOMs as simple Excel sheets. With the equations function, we can perform two types of operations, functions and mathematical operations. Here, we will learn what these include.

Functions

Functions are limited programmed calculations that we can directly use to find specific information and display it in our bill of materials. They include the following:

- **If**: This opens an `if` statement that will allow us to apply a condition.
- **Average**: This finds the average between different values.
- **Count**: This counts the number of cells, regardless of whether the cell is empty or filled.
- **Max**: This finds the maximum value within the selected cells.
- **Min**: This finds the minimum value within the selected cells.
- **Sum**: This finds the sum of values with the selected cells.
- **Total**: This totals all of the values in the column that are located above the selected cell.

Next, we will learn about mathematical operations.

Mathematical operations

In addition to functions, we can also utilize different mathematical operations that are found in a basic calculator. These include the four main operations: addition, subtraction, multiplication, and division. We can use these operations by inputting them using the +, -, *, and / signs from the keyboard.

Now that we know about functions and mathematical operations, we will start using them in our bill.

Inputting equations in a table

Here, we will demonstrate how to add equations in a BOM. We will do this by revisiting our previous BOM. We will add **TOTAL COST PER PART (USD)** and **TOTAL COST** for a full product (a coffee table, in this case). To do this, we will start by applying the mathematical operation known as multiplication. Then, we will apply the *total* function.

Applying a mathematical operation

For this, we will fill the last column of our table with **TOTAL COST PER PART (USD)**, as indicated in the following table. This will show the total cost of purchasing the quantity needed for each part. To calculate the value, we can multiply **COST PER PART (USD)** by **QTY.**:

Add the TOTAL COST PER PART (USD)

ITEM NO.	PART NAME	COST PER PART (USD)	QTY.	
1	Table Top	30	1	
2	Glass Top	15	1	
3	Leg	10	4	
4	Mid surface	7	1	
5	Pin	0.5	4	
6	Floor pad	0.5	4	

Figure 11.20 – We will fill the new column with calculated values for the cost

To do this, we can follow these steps:

1. Highlight the empty column at the far right by clicking the letter **E** at the top. Then, click on the **Equation (Σ)** command, as highlighted in the following screenshot:

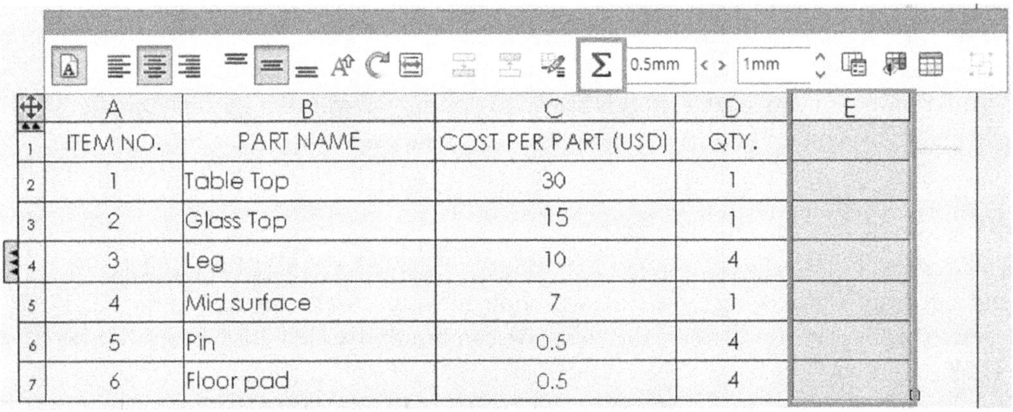

Figure 11.21 – The Equation command allows us to input mathematical operations in the BOM

2. This will open the equation function, as shown in the following screenshot. In the equation space, we need to input the following – COST PER PART (USD)*QTY. Instead of typing all of this in the equation space, we can use the **Columns** drop-down menu. From there, we can select **COST PER PART (USD)**. Then, we will notice that the column name was input in the equation space. After that, we can type * and select **QTY.** from the **Columns** drop-down menu:

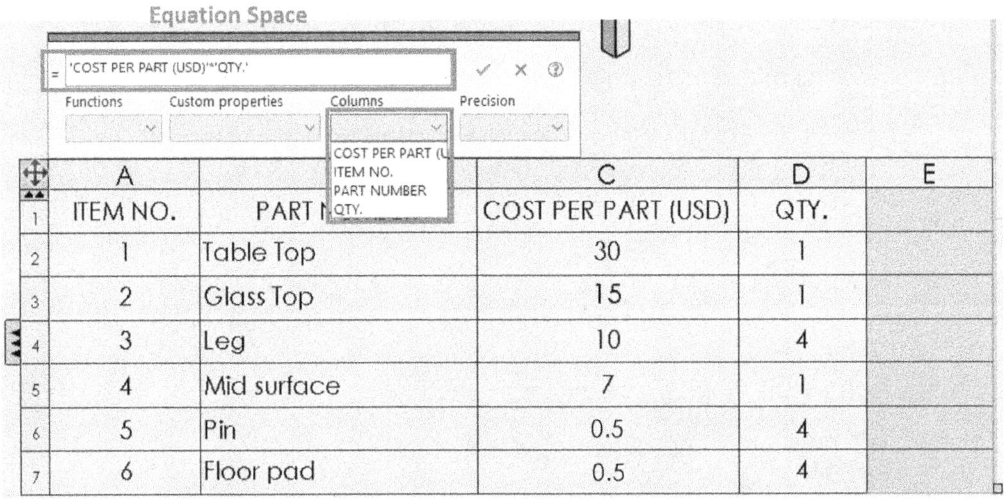

Figure 11.22 – We can input equations to calculate values in the BOM

3. Click on the green check mark to apply the equation. We will notice that the column was filled with numbers that are equal to the number under **COST PER PART (USD)**, multiplied by **QTY.**. To make the table fuller, we can add a title to the column. The title can be **TOTAL COST PER PART (USD)**. The final table will look as follows:

ITEM NO.	PART NUMBER	COST PER PART (USD)	QTY.	/TOTAL COST PER PART (USD)
1	Table Top	30	1	30
2	Glass Top	15	1	15
3	Leg	10	4	40
4	Mid surface	7	1	7
5	Pin	0.5	4	2
6	Floor pad	0.5	4	2

Figure 11.23 – The TOTAL COST column is auto-calculated using a simple multiplication

This concludes using multiplication. All the other operations, such as addition and subtraction, can be applied in the same way. Next, we will learn how to apply a function.

Applying an equation function

Here, we will apply the **Total** function in our BOM. Using this function, we will add all of the values under **TOTAL COST PER PART (USD)**, which we just created. This will give us the total cost per coffee table. To do this, we can follow these steps:

1. Add a new empty row at the bottom of the table. We can do that by right-clicking any cell at the bottom and inserting a row.

2. Select the cell bottom cell under the **TOTAL COST PER PART (USD)** column and select **Equation (Σ)**.

3. In the equations panel, select **TOTAL** under the **Functions** drop-down menu, as highlighted in the following screenshot. Once we select that, the word **TOTAL** will appear in the equation space:

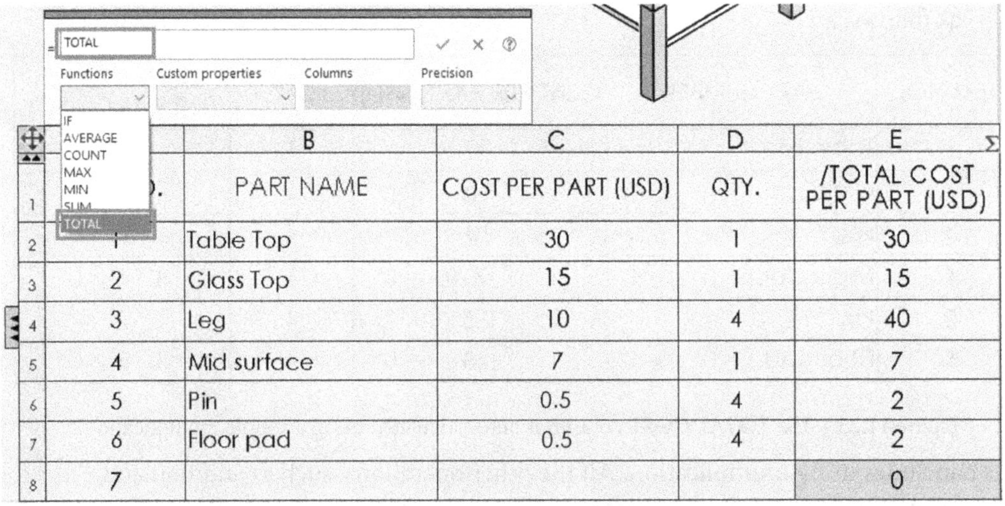

	A	B	C	D	E
1		PART NAME	COST PER PART (USD)	QTY.	/TOTAL COST PER PART (USD)
2	1	Table Top	30	1	30
3	2	Glass Top	15	1	15
4	3	Leg	10	4	40
5	4	Mid surface	7	1	7
6	5	Pin	0.5	4	2
7	6	Floor pad	0.5	4	2
8	7				0

Figure 11.24 – The TOTAL function adds all the values on top of it

4. Click on the *green check mark*. Then, the value of the last cell (**E8**) will change to **96**, which is the sum of all of the numbers in that column. To make the table clearer, we can add the word TOTAL in cell **D8**. The resulting BOM will be as follows. Note that when editing cell **D8**, SOLIDWORKS will ask us whether we want to break the link in the cell and continue editing it; click **Yes** to edit:

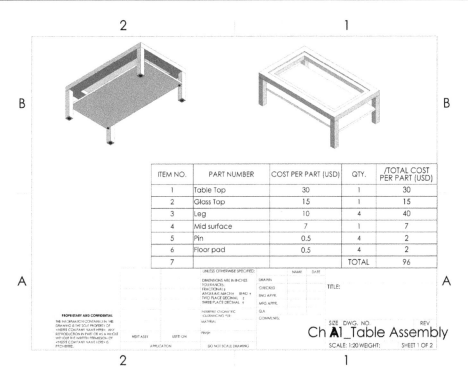

ITEM NO.	PART NUMBER	COST PER PART (USD)	QTY.	/TOTAL COST PER PART (USD)
1	Table Top	30	1	30
2	Glass Top	15	1	15
3	Leg	10	4	40
4	Mid surface	7	1	7
5	Pin	0.5	4	2
6	Floor pad	0.5	4	2
7			TOTAL	96

Figure 11.25 – Our new BOM after calculating the total cost

5. It is good practice to try cleaning our bill as much as possible. For example, in the preceding bill, we can remove item number **7**, as that row only has the total and does not include an item. To hide that item number, we can right-click on that cell and select **Hide Item Number**, as highlighted in the following screenshot:

Figure 11.26 – Hiding the last item number can get us a cleaner BOM

This concludes our work with equations within BOMs. Using equations will allow us to generate new numerical information in our table that is not linked to a particular part. A common application of equations is in relation to the costs of parts. Next, we will learn how to add callouts to the assembly for easier referencing between our bill and the visual display of the assembly.

Utilizing parts callouts

In this section, we will cover how to add callouts to the parts in our BOM. These can help us identify the location of the items in the drawing itself. Here, we will cover auto balloon callouts. To create these callouts, we can do the following:

1. From the **Annotation** tab, select the **Auto Balloon** command, as highlighted in the following screenshot:

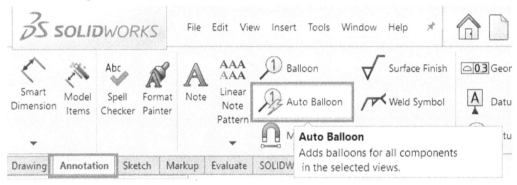

Figure 11.27 – The location of the Auto Balloon command

2. Click on the two views we have on the drawing, one after another. You will notice
    balloons popping up with numbers and arrows pointing toward different parts, as
    shown in the following screenshot. Note that each number in a balloon matches the
    number in the BOM:

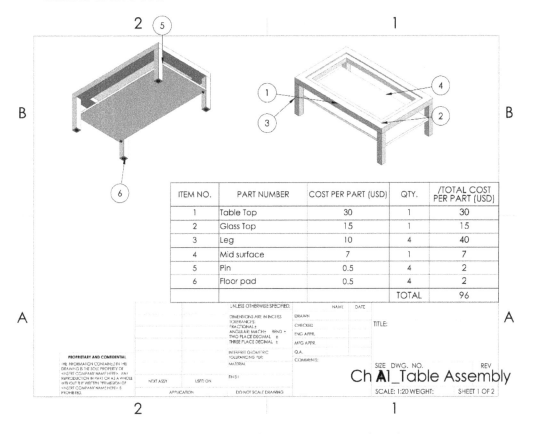

ITEM NO.	PART NUMBER	COST PER PART (USD)	QTY.	/TOTAL COST PER PART (USD)
1	Table Top	30	1	30
2	Glass Top	15	1	15
3	Leg	10	4	40
4	Mid surface	7	1	7
5	Pin	0.5	4	2
6	Floor pad	0.5	4	2
			TOTAL	96

Figure 11.28 – Balloons indicate the item numbers matching the BOM

3. In the PropertyManager, we have multiple options to adjust the callout. For this exercise, adjust the options highlighted in the following screenshot. After adjusting these settings, click on the green check mark to apply the balloon callouts:

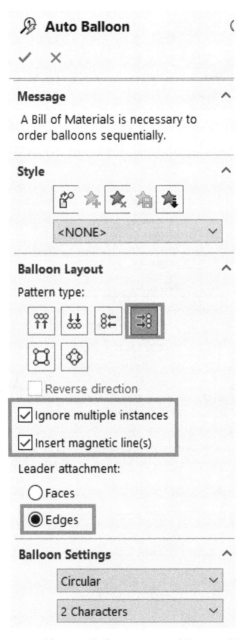

Figure 11.29 – The Auto Balloon command PropertyManager

The final result of our drawing would look as follows. Note that we can manually drag the balloons to change their position as we see fit:

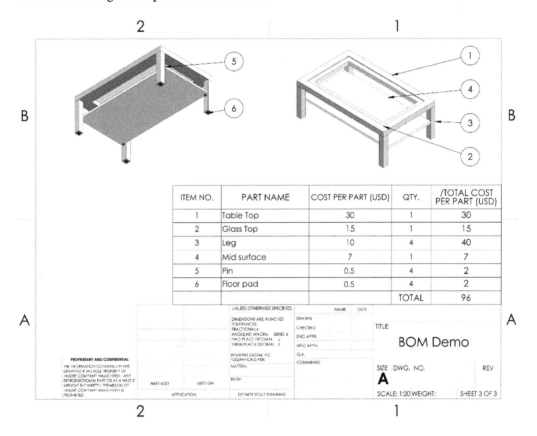

ITEM NO.	PART NAME	COST PER PART (USD)	QTY.	/TOTAL COST PER PART (USD)
1	Table Top	30	1	30
2	Glass Top	15	1	15
3	Leg	10	4	40
4	Mid surface	7	1	7
5	Pin	0.5	4	2
6	Floor pad	0.5	4	2
			TOTAL	96

Figure 11.30 – The final drawing with a clean and communicative BOM

Other than the item numbers displayed in *Figure 11.30*, we can adjust the type of information we want to display. We can see those options in the Auto Balloon PropertyManager, as indicated in the following figure:

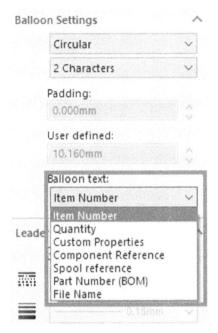

Figure 11.31 – The balloons can display different information other than the item number

This concludes how to create auto balloons for display purposes in our drawing. Note that these balloons will enable anyone who views the drawing to link the information in the BOMs to the displayed assembly.

Manual Balloon command

The Auto Balloon command enables us to quickly display related information to more than one part. However, if we want to include a few balloons with specific information, such as written text notes or any other custom property, then manually adding a balloon can be a more efficient option. Let's examine this by adding a box-shaped balloon linking to the tabletop with the Cut First text. To do this, follow these step:

1. Select the **Balloon** command, as shown in the following figure:

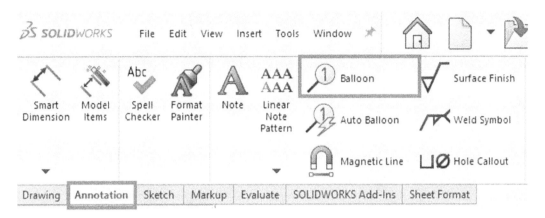

Figure 11.32 – The location of the Balloon command

2. Under **Settings** in the PropertyManager, select the **Box** and **Tight Fit** options, as shown in the following figure:

Figure 11.33 – We can set up the shape and text of the balloon per our needs

3. Under **Balloon text:**, select the **Text** option and type Cut First, as shown in *Figure 11.33*.

4. Using the cursor, go back to the drawing canvas and attach the box-shaped balloon to the *tabletop*. We will end up with the following figure:

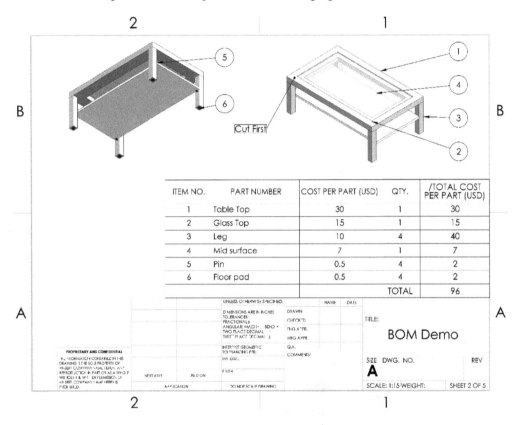

ITEM NO.	PART NUMBER	COST PER PART (USD)	QTY.	/TOTAL COST PER PART (USD)
1	Table Top	30	1	30
2	Glass Top	15	1	15
3	Leg	10	4	40
4	Mid surface	7	1	7
5	Pin	0.5	4	2
6	Floor pad	0.5	4	2
			TOTAL	96

Figure 11.34 – The resulting drawing with a box-shaped balloon

Other than the custom text we did in the previous example, the balloon can also link to the parts' properties and extract information from there in the same way that custom properties work in the BOM. At this point, we can conclude our discussion on utilizing the ballooning options to display specific communicative information.

Summary

Most of the products we work with include more than one part put together. To easily show these parts alongside related details, such as cost, part numbers, materials, and weights, we can use a BOM. Including a BOM is a very common practice when communicating drawings of assemblies. In this chapter, we learned what BOMs are and how to generate a standard one. We then learned how to adjust the information in a standard bill by adding, removing, and regenerating information, rows, and columns. We also learned how to use equations to generate numerical values from our bills. Finally, we learned how to generate callouts to create visual links between the information in our bill and the visual representation of our assembly in the drawing sheet.

Being able to create a BOM is an essential skill in order to communicate products consisting of many different parts. It is also an expected skill of a SOLIDWORKS professional.

In the next chapter, we will start learning about and using another set of advanced features that will enable us to generate more complex 3D models than what we've learned about already. These features will include draft, shell, hole wizard, features mirror, and multi-body parts.

Questions

Answer the following questions to test your knowledge of this chapter:

1. What is a BOM?
2. What information can be found in BOMs?
3. What is linked information in BOMs in SOLIDWORKS?

4. Download the parts and assembly linked to this exercise and generate the following standard BOM:

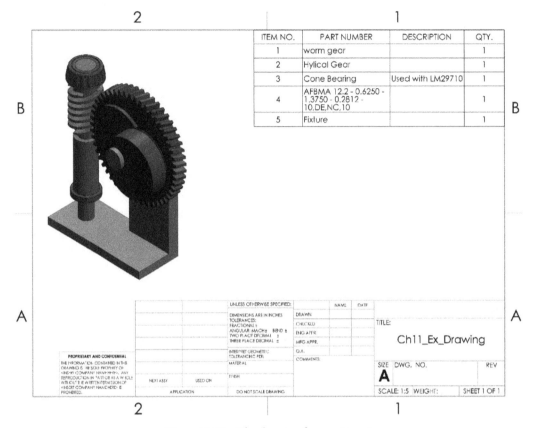

ITEM NO.	PART NUMBER	DESCRIPTION	QTY.
1	worm gear		1
2	Hylical Gear		1
3	Cone Bearing	Used with LM29710	1
4	AFBMA 12.2 - 0.6250 - 1.3750 - 0.2812 - 10,DE,NC,10		1
5	Fixture		1

Figure 11.35 – The drawing for question 4

5. Modify the BOM from the previous question by adding and filling up the cost column (note – cost numbers are not automatically generated; input them manually, as shown in the following screenshot). Also, change the **PART NUMBER** and **DESCRIPTION** columns to **PartNo** and **Vendor**. Your bill will look similar to the one shown here:

ITEM NO.	PartNo	Vendor	/Cost (USD)	QTY.
1	DP-6739	International Gears Limited	800	1
2	DP-5946	International Gears Limited	1200	1
3	DP-9548	International Bearing Limited	50	1
4	DP-3464	International Bearing Limited	60	1
5	DP-4638	Almattar Machine Shop	150	1

Figure 11.36 – The BOM for question 5

6. Use equations to calculate the total cost for all the parts. You will end up with the following bill:

ITEM NO.	PartNo	Vendor	/Cost (USD)	QTY.
1	DP-6739	International Gears Limited	800	1
2	DP-5946	International Gears Limited	1200	1
3	DP-9548	International Bearing Limited	50	1
4	DP-3464	International Bearing Limited	60	1
5	DP-4638	Almattar Machine Shop	150	1
		Total	2260	

Figure 11.37 – The BOM for question 6

7. Use the **Auto Balloon** command to link the item numbers in the drawing views. Your result will look similar to the following figure:

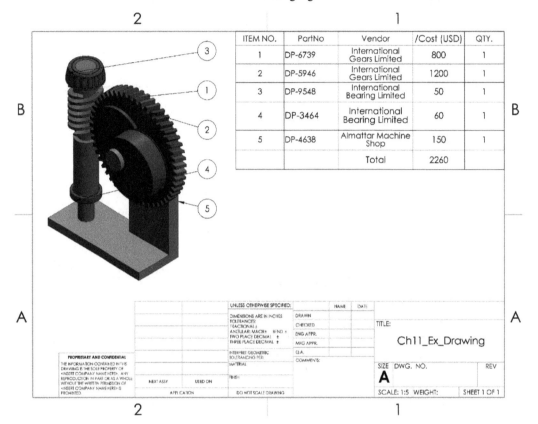

ITEM NO.	PartNo	Vendor	/Cost (USD)	QTY.
1	DP-6739	International Gears Limited	800	1
2	DP-5946	International Gears Limited	1200	1
3	DP-9548	International Bearing Limited	50	1
4	DP-3464	International Bearing Limited	60	1
5	DP-4638	Almattar Machine Shop	150	1
		Total	2260	

Figure 11.38 – The drawing and BOM for question 7

Important Note

The answers to the preceding questions can be found at the end of this book.

Section 6 – Advanced Mechanical Core Features – Professional Level

The professional level is the second level of proficiency for SOLIDWORKS users in core mechanical design applications. This section covers all the features expected of this level. These include features for building 3D models that include draft, shell, hole wizard, mirroring, rib, and feature patterns. Also, you will learn features that can optimize our 3D modeling approach, such as multi-body parts, equations, configurations, and design tables.

This section comprises the following chapters:

- *Chapter 12, Advanced SOLIDWORKS Mechanical Core Features*
- *Chapter 13, Equations, Configurations, and Design Tables*

12
Advanced SOLIDWORKS Mechanical Core Features

In this chapter, we will cover more advanced and less commonly used features in SOLIDWORKS mechanical modeling. These features include the **draft** feature, the **shell** feature, the **Hole Wizard**, **features mirroring**, and the **rib** feature. We will use these features to build models, while also covering multi-body parts. These features will greatly enhance your SOLIDWORKS skills to an advanced level by further simplifying complex model creation and manipulation. They are also essential for passing the SOLIDWORKS professional certification exam.

The following topics will be covered in this chapter:

- Understanding and applying the draft feature
- Understanding and applying the shell feature
- Understanding and utilizing the Hole Wizard

- Understanding and applying features mirroring
- Understanding and applying the rib feature
- Understanding and utilizing multi-body parts
- Understanding and applying the linear, circular, and fill feature patterns

By the end of this chapter, you will be able to generate more complex models with features matching international standards with holes. Also, the draft, shell, mirror, rib, and multi-body parts features will provide us with more means to meet specific design requirements.

Technical requirements

This chapter will require that you have access to the SOLIDWORKS software. The project files for this chapter can be found in this book's GitHub repository: `https://github.com/PacktPublishing/Learn-SOLIDWORKS-Second-Edition/tree/main/Chapter12`.

Check out the following video to see the code in action: `https://bit.ly/323CkEt`

Understanding and applying the draft feature

Drafting refers to changing sharp steps in parts into chamfered ones. Drafting primarily comes from the casting and plastic injection molding industry to make it easier to release parts out of molds. As SOLIDWORKS professionals, we will be expected to apply drafts to a variety of applications where necessary. In this section, we will explore what drafting is and how to use the draft feature.

What are drafts?

Drafts are commonly applied to parts that are made with injection molding. It is a slight tilt between two different surfaces at different levels. In practice, drafts help make parts fit better with the mold and make the parts easier to remove from the mold compared to without it. Also, drafts help increase the success rate of the mold taking effect.

The following diagrams highlight the effect of the draft feature. The first diagram shows the model at hand, as well as the **Cross-section** area we are applying to show the effect of the draft:

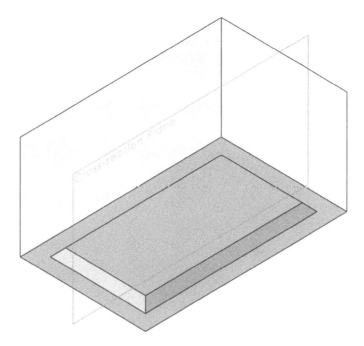

Figure 12.1 – An example of a draft

The second diagram shows the cross-section of the model **With Draft** and **Without Draft**. Note that the draft looks similar to a chamfer:

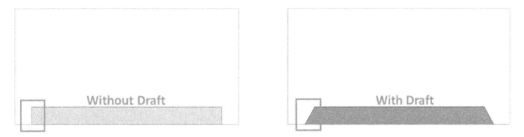

Figure 12.2 – Examples of a design with and without draft

Now that we know what drafts are, we can apply them to create a particular 3D model.

Applying drafts

In this section, we will cover how to apply the draft feature. We will apply a draft to the following model. You can download this model from the package provided for this chapter.

Alternatively, you can create the model from scratch. All of the lengths are in millimeters. Note that the draft is measured by the angle shown in the **Draft DETAIL D** view:

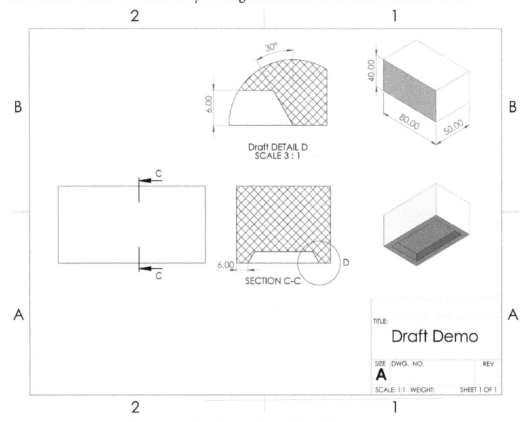

Figure 12.3 – The 3D model we will build in this exercise

To create the draft, do the following:

1.   First, download and open the model for this exercise. The model is as follows:

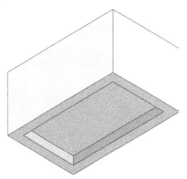

Figure 12.4 – The model to be downloaded for this section

2. Select the **Features** command category, then select the **Draft** command, as highlighted in the following screenshot:

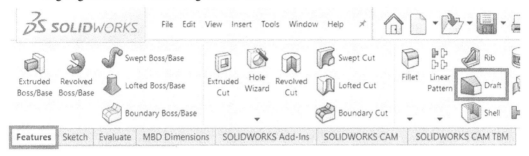

Figure 12.5 – The location of the Draft command

3. Adjust the **PropertyManager** options and selections so that they match those shown in the following screenshot. Here is a brief description of some of the options:

- **Neutral plane**: Here, we can select a surface that will be adjacent to the draft. This neutral surface will remain unchanged during the creation of the draft. Hence, the other adjacent surface will be affected.

- **Faces to Draft**: Here, we can select all of the faces to be drafted. We can select more than one surface as we see fit. In this exercise, we are selecting all four inside faces:

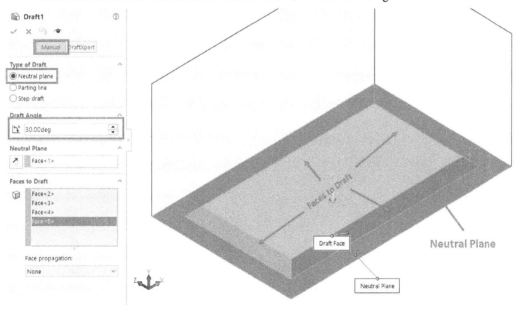

Figure 12.6 – The settings for the draft feature

4. Click on the green checkmark to apply the draft. The resultant draft will be as shown in the screenshot. In the applied draft, note the following:

-   The position of the angle of the draft. In many instances, we might be given the complementary angle in a drawing instead of the draft angle. The complementary angle is 90 minus the given angle.

-   The neutral surface and its surface area remained unchanged compared to all of the other surfaces related to the draft:

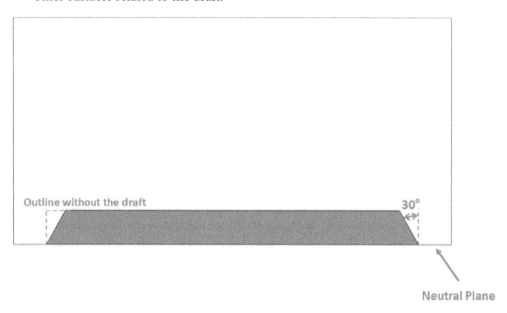

Figure 12.7 – The selected neutral surface remains unchanged once the draft has been applied

Note that, in this exercise, we used the **Neutral plane** draft type. Other types of drafts are available, including **Parting line** and **Step draft**, which can be created similarly. The following diagram highlights an example of a parting line draft application:

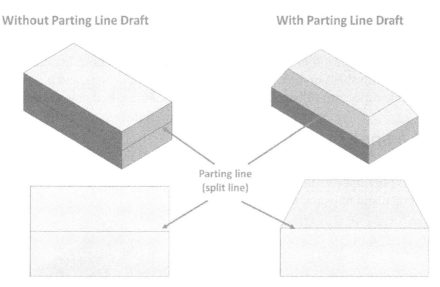

Figure 12.8 – A demo application of the parting line draft

The following diagram highlights an example of a step draft application:

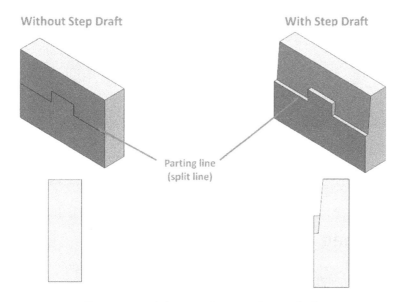

Figure 12.9 – A demo application of a step draft

Both draft options can be found in the draft feature's PropertyManager. This concludes our coverage of the draft feature. We learned what the feature is and how to generate and define it.

Next, we will cover another feature known as the shell feature.

Understanding and applying the shell feature

In this section, we will cover the shell feature. As its name suggests, the shell feature enables us to create a shell out of an existing shape without much effort. In our everyday lives, we interact with shells in multiple products such as cans, laptops, and phone exteriors. In this section, we will explore what the shell feature is and how to apply it.

What is a shell?

The shell feature enables us to make a shell of an existing model. It makes it easy to create objects such as cans and containers. The following screenshot shows a box with the effect of the **Shell** command implemented:

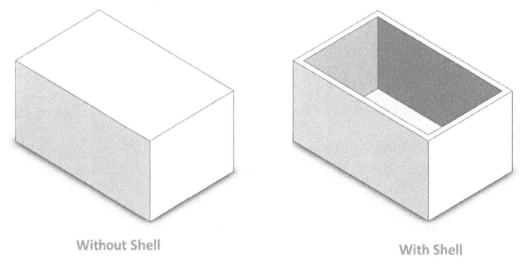

Without Shell

With Shell

Figure 12.10 – The impact of the shell feature

Now that we know what the shell feature does, we can apply it to create a specific model, which we will do next.

Applying a shell

In this section, we will cover how to apply the shell feature. We will apply the **Shell** command to create the following model. You can download this model from the package provided for this chapter. Alternatively, you can create the model from scratch. Note that this is a continuation of the previous model. The thickness of the wall is **3 mm**, as highlighted in the review cloud in the diagram:

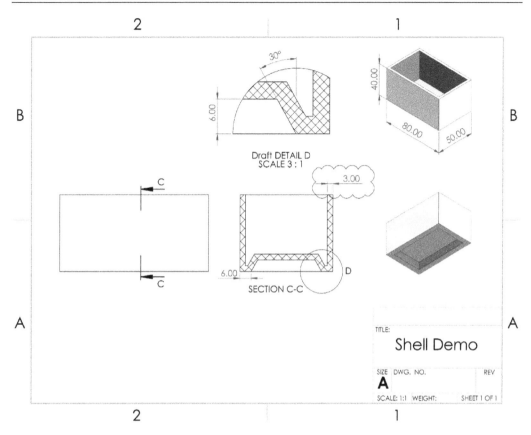

Figure 12.11 – The 3D model we will build in this exercise

To create the highlighted shell, follow these steps:

1. First, download and open the model for this exercise. The model is as follows:

Figure 12.12 – The model that is included with the downloads for this section

2. Select the **Features** command category, then select the **Shell** command, as highlighted in the following screenshot:

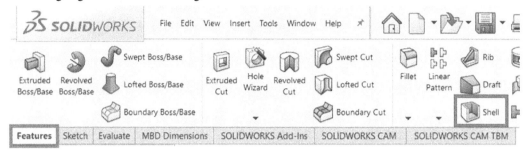

Figure 12.13 – The location of the Shell command

3. Adjust the **PropertyManager** options and selections so that they match the parameters highlighted in the following screenshot:

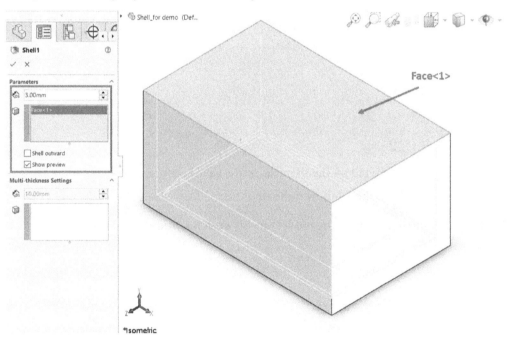

Figure 12.14 – The selected face parameters will be removed after applying the shell

Here is a brief description of some of the options:

- **Faces to Remove**: This is the field below the thickness of the draft. The faces we select here will be removed after we apply the shell. If we leave this field empty, the model will still be shelled from the inside; however, the outer surface will remain unchanged.

- **Shell outward**: This will flip over the shell direction. In other words, the current model will be removed and replaced by a shell starting from its outer surface. You can give it a try to figure out the effect.

- **Multi-thickness Settings**: This allows us to have different wall thicknesses on different sides. We will look at this in more detail in this section.

4. Click on the green checkmark to apply the shell. Our model will look as follows:

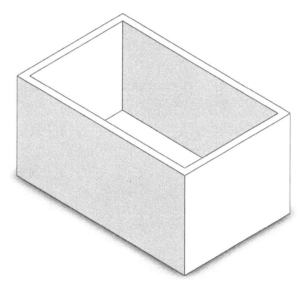

Figure 12.15 – The resulting 3D model after applying the shell feature

This concludes this section on applying a unified shell using the shell feature. Next, we will cover a special condition where we can specify different thicknesses for different sides.

Multi-thickness Settings

In the shell's PropertyManager, we have the option of making a shell with different thicknesses for different walls. To explore this option, we will go back to our model and modify it. This modification is highlighted by the review cloud in the following diagram:

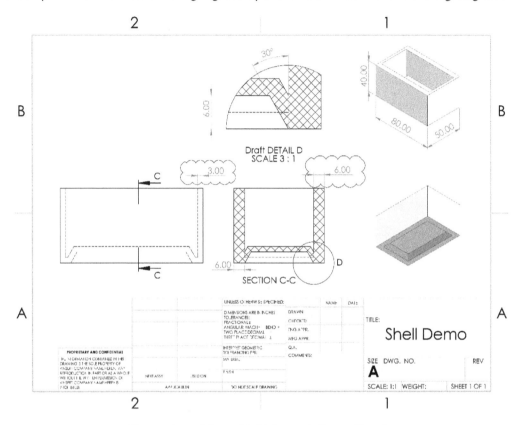

Figure 12.16 – The multi-thickness shell we will apply

Note that the model is still shelled in the same way; however, two of the walls have different thicknesses than the others.

To apply this modification, we can do the following:

1. Right-click on the **Shell** feature from the design tree and select **Edit Feature**.

2. Via **PropertyManager**, we will adjust the fields under **Multi-thickness Settings** by doing the following. The following screenshot highlights **Multi-thickness Settings**:

*   Under the **Multi-thickness Settings** face selection, select one of the faces we want to make thicker. Then, adjust the thickness from 3 mm to 6 mm.

- Select the other face and adjust the thickness again from 3 mm to 6 mm. Note that we have to adjust the thickness of each side separately. The rest of the faces will keep the default thickness we set at top of **PropertyManager**:

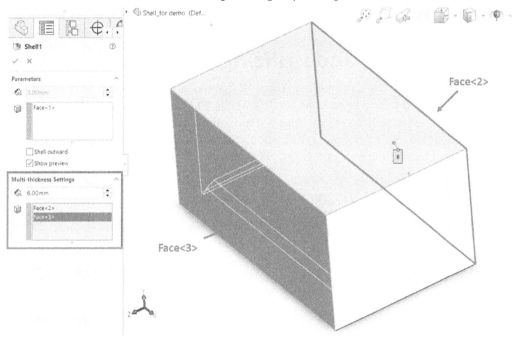

Figure 12.17 – The multi-thickness setting allows us to apply different thicknesses to the walls

3. Click on the green checkmark to apply these **Multi-thickness Settings**. The result of the model will be as follows. Note that two of the side walls are thicker than the others:

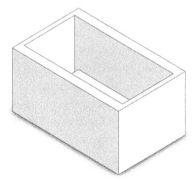

Figure 12.18 – The resulting shell with multiple wall thicknesses

This concludes this exercise on using the shell feature. We covered the shell feature and how to apply it, as well as how to apply both shell thickness and a multi-thickness shell. In the next section, we will cover the Hole Wizard, which will enable us to create different types of holes.

Understanding and utilizing the Hole Wizard

Holes are very common features in most products. If we look at any project, we will likely see screws that hold different parts together. In essence, these are different holes. Usually, these holes are made according to common international standards. The Hole Wizard allows us to create holes as per those standards. In this section, we will explore the Hole Wizard and how to utilize it to create holes.

What is the Hole Wizard and why use it?

The Hole Wizard in SOLIDWORKS enables us to create holes in our model that match international standards for holes. This includes drilling and threading the holes as well. The Hole Wizard makes it easy and convenient to make those holes by selecting the hole standard and type and placing the hole directly on the part.

Identifying a hole in SOLIDWORKS

To identify a hole in the SOLIDWORKS Hole Wizard, we must have the following information about the software:

1. **Hole type**: This includes the following:

    - **Overall shape**: Nine shapes are supported by the Hole Wizard; these include counterbore holes, countersink holes, tapered tap holes, slots, and others. We have a graphical presentation of each hole shape in the SOLIDWORKS interface.

    - **Standard**: This includes internationally recognized standards for defining holes. The most commonly used standards are the ones from the **International Standard Organization (ISO)** and the **American National Standard Institute (ANSI)**. The Hole Wizard also includes more standards such as those from the **British Standard Institute (BSI)**, the **Japanese Industrial Standards (JIS)**, **Korean Standards (KS)**, and many others. These standards mainly differ in holes sizes, hole fittings, and referencing.

    - **Type**: The type of the hole depends on the preceding two options. Each combination of the overall shape and standard will have a different set of types to choose from.

2. **Hole specifications**: This includes the following:

    - **Size**: This allows us to specify the diameter of the hole, as per the standard and type we pick. Note that each standard references sizing differently in terms of naming.

    - **Fit**: For selected hole shapes, we will be able to specify whether we would like to have a normal, loose, or close fit for our hole.

    - **Custom sizing**: This allows us to further customize the size of the hole from the standard size if needed.

3. **End condition**: Similar to the end conditions for the common features we used previously, this allows us to specify the depth of the hole. This will also allow us to specify the depth of the thread if the hole involves that.

We have just learned what the Hole Wizard is and what specifications we need to know about to identify and call out a hole. Next, we will learn how to put all of that into practice by using the Hole Wizard to create holes.

Utilizing the Hole Wizard

In this section, we will use the Hole Wizard to create multiple holes in a box, as shown in the following diagram. You can download the basic box from the models for this chapter. Alternatively, you can create it from scratch. Note that each of the holes is identified with all of the information we need to create it. All of the lengths in the following diagram are in millimeters.

The following diagram highlights the two holes that we will be creating:

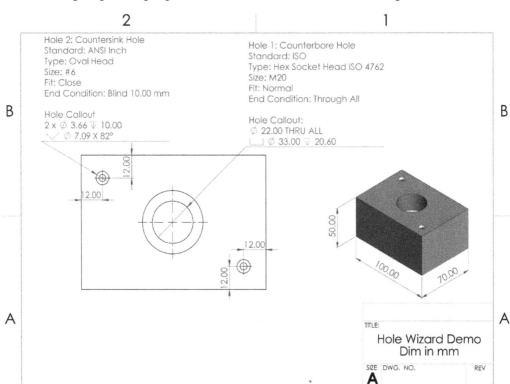

Figure 12.19 – The 3D model we will build in this exercise

To make this, we will create **Hole 1** and then **Hole 2**. Follow these steps to do so:

> **Note**
>
> The preceding diagram shows multiple holes that have been made with different standards. When working with a realistic project, we will only use one standard for the whole product. However, as this is an exercise for demonstration and learning purposes, we are using different standards.

1. First, download and open the model linked to this section.

2. Select the **Hole Wizard** command, as shown in the following screenshot:

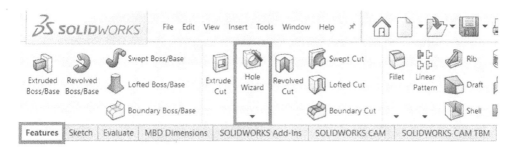

Figure 12.20 – The location of the Hole Wizard command

3. **Making Hole 1**: At this point, we will be making Hole 1, which will be in three stages – choosing the hole specifications, choosing the hole's position, and confirming the hole's position:

    • **Hole Specification**: We can fill out the information shown in the preceding diagram. The settings will be as shown in the following screenshot. After deciding on the hole specifications, we can choose the hole's position. The hole's shape is only shown with figures; however, if we hover our cursor over the icons, the names will appear:

Figure 12.21 – The hole specifications can be applied to match the industry standards

- **Hole Position**: To start working on the positioning of the hole, we can select the **Positions** tap in **PropertyManager**. This will prompt us to select which surface we want the hole on. Select the bluish top surface, as highlighted in the following screenshot:

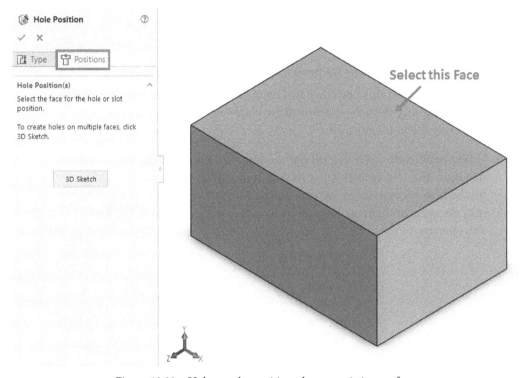

Figure 12.22 – Holes can be positioned on an existing surface

Then, we can place a dot, which will be the center of the hole. This hole is in the center of the shape, which coincides with the origin. Hence, we can place the dot so that it coincides with the origin, as highlighted in the following screenshot. Note that we can see a preview of the hole:

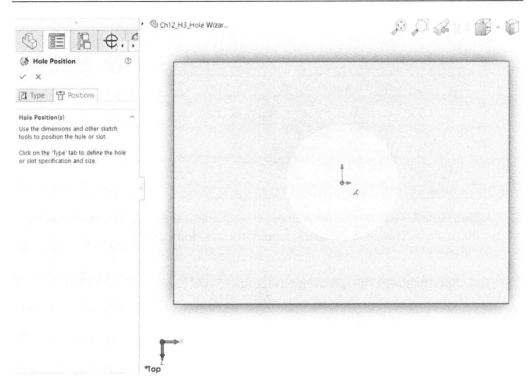

Figure 12.23 – A hole can be positioned using sketching commands

---

Note

If you don't select any surface for the Hole Wizard, a 3D sketch will be started to locate the hole. You can also click on the **3D Sketch** option shown in *Figure 12.22* if you intend to use a 3D sketch for positioning.

Tip

You can pre-select a surface to use with the Hole Wizard.

- **Confirmation**: If we are happy with the hole preview, we can click on the green checkmark to implement the hole. The hole will be added, as shown in the following diagram:

Figure 12.24 – The hole that matches the specified standards will be applied after confirmation

4. **Making Hole 2**: To make the other holes, we will follow the same procedure we did for the first one. However, the specifications, the positions, and the number of holes are different:

- **Hole Specification**: The specifications for the second hole can be set like so:

Figure 12.25 – The specifications of the second hole's types

- **Hole Position**: Now that we have two holes, we can place two dots for each hole. Positioning a hole follows the same sketching commands we used previously. The only exception is that any *point* we place will be interpreted as a center of a hole. Hence, we can position our holes using the **Smart Dimension** command to match what's shown in the following diagram:

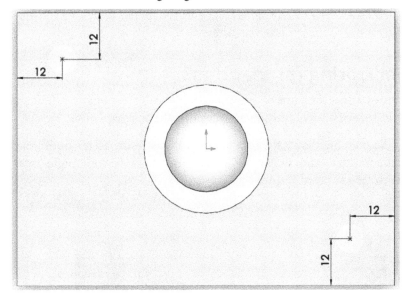

Figure 12.26 – Dimensions can be used to locate a hole

- **Confirmation**: Click on the green checkmark to apply the two holes. The final shape will look as follows:

Figure 12.27 – The final 3D model after confirming the holes

This concludes how to utilize the Hole Wizard. In this section, we learned what the Hole Wizard is and how to identify details for the Hole Wizard. Then, we learned how to use the Hole Wizard's functions in SOLIDWORKS to generate a hole based on different industry standards and identifications. Next, we will cover how to mirror any feature we apply in SOLIDWORKS.

Understanding and applying features mirroring

Sometimes, we need to create a feature or a set of features and then try to duplicate them on the other side of the model. We can accomplish this by mirroring the features. In this section, we will discuss what mirroring is from a feature perspective. We will also learn how to use this feature to mirror other features.

What is mirroring for features?

Mirroring features works the same way as mirroring the entities of a sketch. It enables us to duplicate a feature or a set of features by reflecting them on a plane.

The following diagram highlights the effect of mirroring features. The model on the left highlights a model with a set of features, while the model on the right highlights the model after mirroring selected features:

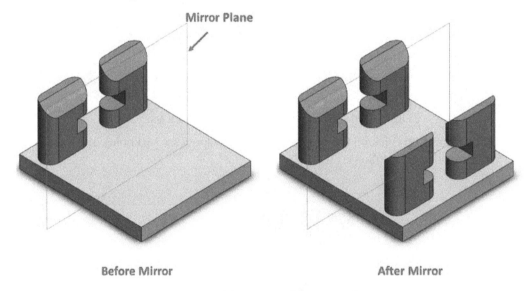

Figure 12.28 – The impact of the mirror feature

Now that we know what is meant by mirroring features, we can start learning how to apply the feature to 3D model creation.

> **Note**
> The mirror plane could be an existing plane or an existing surface. Also, it can be a plane that we create for mirroring purposes.

Utilizing the Mirror command to mirror features

In this section, we will learn how to use the **Mirror** command to mirror features. We will create the model shown in the following diagram. Note that the pillars in the model are mirrors of each other:

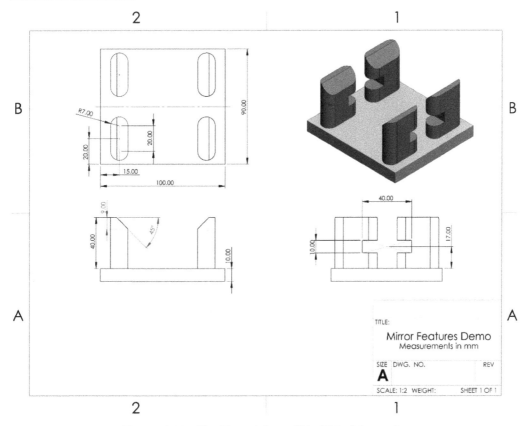

Figure 12.29 – The 3D model we will build in this exercise

To use the **Mirror** command, follow these steps:

1. Download and open the part linked to this section. The model looks as follows. Alternatively, you can create the model from scratch using the information provided in the preceding diagram:

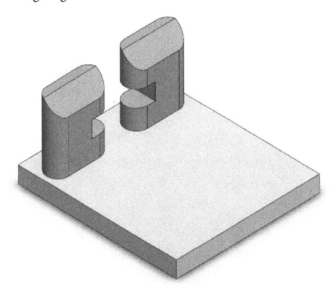

Figure 12.30 – Our starting model, which you can download with this section

2. From the **Features** tab, select the **Mirror** command, as shown in the following screenshot:

Figure 12.31 – The location of the Mirror command

3. The mirror **PropertyManager** will be shown on the left. We can make the following selections:

    - The **Mirror Face/Plane**, in this case, is the same as the default **Right Plane**. Hence, we can select **Right Plane** from the design tree, as shown in the following screenshot:

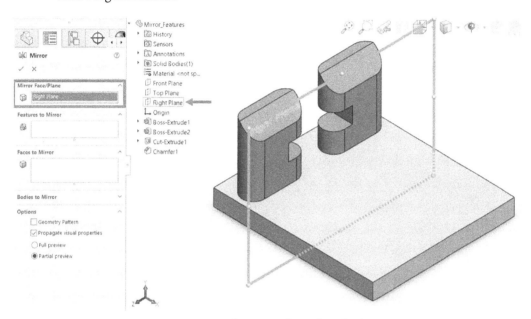

Figure 12.32 – A plane is used to reflect the features

Tip

You can preselect the mirror plane before clicking the **Mirror** command.

- For **Features to Mirror**, we want to mirror a total of three features, which are **Boss-Extrude**, **Cut-Extrude**, and **Chamfer**. We can select all of these features from the design tree, as shown in the following screenshot. Alternatively, we can select the features directly by selecting them from the 3D model shown in the canvas:

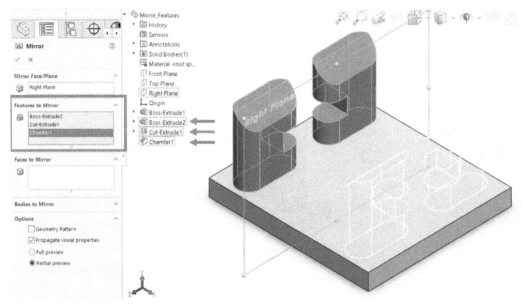

Figure 12.33 – We can select features to mirror from the design tree

4. After confirming the preview, click on the green checkmark to apply the mirror. The final model will look as follows:

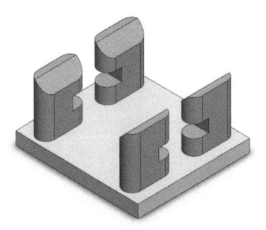

Figure 12.34 – The final 3D model after mirroring the specified features

In this exercise, **Mirror Face/Plane** happened to be the same as the default **Right Plane**. However, it can be any straight face or surface from the model itself. It can also be a new plane that we generate ourselves using reference geometries. We can also mirror any number of features in one go. Note that, similar to mirroring sketches, any modifications we apply to the original features will be reflected in the mirrored features. Before we conclude our discussion of the **Mirror** command, let's explain two notable options shown in *Figure 12.33* – **Geometry Pattern** and **Propagate visual properties**:

- **Geometry Pattern**: Checking this option will mirror a geometrical replica of the original shape while disregarding any logical arrangements that were used, such as features end conditions, if any.

- **Propagate visual properties**: Checking this option will copy the visual textures and colors of the original shape to the mirrored shape.

> **Tip**
>
> Mirroring features is mostly more convenient than mirroring sketches. This is because, with features, we can mirror a combination of features that inherently include sketches. Hence, mirroring end features often results in less modeling time and less time when modifying the model afterward.

This concludes this section on mirroring features. We learned what the features mirroring function is, in addition to how to apply it. Next, we will cover another feature, known as the rib feature.

Understanding and applying the rib feature

Ribs are reinforcement structures that are used to help fix two sides together. In this section, we will learn what ribs are and how to create them using the SOLIDWORKS rib feature. We will also learn how SOLIDWORKS interprets the creation of ribs using the rib feature.

Understanding ribs

Ribs are often welded support structures that are added to link different components or parts together. It is common to find ribs within plastic objects such as toys. They are also commonly found in building structures. The following diagram shows two models, one **Without Ribs** and one **With Ribs**:

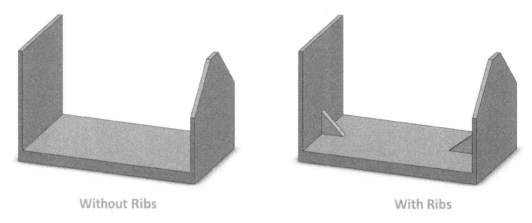

Figure 12.35 – 3D models with and without ribs

Note that we can create ribs out of other features, such as extruded boss and extruded cuts. However, the rib feature provides us with an easier method to both build and define a rib. Now that we know what ribs are and what the rib feature does, we will start learning how to apply it to a SOLIDWORKS model.

Applying the Rib command

In this section, we will learn how to use the **Rib** command to generate ribs in our models. We will create the model shown in the following diagram. The base model we will use can be downloaded from this book's GitHub repository. Note that the ribs are highlighted in detail views **A** and **B**, which we will generate in this exercise:

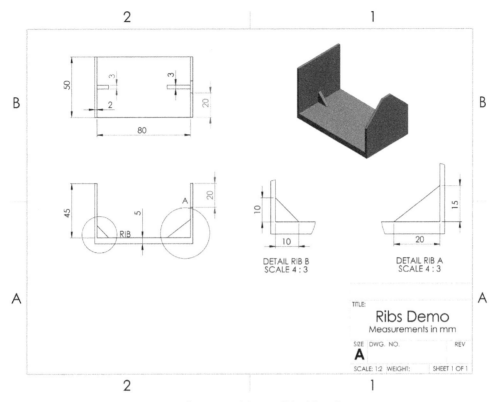

Figure 12.36 – The 3D model we will build in this exercise

To use the rib feature for this exercise, we can follow these steps:

1. First, download and open the SOLIDWORKS part linked to this section concerning ribs. The model will look as follows. Alternatively, you can create the model from scratch using the information provided in the preceding diagram:

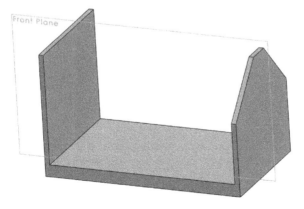

Figure 12.37 – You can download the following model with this section

2. Select **Front Plane** from the design tree. The front plane is shown in *Figure 12.35*.

3. In the **Features** tab, select the **Rib** command, as highlighted in the following screenshot:

Figure 12.38 – The location of the Rib command

4. Sketch a line that outlines the outer boundary of rib A, as follows. This sketch matches the **DETAIL RIB A** view on the drawing. Note that *Steps 2 to 4* are interchangeable; we can choose the feature first, then select the plane and sketch. We can also sketch, then select the **Feature** command:

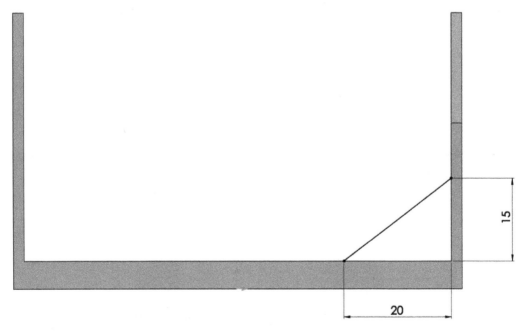

Figure 12.39 – The rib can be built with a single line sketch

5. After we exit the sketch, we will notice the **Rib** command's **PropertyManager** window on the left. For rib A, the options and the preview can be set as shown in the following screenshot. These options will allow us to define our rib in terms of the following:

- The width of the rib.

- The direction of the rib concerning the guiding sketch. Here, we can choose to have all of the width extending toward either direction or to the midway point.

- We can also specify whether we require a draft with the rib and what the draft angle is:

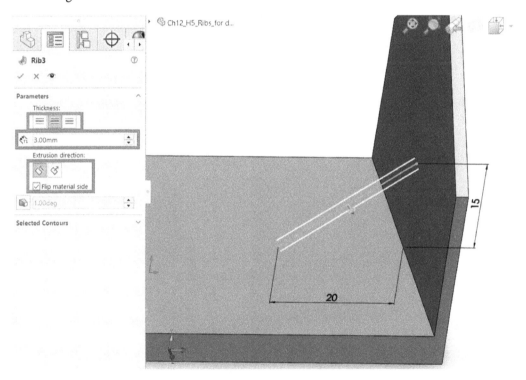

Figure 12.40 – The Rib command's PropertyManager window and its settings

6. Click the green checkmark to apply the rib. The model will look as follows:

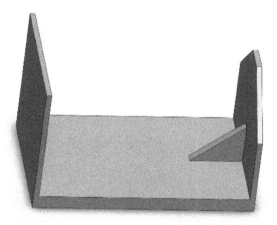

Figure 12.41 – The resulting rib after its application

7. Follow *Steps 2* to *6* again to create rib B. Make sure that you use the dimensions shown in the **DETAIL RIB B** diagram. The final model will look as follows:

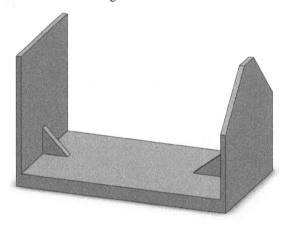

Figure 12.42 – The final 3D model after applying both ribs

We likely have identical rib dimensions in products that we interact with within our day-to-day lives. However, we created ribs with different dimensions to practice how to use the tools.

Drafted rib

One of the key options we can utilize with ribs is adding a draft to them. This option can be used by enabling the **Draft outward** option shown in the feature's **PropertyManager** window, as highlighted in the following screenshot:

Figure 12.43 – The rib feature allows us to make a drafted rib

A drafted rib will have the same look as the drafts we covered in the *Understanding and applying the draft feature*.

This concludes how to use the **Rib** command. We covered what the rib feature is and how to apply it. All of the features we covered earlier in this chapter can be used to construct models directly. Next, we will cover multi-body parts, which is a method for creating parts rather than a feature.

Understanding and utilizing multi-body parts

In all of the applications we have explored in this book, each part file we made consisted of one body. We used assembly files to combine the different parts. In this section, we will explore a different approach with multi-body parts. We will cover what multi-body parts are, how they are created, and what the advantages of multi-body parts are.

Defining multi-body parts and their advantages

Multi-body parts are models made within a SOLIDWORKS part file that contain more than one separate body. Hence, they are called multi-body parts. The following diagram shows the contents of one SOLIDWORKS part file. However, the diagram on the left consists of one solid body, while the one on the right consists of two solid bodies. Note that the difference between these two diagrams is that the right-hand one has an extrusion cut that separates the large triangle (one solid body) into two triangles (two solid bodies):

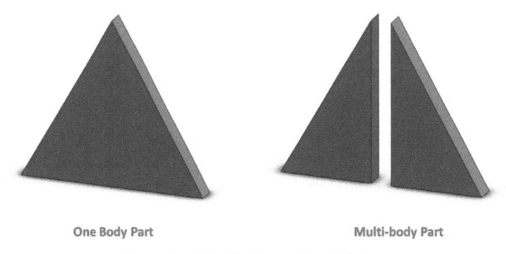

One Body Part   Multi-body Part

Figure 12.44 – A single body part and a multi-body part

However, we should not confuse multi-body parts with assemblies. The different bodies in multi-body parts are not dynamic, as is the case with the different parts in an assembly. This makes multi-body parts appropriate for certain applications that involve static interactions such as frames. This is due to the following advantages:

- The frame and other static elements will all be contained in one file, making them easier to access and modify.

- The work process is faster as we won't need to use more than one SOLIDWORKS file. Also, we won't need to create mates to ensure the different parts fit together, as is the case with assemblies.

However, assemblies also have other advantages over multi-body parts. These include the following:

- We can showcase the moving dynamics within a product.

- There's more flexibility in reusing parts in different assemblies. Also, we have more flexibility in exchanging an independent part.

- Creating separate drawings for each component is more convenient as the parts are set up in different files.

- Having parts in different files makes it easier to have separate part names and numbers for archiving or for inventory references.

There is no right or wrong answer to what approach to choose between assemblies and multi-body parts. As designers and practitioners, we will have to make the choice, weighing up the advantages of both approaches. To do that, we have to be familiar with both approaches. Next, we will create a multi-body frame to put what we just learned into practice.

Generating and dealing with a multi-body part

In this section, we will learn how to generate a part with multiple bodies. To demonstrate this, we will create the frame shown in the following diagram:

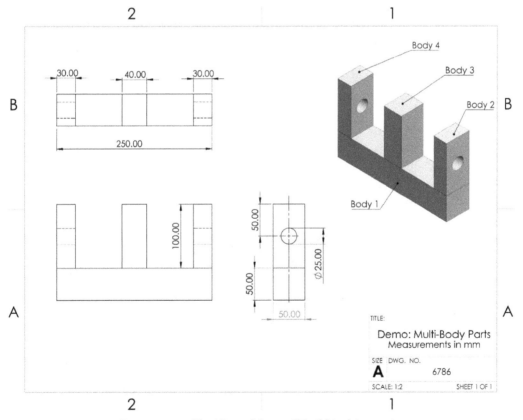

Figure 12.45 – The 3D model we will build in this exercise

Note that each element of the frame is indicated with a different number in the drawing, and the frame consists of four different bodies. To model this frame, we will follow these steps:

1. Create *body 1* as per the dimensions shown in the preceding diagram using the extruded boss feature. It should look as follows:

Figure 12.46 – A one-body part of body 1

2. We will create the three additional bodies with one more extruded boss feature. To do this, we will follow steps similar to those that we have followed previously. However, before applying the extruded boss feature, we will uncheck the **Merge result** box, as highlighted in the following screenshot, which also highlights the sketch we used:

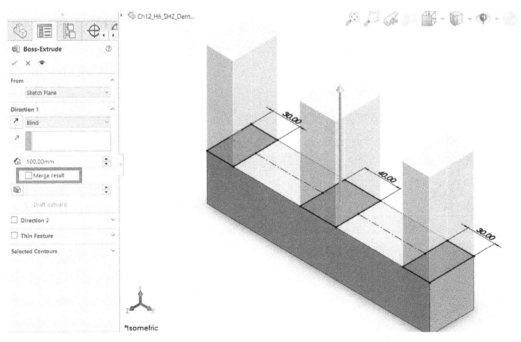

Figure 12.47 – Unchecking the Merge result option will build the extrusions as separate bodies

3. Click on the green checkmark to apply the extrusion. After doing that, we will see two differences compared to our usual extrusion. They are as follows:

    -   We will notice a line separating the different frames, as indicated in the following diagram. These lines indicate a separation between the different bodies in our part:

    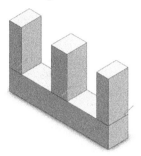

    Figure 12.48 – The lines indicating separate bodies

    -   In the design tree, we will notice a tab with the title **Solid Bodies**, as shown in the highlighted design tree diagram. If we expand this, we will see a list of all of the different bodies we have in our part; clicking on any of them will highlight the body in the canvas. From this list, we can selectively hide, delete, or change the display or assign materials to a specific body by right-clicking on the listing and choosing from the different options:

    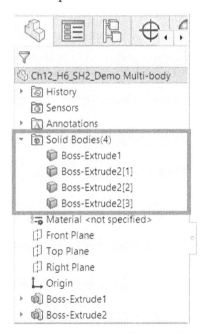

    Figure 12.49 – The design tree will show different bodies

This concludes one of the common ways to generate a multi-body part. We can also intentionally create separate bodies in the canvas, which will automatically result in multi-body parts. Also, whenever we apply features such as an extruded cut, which would result in physically separating bodies, we will have a multi-body part.

Note

One important aspect to note is that SOLIDWORKS, by default, will tend to merge bodies as that is a more common practice. Hence, any feature we apply that physically connects separate bodies will merge them unless we specify otherwise by unchecking the **Merge result** option.

Two important and useful elements concerning multi-body parts are the feature scope and being able to save bodies in different SOLIDWORKS part files. We will cover these two aspects next and apply them to our model.

Feature scope applications

The feature scope refers to the extent to which a feature is applied. For example, in a multi-body part, we can apply a feature such as an extruded cut and specify which body can be included in the cut and which body should not be included. In our exercise, notice in the drawing provided that there is a hole that goes through bodies 2 and 4 and skips body 3.

To utilize the feature scope, we can follow the same steps that we followed for an extruded cut. However, we will notice that the options under the **Feature Scope** tile in our cut extrude PropertyManager. We can see some options highlighted in the following screenshot with both the sketch and the other options for the extruded cut feature. Under the **Feature Scope** options, we can select the **Selected bodies** option and uncheck **Auto-select.**

Then, we can manually select bodies 2 and 4, as highlighted in the following screenshot:

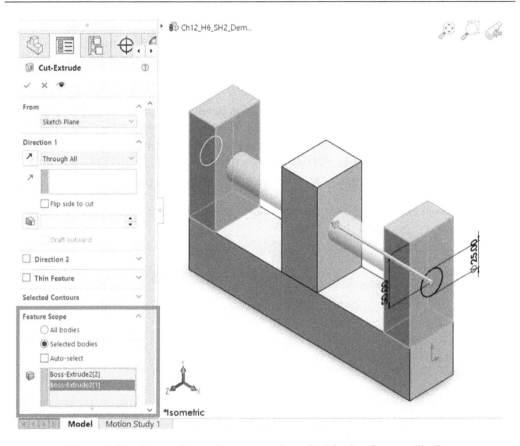

Figure 12.50 – Feature Scope allows us to select which body a feature will affect

As usual, click on the green checkmark to apply the extruded cut feature. Note that the resultant hole is only applied to only the selected bodies, as shown in the following diagram:

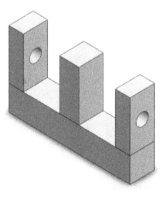

Figure 12.51 – The final multibody 3D model

Being able to scope features enables us to apply our design intent faster and more efficiently as it reduces the number of features we need to apply to reach the same result.

Separating different bodies into different parts

Now that we have the frames, we can face a situation in which we need to have each frame element or body in a separate SOLIDWORKS file. This could be needed for purposes such as generating separate drawings, inputting the separate files into a rapid prototyping machine, or other applications. SOLIDWORKS enables us to separate the different bodies into separate SOLIDWORKS part files. To do this, follow these steps:

1.   Go to **Insert | Features | Save Bodies...**, as highlighted in the following screenshot:

Figure 12.52 – Different bodies can be saved in different SOLIDWORKS part files

2. This will prompt you to select which bodies you want to save separately. You can select the bodies directly from the canvas or the list by checking the box under the save icon. Once you approve the **Save Bodies** command, a new file will be generated for each body. By default, the new files will be located in the same folder as the original file. These command options are shown in the following screenshot:

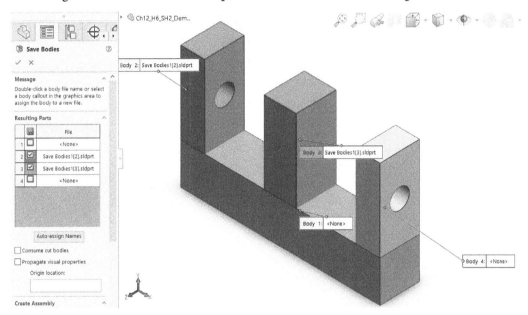

Figure 12.53 – The Save Bodies command allows us to pick which specific bodies we want to save

After applying the **Save Bodies** command, we will notice that this command is listed in the design tree, as highlighted in the following screenshot. The separate files will only reflect the shape from before that feature's listing. Hence, applying more features after using the **Save Bodies** command will not update the already saved bodies. We can drag the features we want to be reflected above the **Save Bodies** command in the design tree:

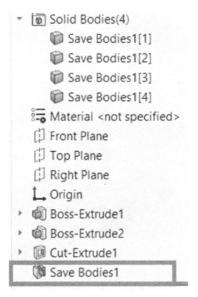

Figure 12.54 – The Save Bodies command will appear in the design tree after its application

This concludes our coverage of multi-body parts. In this section, we learned about the following:

- What multi-body parts are
- The difference between multi-body parts and assemblies and the advantages of each
- How to create a multi-body part
- How to utilize the features scope function
- How to save different bodies into different SOLIDWORKS part files

Knowing how to utilize multi-body parts to our advantage will enable us to optimize the software when targeting different applications, such as static furniture design or beam structure design.

Understanding and applying linear, circular, and fill feature patterns

Feature patterns allow us to duplicate features quickly according to a certain pattern. In this section, we will learn about the linear, circular, and fill patterns. In addition, we will learn how to apply them using the available SOLIDWORKS tools.

Understanding feature patterns

Feature patterns allow us to duplicate features quickly according to a certain pattern. They are similar to sketching patterns, which we covered in *Chapter 4, Special Sketching Commands*. However, with feature patterns, we can build patterns of features and bodies rather than patterns of sketch entities. In this section, we will cover three types of patterns, as follows:

- **Linear pattern**: Duplicates features or bodies in a linear fashion.

- **Circular pattern**: Duplicates features or bodies in a circular fashion.

- **Fill pattern**: Duplicates features within a set boundary by following a specific pattern. Possible patterns include perforation, circular, square, and polygon.

The following table highlights the difference between the three types of patterns:

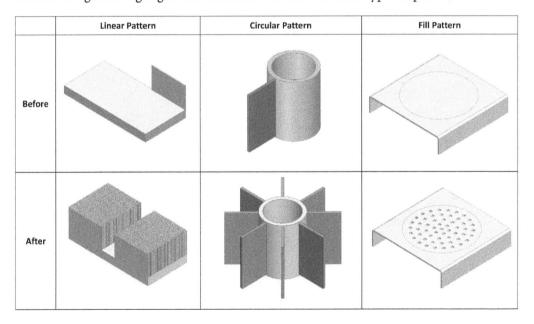

Figure 12.55 – The different types of patterns

Now that we know what to expect of each type of pattern, we can start applying them in the software. Let's get started with linear patterns.

Applying a linear pattern

In this section, we will learn how to use the **Linear Pattern** command to create pattern features. We will create the heat sink shown in the following diagram. The base model we will use can be downloaded from this chapter's GitHub repository. In the following diagram, note the repeated fins, which we will utilize linear patterns to generate:

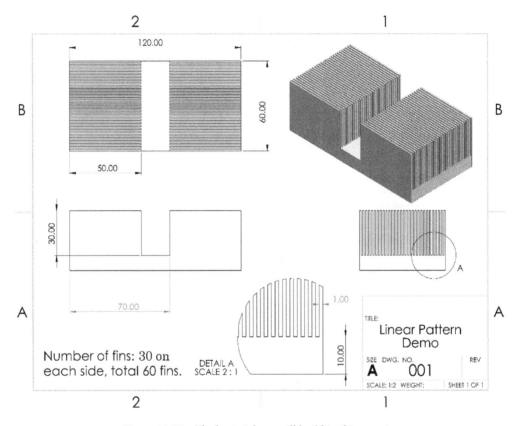

Figure 12.56 – The heat sink we will build in this exercise

To create the linear pattern, follow those steps:

1. First, download and open the SOLIDWORKS part linked to this section concerning feature patterns. The model will look as follows. Alternatively, you can create the model from scratch by using the preceding diagram:

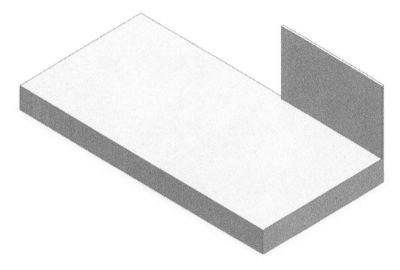

Figure 12.57 – The 3D model that's available for this exercise

2. Select the **Linear Pattern** command, as shown in the following screenshot:

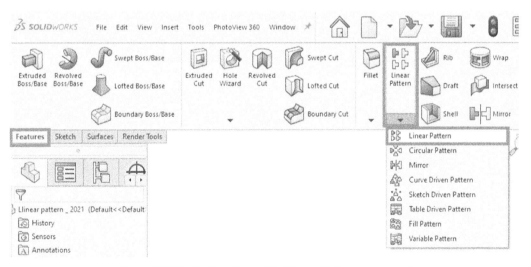

Figure 12.58 – The location of the Linear Pattern command

3. Check the **Features and Faces** box and select the fin, as shown in the following screenshot:

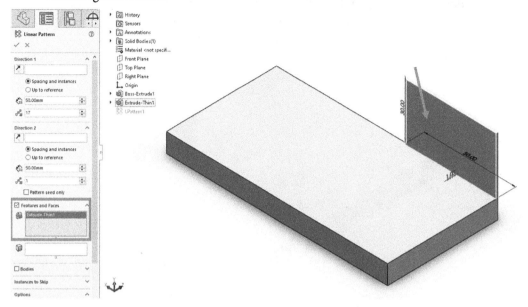

Figure 12.59 – The Features and Faces section of the linear pattern

> **Hint**
> You can use the canvas design tree to select features as well.

4. For **Direction 1**, select the smaller side edge, as highlighted in the following screenshot. A preview will then show the projected pattern:

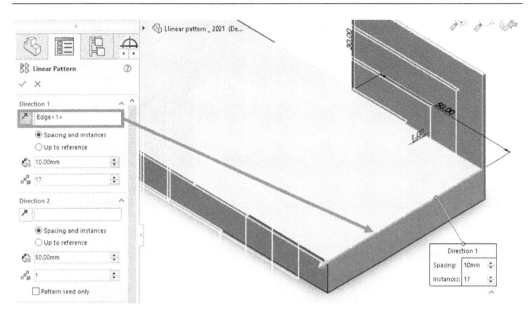

Figure 12.60 – The selection for Direction 1 using an edge

> **Hint**
>
> To indicate the direction of the pattern, we can also select lines from sketches, planes, planar surfaces, axes, and temporary axes.

5. For the end condition, do the following:

    - Select **Up to reference**, then select the end surface of the heat sink base for the first reference geometry, as shown in the following screenshot.

    - Select the **Selected reference** option, then select the indicated face of the fin for the second seed reference, as shown in the following screenshot.

- Define the pattern by the number of instances and input 30 for that, as shown in the following screenshot:

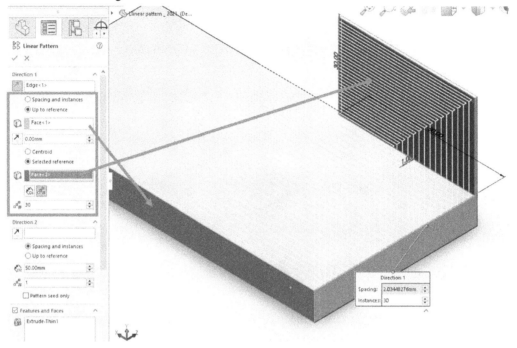

Figure 12.61 – Defining Direction 1 for the linear pattern

With that, we have defined one direction, as you will see from the preview on your screen. This whole pattern is repeated one more time for **Direction 2**. So, let's start defining the second direction in the same way.

6. Repeat *Steps 4* to *5* for **Direction 2**, as follows:

- Select the long edge of the base for the direction, as shown in the following screenshot.

- Select **Up to reference**, then select the end surface of the heat sink base for the first reference geometry, as shown in the following screenshot.

- Select the **Selected reference** option, then select the indicated face of the fin for the second seed reference, as shown in the following screenshot.

- Define the pattern by the number of instances and input 2 for that, as shown in the following screenshot:

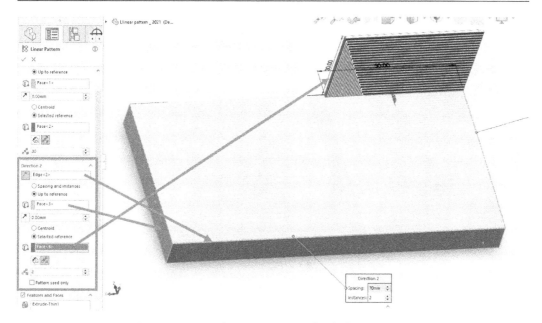

Figure 12.62 – Defining Direction 2 for the linear pattern

7. Click on the green checkmark to apply the pattern. The resulting 3D model will look as follows:

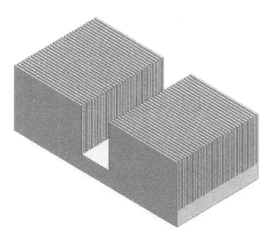

Figure 12.63 – The final 3D model after applying the linear pattern

With that, we have applied the linear pattern. Now, let's define two more important options we did not use in this exercise – **Bodies** and **Instances to Skip**. We can find these options in the linear pattern **PropertyManager** window, as shown in the following screenshot:

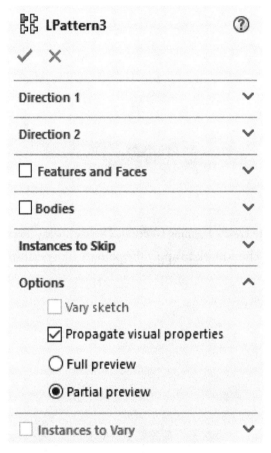

Figure 12.64 – The different options available for a linear pattern

Let's look at these options in more detail:

- **Bodies**: This allows to build a pattern of bodies instead of features. This option can be used when we're working with multi-body parts.

- **Instances to Skip**: This allows us to exclude some of the duplicates that are generated by the pattern manually. This works similarly to the sketch patterns we covered in *Chapter 4, Special Sketching Commands*.

With that, we can conclude our discussion on linear patterns. Next, we will take a closer look at circular patterns.

Applying a circular pattern

In this section, we will learn how to use the **Circular Pattern** command to pattern features. Note that setting up a circular pattern follows a similar procedure to setting up a linear pattern. We will create a simple waterwheel design, as shown in the following diagram. Note the repeated blades that we will use the circular pattern to generate:

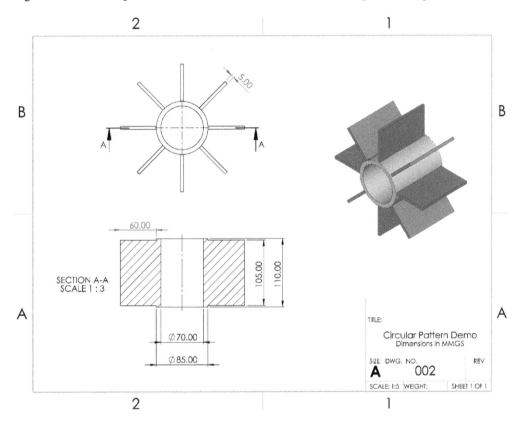

Figure 12.65 – The water wheel we will create in this exercise

To complete the exercise, follow those steps:

1. First, download and open the SOLIDWORKS part linked to this section regarding feature patterns. The model looks as follows. Alternatively, you can create the model from scratch using the preceding diagram:

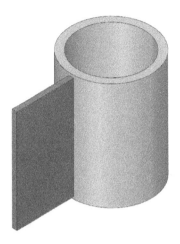

Figure 12.66 – The 3D model for this exercise

2. Select the **Circular Pattern** command, as shown in the following screenshot:

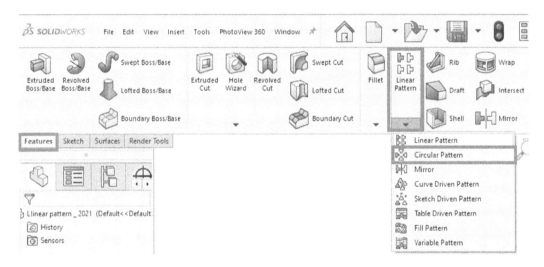

Figure 12.67 – The location of the Circular Pattern command

3. Check the **Features and Faces** box and select the blade, as shown in the following screenshot:

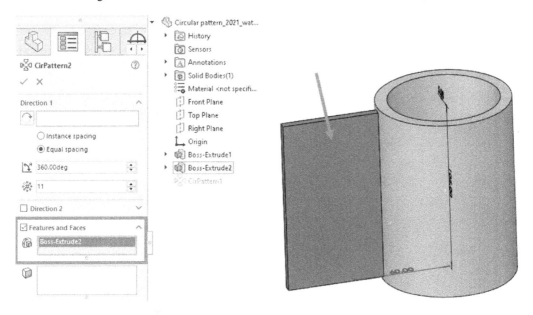

Figure 12.68 – Features section for the linear pattern

Hint

You can use the canvas design tree to select features as well.

---

4. For **Direction 1**, select the circular face, as shown here. Also, set the pattern to **equal spacing**, the angle to 360, and the instances to 8. Note that the number of instances includes the base feature:

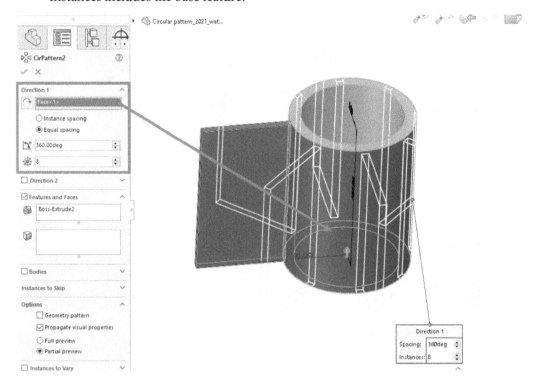

Figure 12.69 – The selection for Direction 1 using a circular face

> **Tip**
> For the direction, we can use faces, circular edges, axes, and temporary axes.

5. Click on the green checkmark to apply the pattern. The resulting 3D model will be as follows:

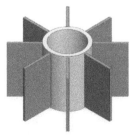

Figure 12.70 – The final 3D model after applying the linear pattern

With that, we can conclude our application of the circular pattern. Other options, such as **Bodies** and **Instances to Skip**, have the same functionality as their counterparts for linear patterns. Next, we will discuss the fill pattern feature.

Applying a fill pattern

In this section, we will learn how to use the **Fill Pattern** command to pattern features. We will create the simple holes grill shown in the following diagram. Note the repeated square holes on the top surface; we will use these to the fill pattern feature to generate. Fill patterns are commonly used for grills that are used in sound systems, in ventilation for electronics, and for weight reduction purposes:

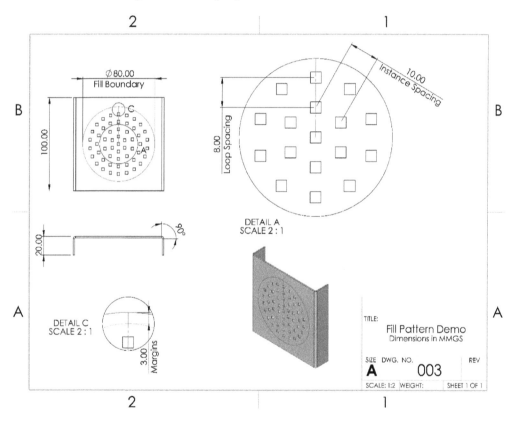

Figure 12.71 – The 3D model we will create in this exercise

Before applying the fill pattern, we must know that a fill pattern requires a boundary and a direction to be applied. Thus, we will be defining them before applying the pattern.

To complete this exercise, follow those steps:

1. First, download and open the SOLIDWORKS part linked to this section regarding feature patterns. The model looks as follows. Alternatively, you can create the model from scratch by using the preceding diagram:

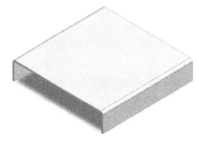

Figure 12.72 – The base model for this exercise

2. Sketch the 80 mm circular boundary on the top surface using a solid line. Then, sketch a construction line for the direction, as shown in the following screenshot:

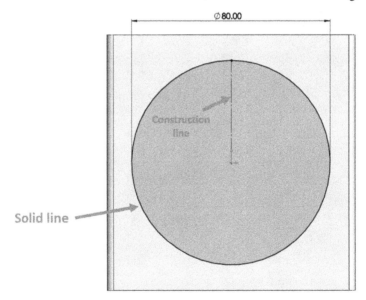

Figure 12.73 – The fill pattern requires a boundary and direction to apply

---

Tip

The boundary and direction do not have to be sketched separately. The existing linear surface can be used as a boundary, and existing edges can be used for direction. However, making new sketches allows us to build custom boundaries.

---

3.   Select the **Fill Pattern** command, as shown in the following screenshot:

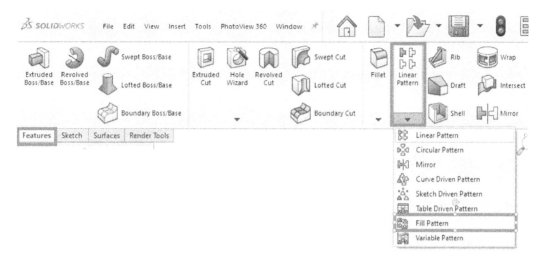

Figure 12.74 – The location of the Fill Pattern command

4.   Under **Fill Boundary**, select the circular sketch, as shown in the following screenshot:

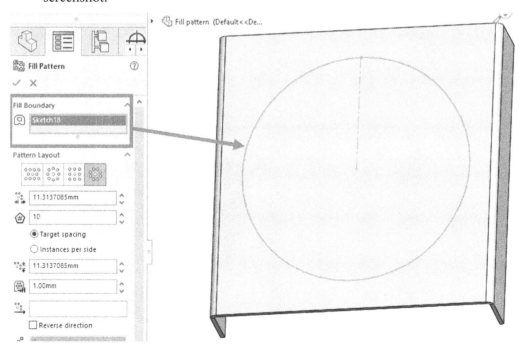

Figure 12.75 – A fill boundary is required for the fill pattern

5.   For **Pattern Layout**, select the **Circular** pattern and select the construction line for
     the direction, as shown in the following screenshot:

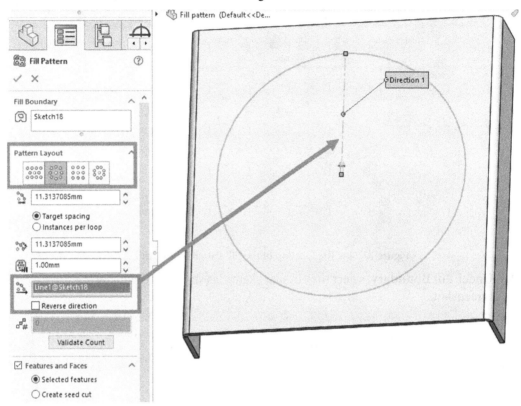

Figure 12.76 – Selecting the layout and direction

6. Check the **Features and Faces** box, select **Create seed cut**, and pick the square option. We can specify 3 mm as the side length. The **Vertex** selection indicates the origin point of the pattern. For example, for a circular pattern layout, the vertex will indicate the center of the square. These settings are highlighted in the following screenshot:

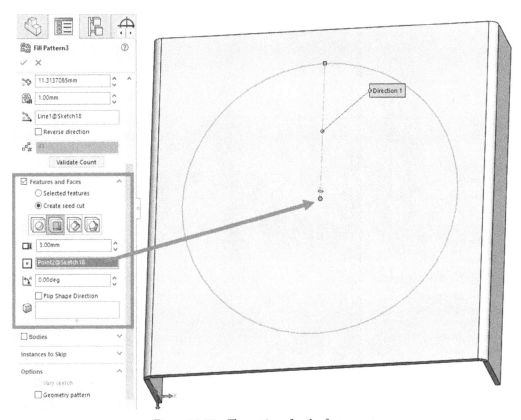

Figure 12.77 – The settings for the feature cut

> **Note**
>
> The **Create seed cut** option allows us to create common cuts that are associated with fill patterns. However, we can create fill patterns for any other feature as well.

7. Now, we can fine-tune the pattern. Make sure that the **Target spacing** option is selected. Then, set **Loop Spacing** to 8 mm, **Instance Spacing** to 10 mm, and **Margins** to 3 mm. Note that the on-screen preview will change as we are adjusting those parameters. Also, note that **Instance Count** will change according to the number of repeated instances. The following screenshot highlights these settings:

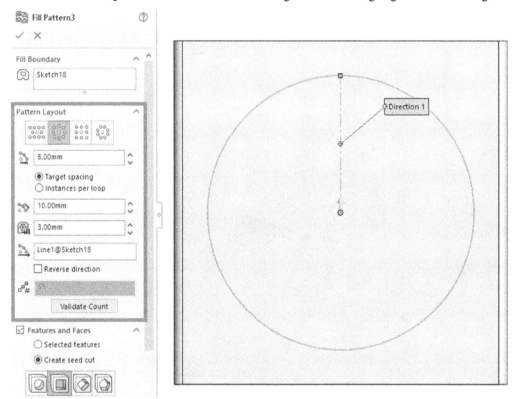

Figure 12.78 – The pattern layout instance's settings

> **Tip**
>
> It is common to keep adjusting the instance settings in a trial-and-error fashion until we get the desired result.

8. Click on the green checkmark to apply the fill pattern. The resulting 3D model will look as follows:

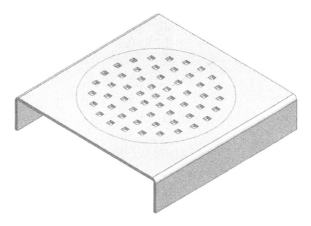

Figure 12.79 – The final 3D model after applying the fill pattern

With that, we have implemented the fill pattern feature. In this exercise, we applied the fill pattern within one boundary. However, it is also possible to apply a fill pattern that covers more than one boundary at a time. We'll look at this in the next section.

Applying the fill pattern to more than one boundary

In the previous exercise, we applied a simple fill pattern to one boundary area. However, the feature can apply one pattern that extends more than one boundary. This allows us to create harmonious-looking patterns with an elegant look and feel. To highlight this, let's look at the following pattern:

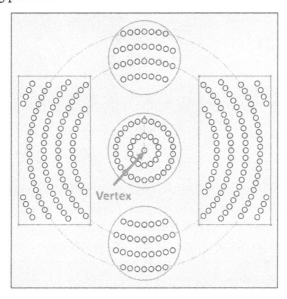

Figure 12.80 – Mutiboundary circular fill pattern with a central vertex

This fill pattern follows a circular pattern originating from the rectangular piece's center, as indicated by the *Vertex*. Here, we can see the formation of the circular fill pattern with the selected vertex as its center.

> **Note**
>
> The different boundaries can be in one sketch, or they can be in more than one sketch.

The vertex can also be located outside the areas of the boundaries as we see fit for our design. For example, the following circular fill boundary has the same specifications as the preceding one, with the vertex located differently. Here, the overall circular alignment of the pattern is preserved. However, the center of the base circle is located differently:

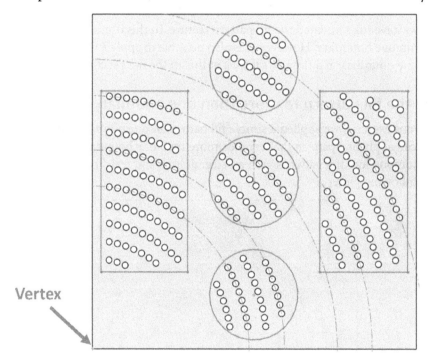

Figure 12.81 – Mutiboundary circular fill pattern with a corner vertex

With that, we have finished looking at the major types of feature patterns. We covered the linear, circular, and fill patterns. The linear and circular feature patterns are very similar to the patterns available for sketching, while the fill pattern allows us to apply patterns within selected boundaries easily. All these feature boundaries allow us to build patterns for specific features or whole bodies.

Summary

In this chapter, we learned about a variety of relatively advanced features for building more complex models. We covered the draft, shell, and rib features for creating specific geometries faster. We also learned about using the Hole Wizard to create industry-standard holes and covered how to mirror features to save us time that would otherwise be spent remaking features. We also learned about multi-body parts, their advantages, and how to utilize them. At the end of this chapter, we learned how to apply linear, circular, and fill patterns for features and bodies, which allow us to duplicate features or bodies in specific formations.

Knowing about the topics that were covered in this chapter is what separates professional users of SOLIDWORKS from amateurs. Mastering this chapter's topics will help you save time and create complex shapes faster while capturing more specific design intents.

In the next chapter, we will cover equations, configurations, and design tables. These skills will allow us to create more connected models and allow us to have multiple variations of a part within one part file.

Questions

Answer the following questions to test your knowledge of this chapter:

> **Note**
> The following questions will reinforce the main topics we learned in this chapter. However, it is also good practice to pick random objects and model them in SOLIDWORKS to improve your skills.

1. Describe the functions of the draft, shell, and rib features.

2. What is the Hole Wizard, and why is it useful?

3. Create the following part in SOLIDWORKS. Hint: Use the draft and shell features:

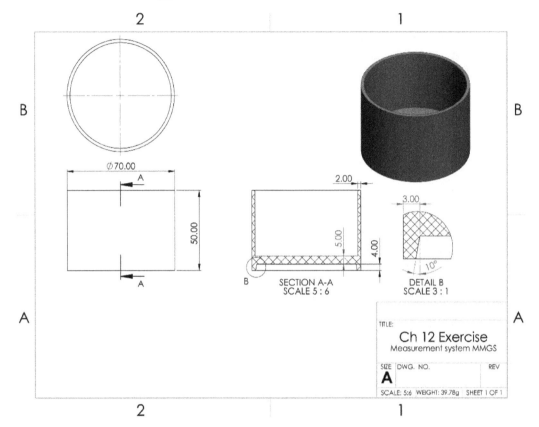

Figure 12.82 – The drawing for Question 3

4. Create the following part in SOLIDWORKS. Hint: Use the rib feature and the Hole Wizard:

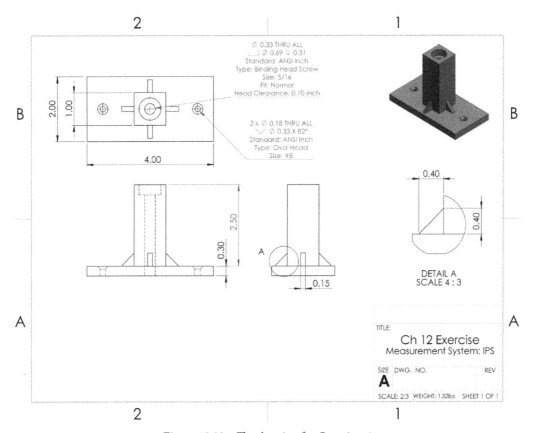

Figure 12.83 – The drawing for Question 4

5. Create the following part in SOLIDWORKS. Hint: Use the rib and draft features and the Hole Wizard. You may also use reference geometries and the swept boss. Due to the amount of information that this part contains, the drawing has been split into two diagrams. Both diagrams are for this one question:

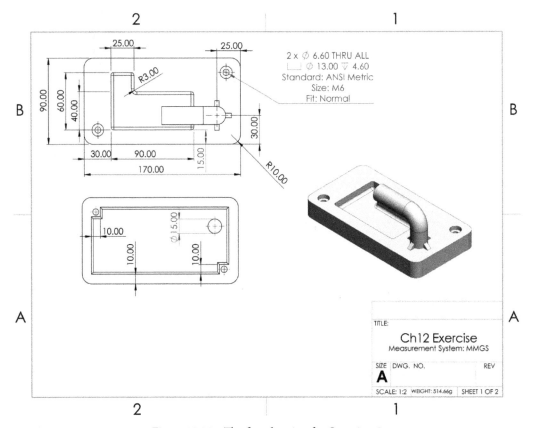

Figure 12.84 – The first drawing for Question 5

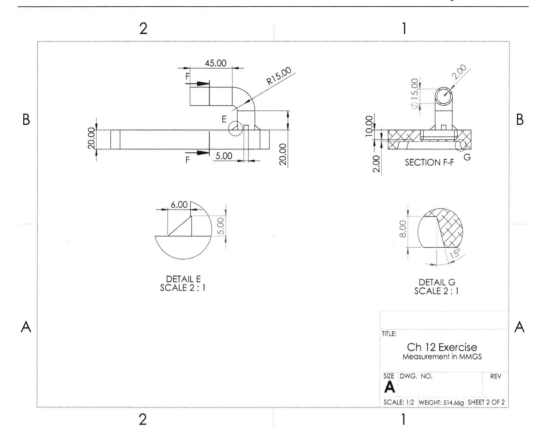

Figure 12.85 – The second drawing for Question 5

6.   What are multi-body parts, and what are some advantages of using them?

7. Create the following frame in a multi-body format. Use the annotations in the drawing for the different bodies. Hint: Use features scope to extrude cut the slot shown on bodies 2 and 4:

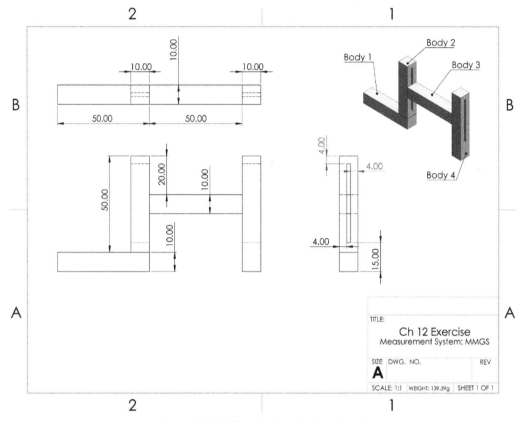

Figure 12.86 – The drawing for Question 7

Important Note

The answers to the preceding questions can be found at the end of this book.

13
Equations, Configurations, and Design Tables

In this chapter, we will cover how to use equations, configurations, and design tables. These functionalities are not used to directly add or remove materials or features, such as an extruded boss, ribs, and sweeps. Rather, we will cover how to link different lengths with equations and how to generate multiple versions of a part using configurations and design tables. Mastering these functionalities will enable us to generate more interlinked models that are more robust and easier to modify. Also, it will enable us to generate multiple model versions for testing and evaluation.

The following topics will be covered in this chapter:

- Understanding and applying equations in parts
- Understanding and utilizing configurations
- Understanding and utilizing design tables

By the end of this chapter, we will be able to link different dimensions with equations and create multiple variations of a part within one SOLIDWORKS file. This will enable you to both optimize and accelerate your design process. Note that, in this chapter, we will focus on applications within part files. Similar functions are also available within assemblies.

Technical requirements

This chapter will require access to SOLIDWORKS and Microsoft Excel software on the same computer.

Check out the following video to see the code in action: `https://bit.ly/3ytbT7j`

Understanding and applying equations in parts

When creating 3D models, we often use a variety of dimensions to define sketch entities, such as squares and arcs. We also use dimensions to define features such as an extruded boss, an extruded cut, a revolved boss, and a revolved cut. In many applications, these dimensions are not isolated from each other. Rather, they are connected with mathematical relations. For example, the length of a rectangle should be 75% of its width or the height of a cylinder should be double the length.

In this part, we will learn how to set up these relations with equations. First, we will explain the equations, and then we will apply equations in the modeling of a SOLIDWORKS part.

Understanding equations

Equations within parts allow us to both define and link different dimensions together within the parts. By defining variables and equations, we will be able to build a more interconnected 3D model. This will give us certain advantages, such as the following:

- There will be ease in accessing specific defined dimensions, making them easier to adjust, as we can access those dimensions from the equations panel rather than by looking up the dimension from the design tree.

- There will be ease in modifying connected dimensions as equations will enable us to modify one dimension and have all other linked dimensions updated accordingly rather than modifying each dimension separately.

To theoretically demonstrate equations in the SOLIDWORKS part context, we will look at the rectangular cuboid scenario highlighted in *Figure 13.1*. Let's assume we were required to model a rectangular cuboid with the following specifications:

- A width equal to 5 millimeters

- A depth that is double the width (which equals 10 millimeters)

- A height that is 3 millimeters longer than the width (which equals 8 millimeters)

According to the given specifications, we can look at defining the rectangular cuboid dimensions in two different ways. The first is what we did previously using numerical values, and the other is using equations. The following screenshot highlights both methods; the rectangular cuboid on the left highlights directly inputting **Numerical Values**, while the one on the right highlights inputting **Equations**:

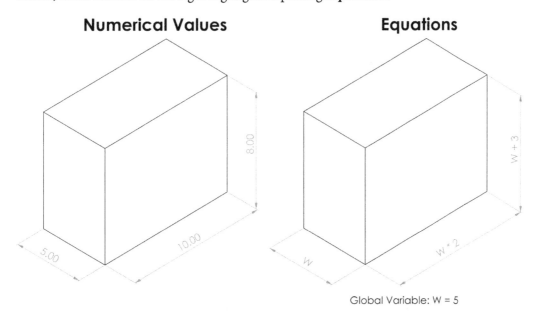

Figure 13.1 – A rectangular cuboid built in two methods

The initial final result of both methods gives the exact same cuboid. However, the way the dimensions are defined is different. Let's assume that, after creating this cuboid, we were required to adjust the width from 5 mm to 8 mm. With equations, all we will need to do is to change the W global variable from 5 to 8, and then all of the other dimensions for depth and height will change accordingly. However, if we input numerical values, we will have to change all three dimensions individually. This is why mastering equations is key when building more connected models.

Now that we know what equations are in a part modeling context, their advantages, and how they work, we can start learning how to use equations to enhance our 3D-modeling process.

Applying equations in parts

To demonstrate how to use equations when modeling parts, we will create the following simple rectangular cuboid with the variables shown in the screenshot:

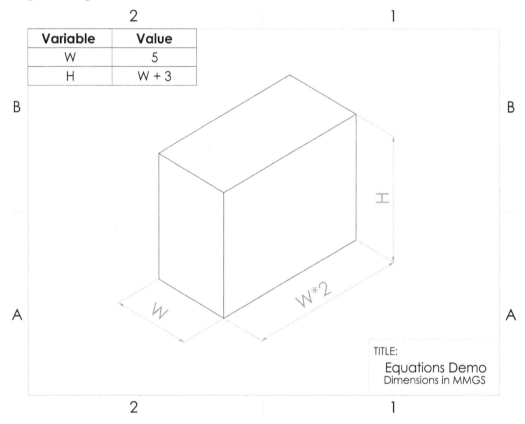

Variable	Value
W	5
H	W + 3

Figure 13.2 – The rectangular cuboid we will create in this exercise

To start, we will define our global variables. To do this, follow these steps:

1. Open the equations manager by right-clicking on the part name on the top of the design tree, and then choose **Hidden Tree Items | Equations | Manage Equations...**, as highlighted in the screenshot:

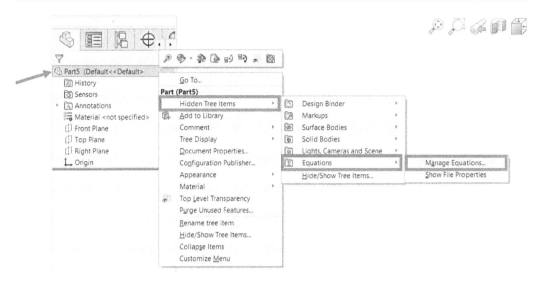

Figure 13.3 – The location of the equations manager

> **Tip**
>
> You can also go to equations by going to **Tools** and then **Equations**.

2. Input the **W** and **H** variables under **Global Variables** and input the values, as highlighted in the screenshot. Click **OK** after defining the variables:

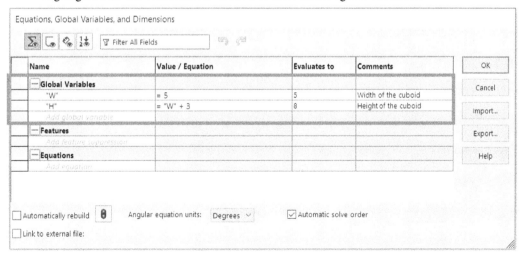

Figure 13.4 – Global variables are active through the model

3. Verify that the variables are already defined by checking the new folder in the design tree under the **Equations** name, which will show all of our variables. The following screenshot highlights the new **Equations** folder:

Figure 13.5 – The defined variables are listed in the design tree

4. Start creating our rectangular cuboid by inputting the variables instead of the lengths. When dimensioning the width, we can input = "W" instead of the number 5, as shown in the screenshot. Note that the dimension will then show as Σ **5.00** to indicate the involvement of the function of the equation:

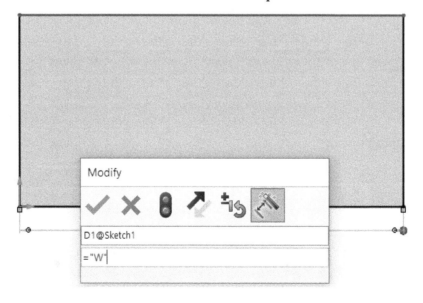

Figure 13.6 – Inputting the variable in the smart dimension field instead of the dimension

5. Dimension the depth by inputting =`"W"*2`, as shown in the screenshot. Similar to the width, the depth dimension will be displayed as **Σ 10** to indicate the involvement of the equation function:

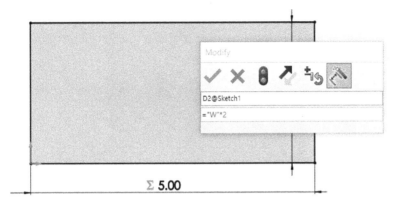

Figure 13.7 – Inputting the multiplication with a variable in place of the dimension value

6. Apply the extruded boss feature; we can follow the same technique. When specifying the dimension, we can input =`"H"`, which will result in the value **8**, as highlighted in the screenshot:

Figure 13.8 – Inputting the defined variable in the PropertyManager feature instead of the value

After applying the Extruded Boss feature, we will have the complete rectangular cuboid, as shown in the following screenshot:

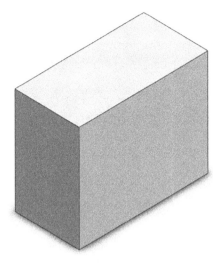

Figure 13.9 – The resulting rectangular cuboid

Note that we can define our variables using any term we want. It can be a single letter or a word. A good practice is to use a term that would make it easier for us to recall while creating the model.

Now that we know how to apply equations, let's examine how easy this will make implementing modifications.

Modifying dimensions with equations

To illustrate how to modify dimensions, let's change the width (**W**) from 5 millimetres to 8 millimetres. To do this, we can follow these steps:

1. Open the Equations Manager, which we will be able to find by right-clicking on the Equations folder in the design tree.

2. Change the value of the **W** global variable from 5 to 8, as shown in the following screenshot, and then click **OK**:

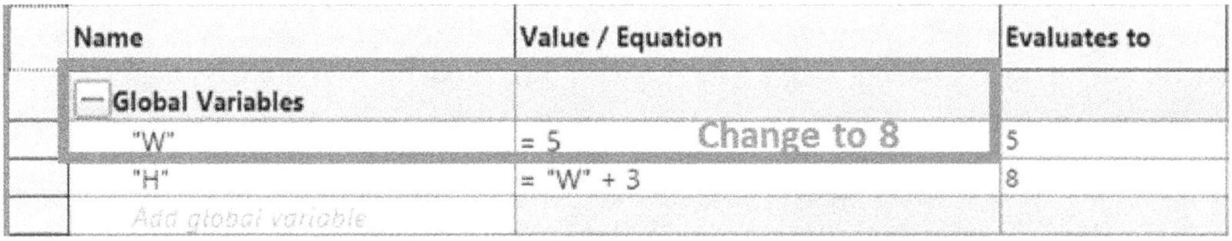

Name	Value / Equation	Evaluates to
─ Global Variables		
"W"	= 5 Change to 8	5
"H"	= "W" + 3	8
Add global variable		

Figure 13.10: Changing the variable's value from the Equations Manager

After implementing this change, we will notice that all of the dimensions linked to the **W** variable have changed as well. In this, the height, **H**, changed from 8 to 11 and the depth changed from 10 to 16. Not only do equations allow us to change our dimensions faster, but they also allow us to keep the design intent while doing so.

Before we cover more in the next topic, let's examine a couple of more notes when dealing with equations. One is regarding the Equations Manager and the other is regarding design intents with equations.

Equations within the Equations Manager

If we go back to the Equations Manager, we will notice a tab at the bottom titled **Equations**, as highlighted in the following screenshot:

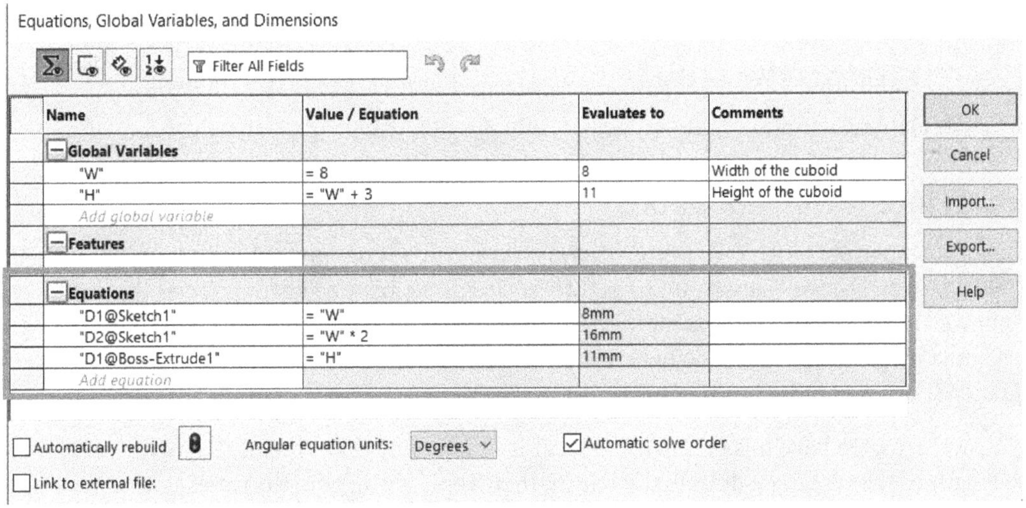

Figure 13.11 – Used equations are listed in the Equations Manager

This shows all of the equation applications we have in the model. It also gives us quick access to all of them in case we need to apply any change without needing to look up the actual feature in the design tree. The first column shows the names of the dimensions. For example, **D1@Sketch1** refers to **dimension 1** from **Sketch1**, which is listed in the design tree.

You may think that it is not convenient for us to recognize these codes (for example, **D1@ Sketch1**), especially when we build models that contain many different measurements and sketches. To make this process easier, we can change these names as we are inputting the dimensions. The following screenshot shows the dialog box we get when we enter a specific dimension, highlighting where we can change the name of that specific dimension. The following highlighted box will enable us to change **D1**. To change **@Sketch1**, we can rename the sketch entry found in the design tree:

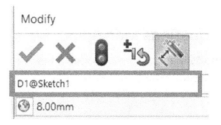

Figure 13.12 – We can change the name of the dimension as we are inputting it

Now, let's elaborate a bit more about design intent when using equations.

Design intent with equations

One important note to keep in mind when working with equations is the design intent. Whenever we 3D-model anything in SOLIDWORKS, we have to keep in mind the design intent we are aiming for. For example, if we are to sketch a rectangle with a width of 5 mm and a length of 10 mm, one question is, should we link the two dimensions with an equation?

To answer this, we have to ask ourselves what is important. If we intend to always have the length of the rectangle double the width, then applying an equation stating that would be the better practice. However, if we intend to have the length as 10 mm regardless of the width's value, then entering a direct numerical value would be the better practice. Referring to *Figure 13.13*, the rectangle on the left shows the dimensions input if our priority is to keep the length as double the width, while the rectangle on the right shows the dimensions input if our priority is to keep the length as a constant value:

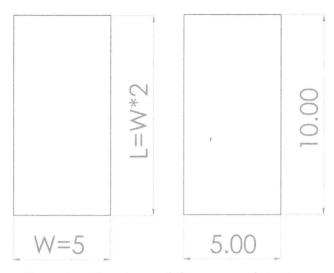

Figure 13.13 – Equations can help us preserve design intent

The rectangle example is a very simple one. However, the same principle is applicable for more complex parts or multiple parts linked together.

This concludes the topic of equations. We learned what equations are, how to apply and modify them, and, finally, important considerations with design intent and equations. Now, we can move on to exploring what configurations are within SOLIDWORKS parts modeling.

Understanding and utilizing configurations

Oftentimes, when creating a product, we will create multiple versions of it, with each of the versions having a slight variation from the others. SOLIDWORKS provides special tools for such configurations. In this part, we will learn all about configurations and how to use them to create different variations of a certain product or object.

What are configurations?

Configurations are different variations of a particular product or object. These variations would have small differences when compared to each other. For example, the four drawings in the following figure show different configurations of the same object:

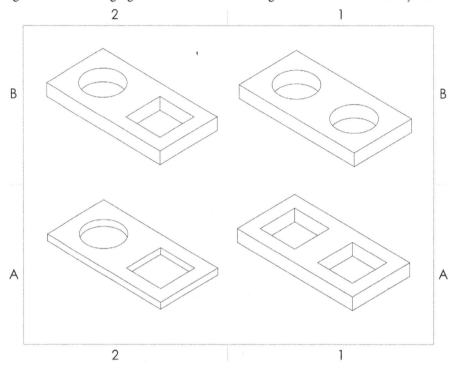

Figure 13.14 – Different configurations of the same object

Note that the four configurations do not have major differences from each other. Because of this similarity, it is an advantage for us to be able to create all of the different configurations in one SOLIDWORKS file rather than having a separate file for each configuration.

Now that we have a better idea of configurations, let's start applying them in SOLIDWORKS.

Applying configurations

Whenever dealing with configurations, we can start by creating a base model, and then we can create configurations of the base model. To highlight the application of configurations, we will create the model and the configurations highlighted in *Figure 13.15*. To make the exercise easier to follow, the dimensions in bold refer to dimensions that are different from one configuration to another:

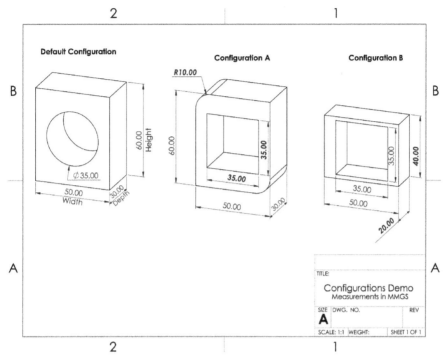

Figure 13.15 – The 3D model we are creating in this exercise

To complete this exercise, we will follow these steps:

1. Create a model titled **Default Configuration**, as highlighted in the preceding drawing. You can follow the procedure highlighted in the following screenshot for your first extrusion. Note that the circle is at the center of the rectangular face:

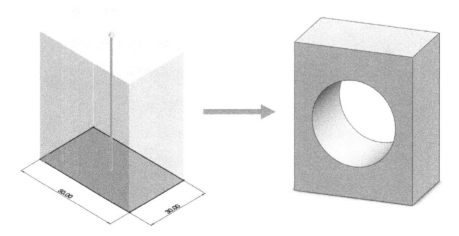

Figure 13.16 – The steps for creating the default configuration

2. To start creating the first configuration, we have to add it by going to the **ConfigurationsManager** at the top of the design tree, and then *right-clicking* on **Configurations** and selecting **Add Configuration...**, as shown in the following screenshot:

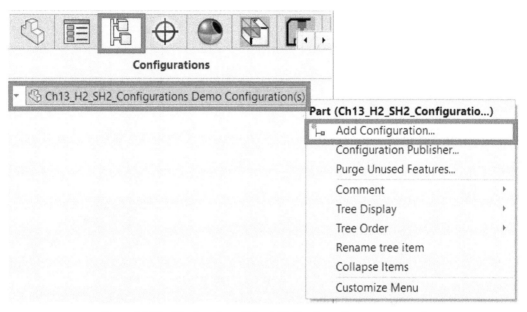

Figure 13.17 – The location of the Add Configuration... command

3. After doing that, we will be prompted to enter a name in the **Configuration name** field as well as a small description in the **Description** field, as highlighted in the screenshot. We are free to choose a name and description that would help us to identify the configuration. Click on the green check mark to introduce the configuration:

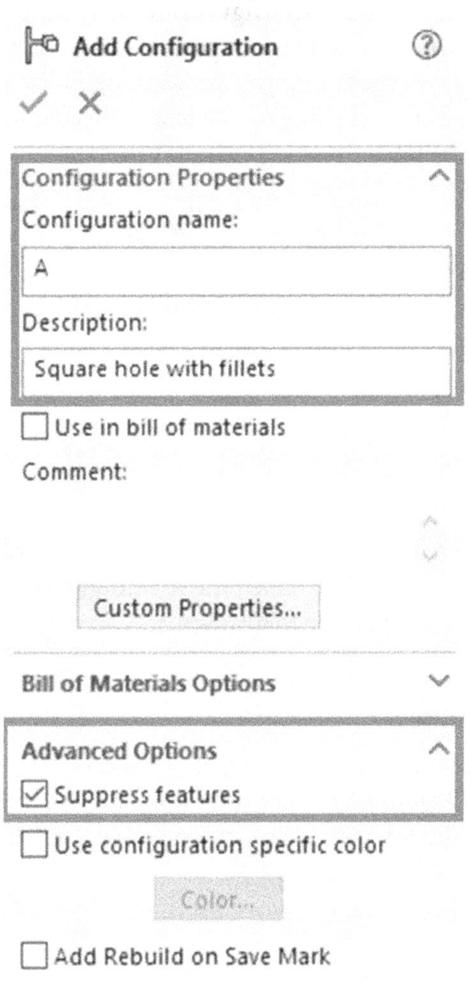

Figure 13.18 – Naming and describing the configuration can help us identify it later on

Important Note

The indicated **Suppress features** advanced option will have all-new features in the selected configuration that are suppressed in all other configurations.

4. Note the different coloring on the **Configurations** list in the
    ConfigurationsManager. The active configuration will not be grayed out, as shown
    in the screenshot. Double-check that you are working in the **A** configuration. Then,
    go back to the design tree by clicking on the icon indicated with a square, as shown
    in the following screenshot:

Configurations

    ▾ ◌ Ch13_H2_SH2_Configurations Demo Configuration(s) (A)
        ⊩ ✓ A [Ch13_H2_SH2_Configurations Demo]
        ⊩ ✓ Default [Ch13_H2_SH2_Configurations Demo]

Figure 13.19 – Available configurations are listed in the ConfigurationsManager

5. Now, we can modify our model to match the **A** configuration given on the initial
    drawing. To do that, we will do the following:

•   Suppress the Extruded Cut feature by right-clicking on the feature from the design
    tree and selecting **Suppress**.

•   Implement a new extruded cut and the Fillet features to create the square extruded
    cut, as highlighted in *Figure 13.15*. Note that the square is at the center of the
    rectangular face. We should have the model shown in the following screenshot,
    which presents the **A** configuration:

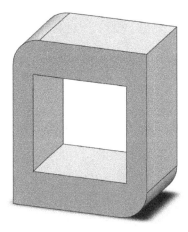

Figure 13.20 – The 3D model for the A configuration

Tip

After creating new configurations, a good practice is to go back to the Configuration Manager and double-check the status of other configurations. In this case, in the ConfigurationManager, double-click on the **Default** configuration to check the difference and ensure that we created the new configuration successfully.

6. Repeat *steps 2–4* to generate the B configuration.

7. Modify the model to match the B configuration, given in the initial drawing. To do that, we can do the following:

- Suppress the Fillet feature that we applied previously from the design tree.

- Edit the Extruded Boss feature (height) by changing the dimension to 40 mm. Also, ensure the change is applied only to **This configuration** by selecting the option from the PropertyManager, as shown in the screenshot:

Figure 13.21 – Specifying the related configuration when changing an existing feature's dimension

- Modify the depth by modifying the dimension in the initial sketch from **30.00** mm to **20.00** mm. Make sure this modification only applies to **This Configuration**, as shown in the following screenshot:

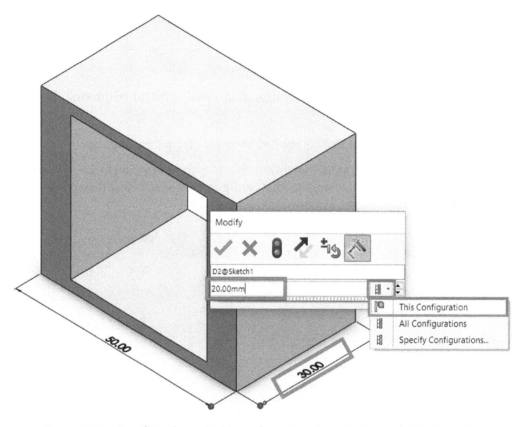

Figure 13.22 – Specifying the applicable configuration when adjusting a sketch dimension

8. Go back to the ConfigurationManager and double-click on each of the configurations to double-check that they were applied correctly. At this point, our three configurations should be as they were in the initial drawing.

> **Important Note**
>
> When editing a dimension within an existing feature or a sketch, SOLIDWORKS allows us to pick which configuration this change should apply to. For example, in the exercise, we applied all of our edits to only the active configuration by selecting the **This Configuration** option. We can also choose to apply the edits to **All Configurations** or specify which configurations we want the edits to apply to.

This concludes this exercise in creating different configurations for a specific model. Remember that it is a good practice to go back to the ConfigurationManager and double-check that all of our configurations are accurate by double-clicking on each of them. Here are some important takeaways from this section:

- We can suppress and un-suppress features to set them to different configurations.
- When modifying dimensions on existing features or sketches, we have to select the scope in terms of which configuration(s) the adjustment will apply to.

Now, we can start working with design tables, which is another method that will enable us to create different configurations.

Understanding and utilizing design tables

Through configurations, we were able to create different variations of a particular 3D model. Design tables will also allow us to create different variations of a 3D model. However, unlike directly setting up configurations, design tables will enable us to generate more than one variation at the same time instead of generating them one after the other. In this section, we will cover how to set up design tables and some scenarios in which design tables will give us an advantage over directly setting up configurations.

What are design tables?

Design tables are one method that will enable us to create multiple variations of a specific 3D model at once. Design tables make it easy and efficient to set up different dimensions for lengths and angles. They also allow easy manipulation for suppressing certain features in specific model variations or configurations. The following drawing highlights a simple application of a design table with the multiple configurations it generates. Note that, with design tables, we do not enter the dimensions for the different configurations manually one by one.

Rather, we just enter a table highlighting how the different configurations will differ, and SOLIDWORKS will then generate all of the configurations at once. In the following screenshot, all the configurations are different, based on the three **Length**, **Width**, and **Thickness** parameters only:

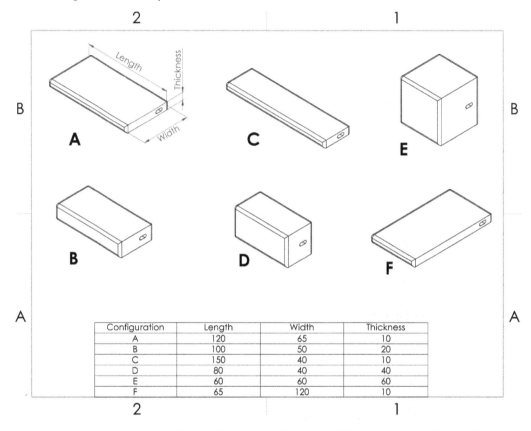

Configuration	Length	Width	Thickness
A	120	65	10
B	100	50	20
C	150	40	10
D	80	40	40
E	60	60	60
F	65	120	10

Figure 13.23 – Design tables enable us to quickly create different variations of a model

When working with design tables, a good practice is to start by creating a base model that includes all of the dimensions and features we will vary on other configurations. It is also a good practice to make custom names for all of the dimensions and features we want to vary for easy identification. Now that we have an idea of design tables, we can start applying them when creating models.

Setting up a design table

When dealing with design tables, we can start by creating a base model. After that, we can use a design table to create multiple configurations. This will be the process we follow in this exercise: creating the base model and using the configurations highlighted in the following diagram. SOLIDWORKS uses Microsoft Excel to generate design tables, so you must have Microsoft Excel installed on your computer to use design tables:

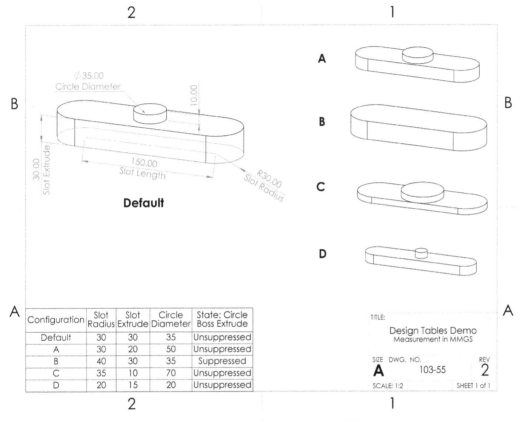

Configuration	Slot Radius	Slot Extrude	Circle Diameter	State: Circle Boss Extrude
Default	30	30	35	Unsuppressed
A	30	20	50	Unsuppressed
B	40	30	35	Suppressed
C	35	10	70	Unsuppressed
D	20	15	20	Unsuppressed

Figure 13.24 – The drawing for the design table's exercise

To create the following model utilizing design tables, follow these steps:

1. Create the base model, as highlighted in *Figure 13.25*. When doing that, you can use the following good practices. They will minimize confusion when generating and using a design table:

2. Give names to your dimensions as you are inputting them. The following screenshot shows inputting the name for `Slot Radius`. You can follow the naming, as shown in *Figure 13.26*:

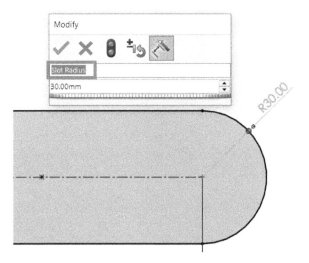

Figure 13.25 – Naming the dimensions can help identify them easie

*   Rename your features and sketches listed in the design tree to make them easy to identify. The following screenshot highlights a sample of renamed features and sketches for use in this exercise. Note the names are as shown in *Figure 13.27*:

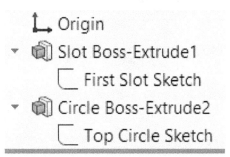

Figure 13.26 – Naming the features can make them easier to identify

3. To add a design table, go to **Insert**, then **Tables**, and select **Design Table...**, as highlighted in the following screenshot:

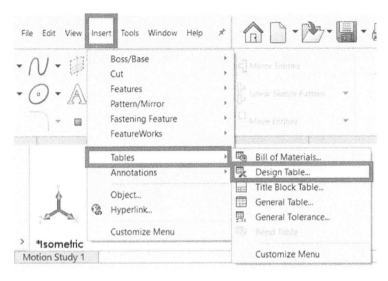

Figure 13.27 – Inserting a design table

4. Select the source as **Auto-create,** as in the following screenshot. **Edit Control** is set by default to **Allow model edits to update the design table**; we will leave that selected. We will discuss this option later in this section. Click on the *green check mark* to initiate the design table:

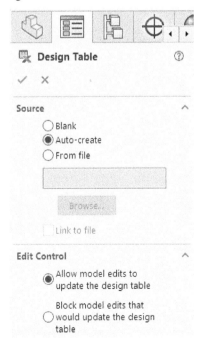

Figure 13.28 – The initial options for generating a design table

5. Select the dimensions we want to vary, as shown in the following screenshot. Note that this selection only includes dimensions related to sketches and features, not whether a feature is applied or suppressed. Also, note how naming the dimensions and features make them easier to identify:

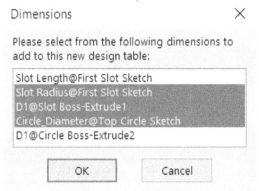

Figure 13.29 – We can select the dimensions to be inserted into the design tree

6. Fill in the design table, as highlighted in the following screenshot. The first column in the table shows the name of the configuration, while the others show dimensions unique to that particular configuration. Note that empty cells will automatically take the values of the active default configuration:

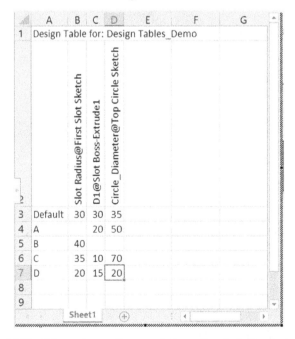

Figure 13.30 – We can input all our variations directly in the design table

7. The table only has dimensions so far. However, the **B** configuration has the Circle-Extruded Boss features suppressed. To add the state of the feature to the design table, select the first empty title cell, and then *double-click* on the feature on the design tree, as highlighted in *Figure 13.32*. This will automatically add the state of the feature to the design table.

8. To adjust the status of the feature, we can type the letter U for **unsuppressed** or the letter S for **suppressed**. Alternatively, we can type the full words (unsuppressed or suppressed) or 0 and 1, which stand for unsuppressed and suppressed respectively. Adjust the feature status, as shown in the following screenshot:

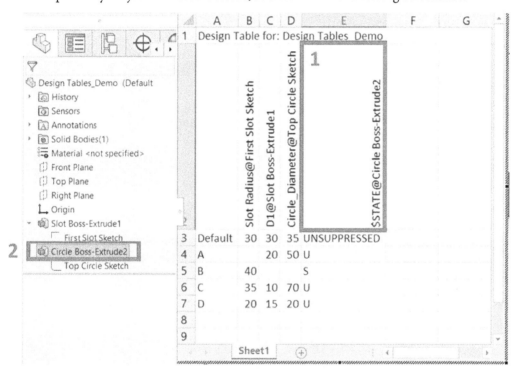

Figure 13.31 – We can indicate the state of a feature in the design tree

> **Tip**
> We can also add any dimension to the design table by directly clicking on it
> from the sketch on the canvas, similar to adding the feature status.

9. Click anywhere on the canvas outside of the table to confirm the table. We will see
    the following message, confirming the new configurations:

Figure 13.32 – A conformation message after generating the design table

After generating the different configurations, it is a good practice to double-check them.
So, go to the ConfigurationManager and check all of the configurations that we just
generated, and note how they are different from each other.

This concludes our coverage of how to generate a design table to create multiple
configurations. However, now that we know how to initiate a design table, we will also
learn how to edit it.

Editing a design table

There are two ways in which we can edit or update our design table. The first one is
through the design table itself and the other is through directly modifying the dimensions
in the model. We will examine both ways.

Editing directly from the design table

To edit the design table, we can find it on the ConfigurationManager. Then, right-click on the table and select **Edit Table**, as shown in the following screenshot:

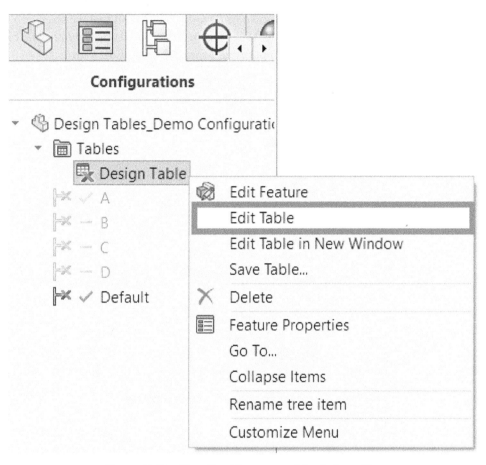

Figure 13.33 – The edit table command for design tables

After selecting the **Edit Table** command, the table will open for us to modify as we see fit. After modification, we can simply click anywhere on the canvas for all of the modifications to be applied.

Editing the design table by modifying the model

Another way of editing the design table is by directly editing the model by editing sketches or features from the design tree. This will update the corresponding cells in the design tree. However, this will only happen if we select the **Allow model edits to update the design table** option for **Edit Control**, as shown in *Figure 13.35*. Recall that we selected this option when we were creating the table. We can adjust this option by *right-clicking* on the design table from the ConfigurationManager and then selecting **Edit Feature**:

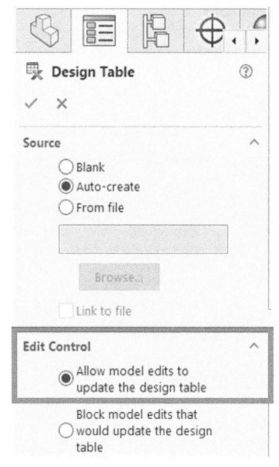

Figure 13.34 – Manual 3D model edits can automatically update the design table

Note that applying a design table does not prevent us from adding additional configurations, as we explored in a previous section when discussing configurations. We can use both methods together as we see fit.

This concludes this section about design tables. We covered what design tables are and how to apply and edit them. Design tables are a very efficient method that enables us to generate multiple configurations at once and are an essential tool for SOLIDWORKS professionals.

Summary

In this chapter, we learned skills that will enable us to create more robust and agile models. We covered equations that will enable us to create more connected models to help us to deliver our design intents. We also learned how to create different configurations of a specific model. We learned how to do that by directly and manually adding and adjusting configuration, or by using design tables to accomplish a similar objective.

The new skills in this chapter will enable us to generate more connected models. They will also enable us to generate many different variations of a model in a single SOLIDWORKS file. These will enable us to more efficiently conduct variation testing and quicker adjustments, which were the goals of this chapter.

In the next chapter, we will cover advanced mates within assemblies, which will help us to create assemblies with parts that have more complex interactions between each other.

Questions

The following questions will help to emphasize the main points we have learned in this chapter. However, in terms of practical exercises, do not limit yourself to what we provide you with here. Try modeling random objects around you or come up with your own innovative mode to increase your fluency using the software:

1. What are the equations when modeling parts? Why do we use them?
2. What are the configurations for a specific part?
3. What are design tables?

4. Create the model shown in the following screenshot, utilizing global variables and equations:

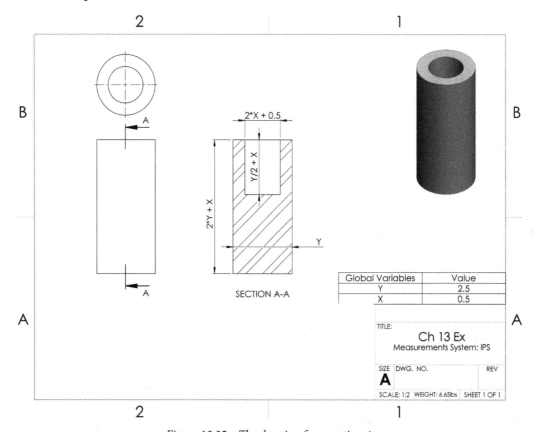

Figure 13.35 – The drawing for question 4

5. Create the base model shown, including the different configurations it highlights. Note that the two squares and the two circles in the **A** and **B** configurations are mirrors of each other:

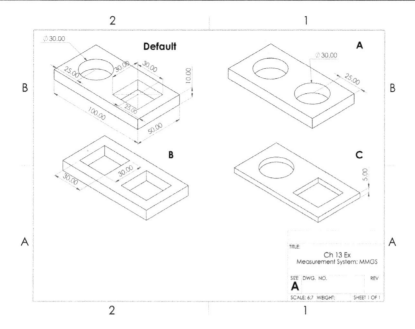

Figure 13.36 – The drawing for question 5

6. Create the following model and the different configurations shown using design tables. Use the MMGS measurement system when completing this exercise:

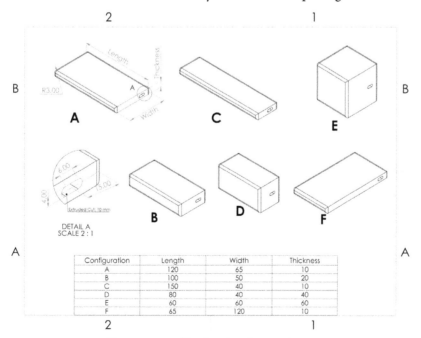

Configuration	Length	Width	Thickness
A	120	65	10
B	100	50	20
C	150	40	10
D	80	40	40
E	60	60	60
F	65	120	10

Figure 13.37 – The drawing for question 6

7. Create the following model and the different highlighted configurations. The model is displayed in two different screenshots to cover all of the requirements. Hint – use the mass values to double-check your model's accuracy:

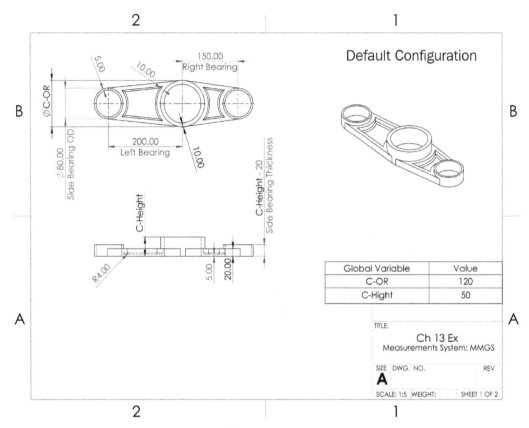

Figure 13.38 – The first drawing for question 7

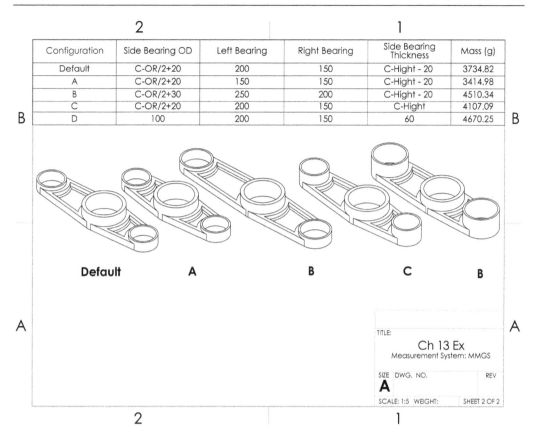

Configuration	Side Bearing OD	Left Bearing	Right Bearing	Side Bearing Thickness	Mass (g)
Default	C-OR/2+20	200	150	C-Hight - 20	3734.82
A	C-OR/2+20	150	150	C-Hight - 20	3414.98
B	C-OR/2+30	250	200	C-Hight - 20	4510.34
C	C-OR/2+20	200	150	C-Hight	4107.09
D	100	200	150	60	4670.25

TITLE:
Ch 13 Ex
Measurement System: MMGS

SIZE DWG. NO.   REV
A

SCALE: 1:5 WEIGHT: SHEET 2 OF 2

Figure 13.39 – The second drawing for question 7

Important Note

The answers to the preceding questions can be found at the end of this book.

Section 7 – Advanced Assemblies – Professional Level

This section will take our assemblies skills to a higher level than what we covered at the associate level. Here, you will learn about advanced mates and other assembly functions, including interference and collection detection, assembly features, design tables, and configurations for assemblies. It also includes a comprehensive project, covering topics from across the book toward the 3D modeling of a remote-control helicopter.

This section comprises the following chapters:

- *Chapter 14, SOLIDWORKS Assemblies and Advanced Mates*
- *Chapter 15, Advanced SOLIDWORKS Assemblies Competencies*
- *Project 2, 3D - Modeling an RC Helicopter*

14

SOLIDWORKS Assemblies and Advanced Mates

Mates within SOLIDWORKS assemblies fall into three major types – standard, advanced, and mechanical. This chapter will cover the usage of advanced mates within the SOLIDWORKS assembly's environment. These include profile center, symmetric, width, path mate, linear/linear coupler, distance range, and angle range. For each of the mates, we will learn what they do and how to use them. These advanced mates will enable us to generate assemblies with more complex part-to-part interactions than when only using standard mates.

The following topics will be covered in this chapter:

- Understanding and using the profile center mate
- Understanding and using the symmetric and width mates
- Understanding and using the distance range and angle range mates
- Understanding and using the path mate and linear/linear coupler mates

By the end of this chapter, you will be able to generate more complex interactions within different parts of an assembly. Those include both dynamic and static interactions.

Technical requirements

This chapter will require access to SOLIDWORKS software.

The project files for this chapter can be found in the following GitHub repository: `https://github.com/PacktPublishing/Learn-SOLIDWORKS-Second-Edition/tree/main/Chapter14`.

Check out the following video to see the code in action: `https://bit.ly/3EXToub`

Understanding and using the profile center mate

The **profile center** advanced mate allows us to create a centered relationship between two straight profiles in one step. This can save us time as well as deliver a precise centered design intent in one go. However, the mate also has its own limitations. The profile center mate is an essential part of our assembly toolkit. In this section, we will learn about the profile center and how to apply this mate to two different parts.

Defining the profile center advanced mate

The profile center advanced mates option helps us to center two surfaces in relation to each other in an assembly. The following screenshot highlights the effect of the profile center mate. The assembly on the left is before applying the mate, while the one on the right is after applying the mate:

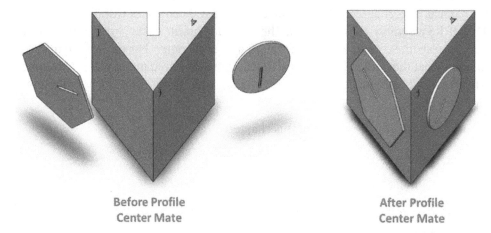

Before Profile
Center Mate

After Profile
Center Mate

Figure 14.1 – The profile center mate can position one part at the center of another

Note that SOLIDWORKS interprets the profile of the shape by its outer outline only. Also, we can only use the profile center mate with limited profiles that include circles and polygons. Polygons can include triangles, rectangles, hexagons, and so on. Any other irregular profiles cannot be used with the profile center mate. Now that we know what the profile center advanced mate does, we can start applying the mate to different parts, which we will do next.

Applying the profile center mate

In this section, we will apply the profile center advanced mate to generate the assembly shown in the following figure. All of the parts needed for this assembly are available for you to download with this chapter. Note that in the screenshot, both the circular and hexagonal parts are in the center of the faces, indicated by the numbers **1** and **3**:

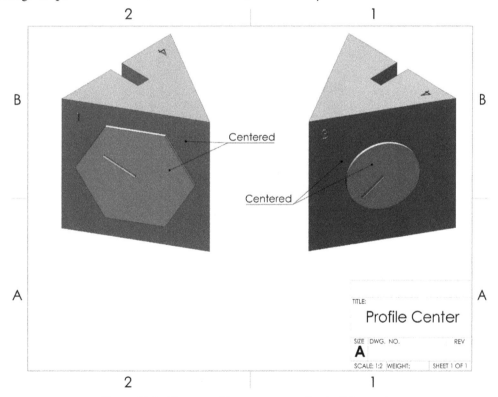

Figure 14.2 – The assembly we are generating in this exercise

To generate the assembly, we can perform the following steps:

1. Download all of the parts linked to this part and input them in an assembly file. Have the triangular prism as the fixed base part.

2. We will first start by generating the mate between the hexagonal and side **1** in the triangular prism. We will first select two profiles from each part, as indicated in the screenshot. Then, click on **Mate**:

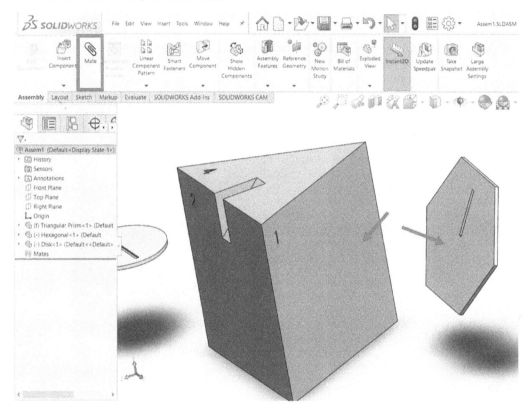

Figure 14.3 – The first selections for the mate

3. This will apply the standard mate, **Coincident**, by default. Change it by selecting the **Advanced** tab option and then the **Profile Center** mate, as shown in *Figure 14.4*.

4. This will drive the parts together for a preview with the hexagon centered in face **1**. Options for **Orientation** and **Mate alignment** can be used to adjust the positioning of the centered part as needed. Both options are highlighted in *Figure 14.4*.

This will show us a preview of the mate. Note that the hexagonal became stationed at the center of side **1** of the triangular prism. Note that there are more options under the mate that we can use to orient the parts with each other:

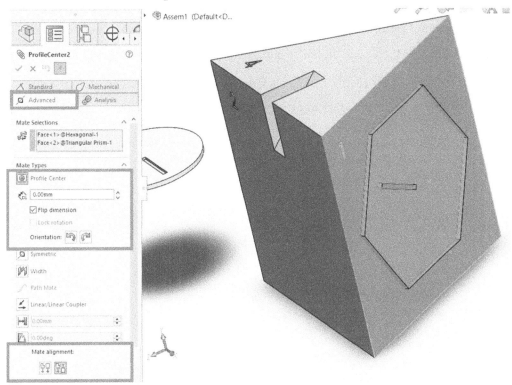

Figure 14.4 – The PropertyManager and setting for the profile center mate

Here is an explanation for all of the options available in this case:

- **Offset distance**: The distance field allow us to have a certain distance to separate the two profiles. You may enter a random number in the field to see a preview of the effect.

- **Flip dimension**: This relates to the distance field explained in the previous point. It will flip the distance measurement to the other side. You can try checking the box to see the effect of the option. In this exercise, the offset distance is zero, so the option will not make a difference.

- **Lock rotation**: This option only applies when the matted face is circular. Having this option checked will stop the circular profile from rotating.

- **Orientation**: This option allows us to change the orientation of the centered profile. This option is activated where there is a countable number of orientations in which our profiles can be centered in relation to each other. In this exercise of a hexagonal and a square profile, the profiles can be centered in four different rotational orientations. Click on the arrows until you get the orientation position you require.

- **Mate alignment**: This will flip the two mated surfaces to opposite sides of each other.

5. After fixing all of the requirements, click on the *green check mark* to apply the mate.

6. We can follow *steps 2–5* again to center the circular profiles for the disk part with face **3** of the triangular prism. Note that, in the PropertyManager, the **Lock rotation** option will now be available. After this step, our assembly will look similar to the following figure, which matches the initial assembly drawing:

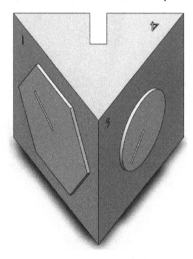

Figure 14.5 – The final result of the assembly

One aspect to note is that the profile center mate will only work on circular or polygonal profiles. In the preceding exercise, faces **2** and **4** do not have a regular polygonal profile because the profiles are not continuous. As a result, we will not be able to use the profile center command with those sides. To address these restrictions, we can instead use a reference sketch with the profile center mate. For example, we can create a new sketch on face **4** of a regular triangle and then use that to represent face **4** in applying the profile center mate.

> **Tip**
> Sketches can be used as a reference for mate selection the same way that planner surfaces work.

This concludes this section on the advanced mate profile center. We covered its definition and how to use it. Next, we will cover two advanced mates, symmetric and width.

Understanding and using the width and symmetric mates

Symmetric and **width** are two advanced mates that we can use within SOLIDWORKS assemblies. They enable us to have more flexible control in terms of part movement in relation to both symmetrical movements and setting width-related adjustments. In this section, we will cover the width and symmetric mates and how to use them for assemblies.

Defining the width advanced mate

The width advanced mate relates to the adjustment of surfaces in relation to each other in what is commonly known as the width dimension. A common application of the width mate is in mechanical joints and mechanical slots. For example, take note of the mechanical joint shown in *Figure 14.6*. The width advanced mate will help adjust the location of the *inner joint* in relation to the *outer joint*.

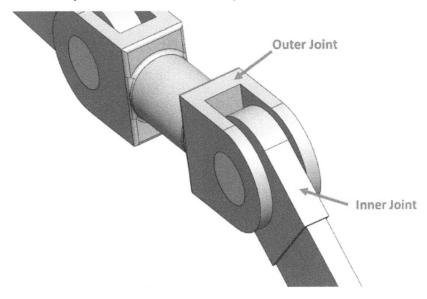

Figure 14.6 – The width mate allow us to adjust the location of the inner joint in relation to the outer one

These relationships include the following:

- The *Inner Joint* moves freely between the space allowed by the *Outer Joint* boundary.

- The *Inner Joint* is centered in the space allowed between the *Outer Joint*.

- The *Inner Joint* is fixed to a certain distance or percentage within the *Outer Joint* boundary.

Now that we know what the width relation does, we can start applying it next.

Applying the width advanced mate

In this section, we will learn how to apply the width advanced mate. We will do that by fixing the width of the *Inner Joint* to the *Outer Joint*, as highlighted in the following figure. All of the parts needed for this exercise are available for you to download with this section:

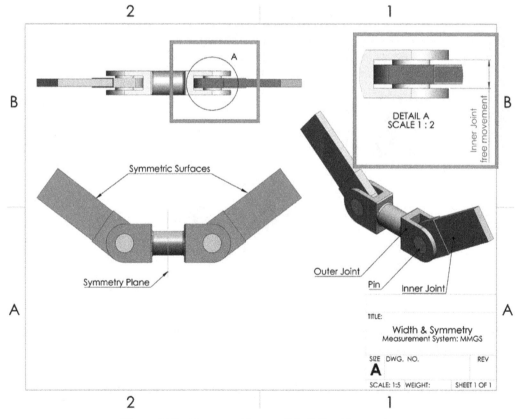

Figure 14.7 – The assembly we will build in this exercise

To generate this mate, follow these steps:

1. Download all of the parts and the assembly file linked with this part. Then, open the assembly file as a starting point.

2. Select the **Width** mate by selecting the **Mate** command and then the **Advanced** tab option. Then, select the **Width** mate, as highlighted in the following screenshot:

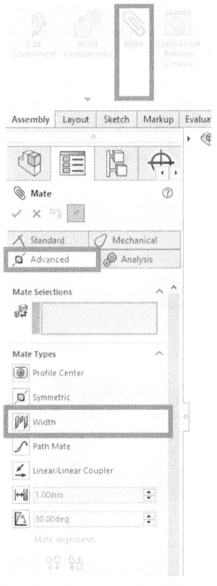

Figure 14.8 – The location of the Width mate

3. Once we select the mate, we will get more options to set it up. We can follow the selections highlighted in the following screenshot, which are as follows:

- **Width selections**: For this, we can select the outer boundaries for our width constraints. In this case, they are the inner surfaces of our outer joint.

- **Tab selections**: For this, we can select the inner boundaries for our width constraints. In this case, they are the outer surfaces of our inner joint.

- **Constraint**: This allows us to determine the relationship between our **Width selections** and our **Tab selections**. There are four options that we can put in the context of this example. **Centered** will make the inner joint centered in the space available between the outer joint boundaries. **Free** will allow the inner joint to move both sides within the boundaries of the outer joint. **Dimension** will allow us to set the inner joint to a certain distance from the boundaries of the outer joint. **Percent** is similar to **Dimension**; however, instead of setting a distance dimension, we can set a percentage. In this exercise, we are required to set the constraints to **Free**, as per the initial drawing. However, while you are here, you can also experiment with the other constraints:

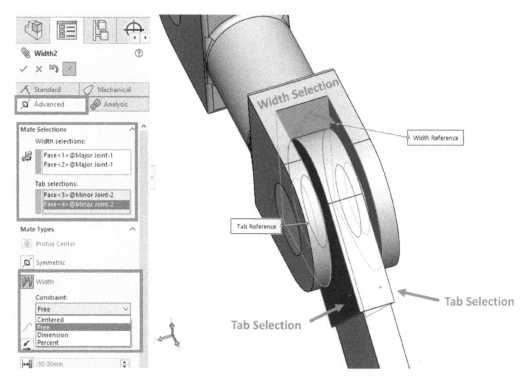

Figure 14.9 – The selection of the width and tab

4. Click on the *green check mark* to confirm the mate. Once you confirm the mate, it is a good practice to check its effect. So, at this point, you can try moving the inner joint to different sides to see the effect of the new width advanced mate. You will notice that the inner joint will move to the sides only within the space available before it hits the outer joint.

5. Repeat *steps 2–4* for the other joints.

> **Important Note**
>
> We can also switch **Width selections** with **Tab selections**. However, this will result in having only one constraint setting, which is **Centered**.

This concludes this exercise on the width advanced mate. We covered what a mate is as well as how to use it. Now, we will move on to another advanced mate, which is symmetric.

Defining the symmetric advanced mate

The symmetric mate allows us to set two different surfaces to have a symmetric dynamic relation to each other around a symmetric plane. A key thing to note with the symmetric advanced mate is that it builds a relationship between surfaces, not the parts themselves. Hence, it establishes symmetric surfaces rather than symmetric parts. The following screenshot highlights the functionality of the symmetric mate. The symmetric surfaces are symmetric around the symmetry plane. Hence, if we rotate any of the surfaces, clockwise or anticlockwise, the other surface will rotate as well.

The following screenshot highlights the components involved in constructing symmetric surfaces:

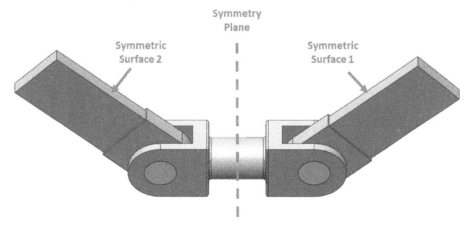

Figure 14.10 – The symmetric mates allow us to build movement symmetry

However, if we look at the assembly from a different angle, as shown in the following screenshot, we can see the two parts (inner joints) are not fully symmetrical, as is the case around **Plane 1**:

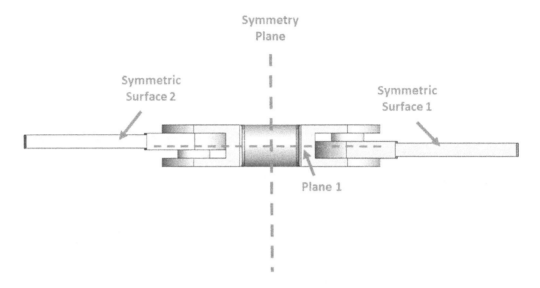

Figure 14.11 – The symmetric build with the symmetric mate is not absolute

We can also apply the symmetric mate toward edges or points. One important note to know is this symmetrical relation is also directional, according to our mate selections. For example, if we select two surfaces, the symmetry will be applied toward movements that are normal to those surfaces. This is a similar scenario when selecting edges for symmetry.

Now that we know what the symmetric mate does, we can start applying it to our assembly.

Applying the symmetric advanced mate

In this section, we will learn how to apply the symmetric advanced mate. We will do that by applying the symmetric mate to the indicated surfaces around the shown symmetry plane in the following figure. In this exercise, we are continuing our work on the assembly that we started with the width mate:

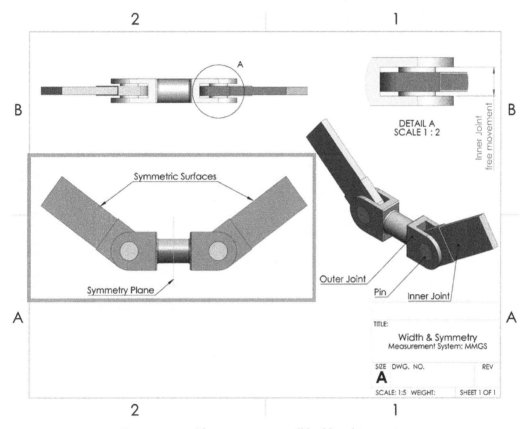

Figure 14.12 – The symmetry we will build in this exercise

To set up the symmetric mate, we need to meet two requirements, a symmetry plane and surfaces to make it symmetrical. To apply the symmetry mate, follow these steps:

1. Check whether any of the default planes with the assembly match our required plane. In this case, there are none. Hence, we can create a new plane using the reference geometry function. The symmetry plane should be in the middle of the outer joint, as highlighted in the preceding figure. The following screenshot highlights the settings for our new reference geometry plane:

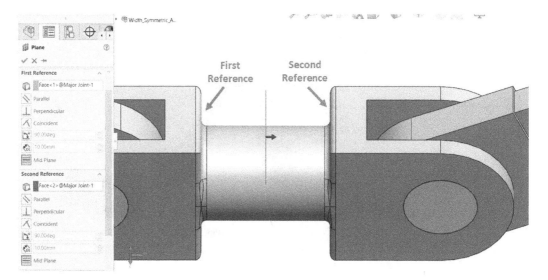

Figure 14.13 – Defining a new reference plane might be needed for our symmetry

> **Important Note**
>
> SOLIDWORKS automatically switches to mid plane when two face/planner surfaces are selected as references.

2. Select the **Symmetric** mate by selecting the **Mate** command, the **Advanced** tab option, and then the **Symmetric** mate, as highlighted in the following screenshot:

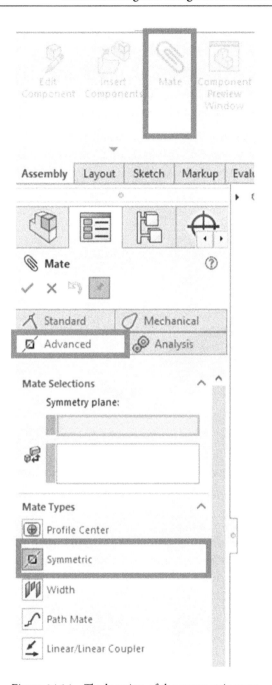

Figure 14.14 – The location of the symmetric mate

3. Now, we will be prompted to set up the mate. We can set it up as highlighted in the following screenshot. Here is more explanation about the selection:

- **Symmetry plane**: Here, we can select the plane we created in *step 1*. Note that if the plane is not visible in the canvas, we can find it and select it by expanding the design tree available in the canvas, as indicated in the following screenshot.

- **Entities to mate**: This is the field under **Symmetry plane**. Here, we can select the two surfaces we want to mate. In this case, they are the two surfaces highlighted in the following screenshot:

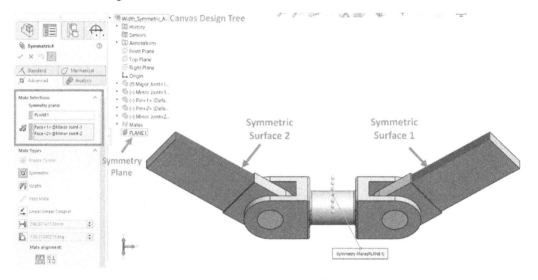

Figure 14.15 – The selection and PropertyManager for our symmetric mate

4. Click on the *green check mark* to apply the mate. Once you confirm the mate, try moving the different parts to see the effect of the new mate.

This concludes how to use the symmetric mate. We covered what it is and how to apply it by making two surfaces dynamically symmetrical about a plane. Next, we will explore the distance range and angle range mates.

Understanding and using the distance range and angle range mates

In this section, we will explore the advanced mates, **distance range** and **angle range**, also known as **limit mates**. With these mates, we will be able to limit the movement of a specific component to be within a defined range rather than a set distance or an angle. We will first define what these mates are and then we will use them by following a practical exercise.

Defining the distance range and angle range

Distance range and angle range will allow us to set up an acceptable range of movement. These include angular movements as well as linear movements. Common scenarios in which these mates are used are in joints where we want to restrict the angular movements of an arm. Gas springs and distance-based switches are examples in which we require the linear movement to be restricted to a specific range. The following diagram highlights an assembly as well as what we can accomplish with the distance range and angle range advanced mates. With set angle and distance ranges, we can allow the lever and switch to only move freely within that certain range:

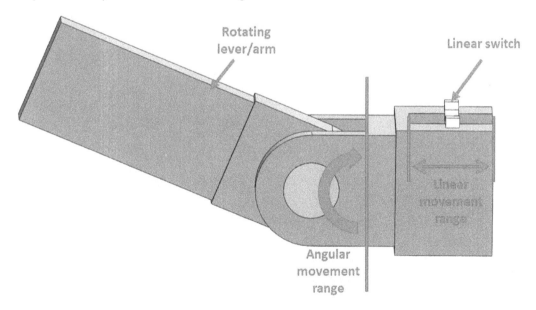

Figure 14.16 – The ranges mate allows us to restrain the movement of our parts to limits

After knowing about the distance range and angle range mates, we will cover how to apply the mates in an assembly.

Applying the distance range mate

In this section, we will learn how to apply the distance range advanced mate. We will do that by setting a linear movement range of 0 to 50 mm for the switch, as indicated in the following figure. All of the part files, as well as the initial assembly file, are available for you to download with this chapter:

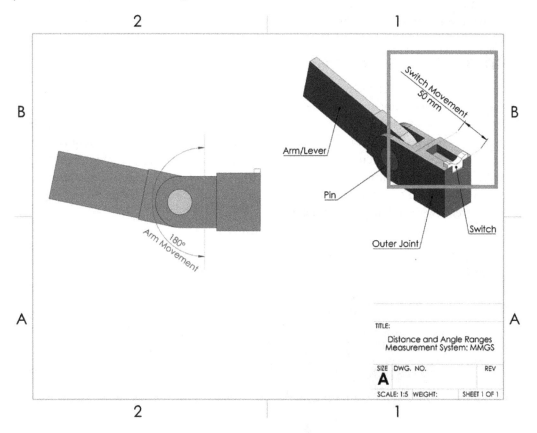

Figure 14.17 – The distance range mate we will apply in this exercise

To set up the distance range mate, we are only required to have two surfaces, edges, or vertexes for two different parts. In this example, we can follow these steps to apply it:

1. Download the linked SOLIDWORKS files and open the provided assembly, as we will use it as a starting point.

2. Select the **Distance** mate by selecting the **Mate** command, the **Advanced** tab option, and then the **Distance** mate, as highlighted in the following screenshot:

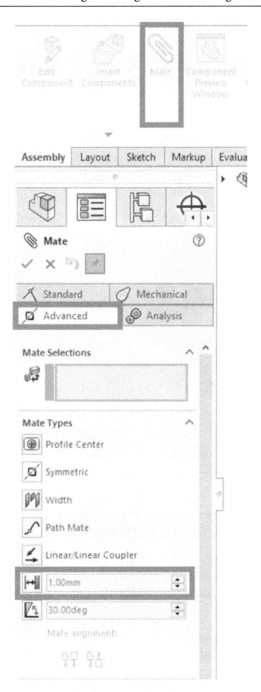

Figure 14.18 – The location of the distance range mate

3. Once we select the mate, we will be prompted to set it up, as shown in the following screenshot:

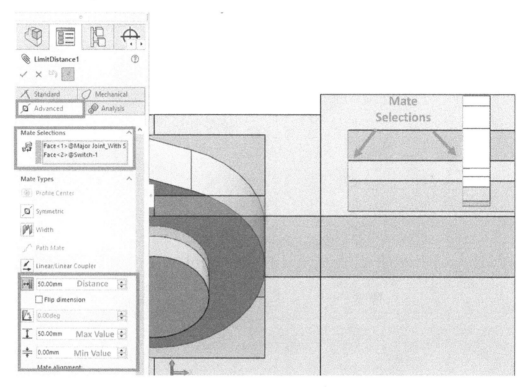

Figure 14.19 – The selections and PropertyManager for our mates

Here is an overview of the settings indicated:

*   **Mate Selections**: Here, we should select the entitles we want for the mate. This can include surfaces, edges, or vertexes. In this exercise, we can select the two faces indicated in the screenshot between **Major Joint** and **Switch**.

*   **Distance**: This acts as an overview of what a specific distance would look like. In the screenshot, the distance is set to 50 mm, which is the actual distance shown on the canvas. Also, the distance listed here will be the initial distance applied to the parts after confirming the mate.

*   **Flip dimension**: Checking and unchecking this option will flip the direction in which the distance is measured. We can always check this box to match our requirements.

*   **Max Value**: Here, we can input the maximum limit of the distance range. We can fill this with 50 mm, as highlighted in the screenshot. This field can be filled with both positive and negative values.

- **Min Value**: This is the opposite of the maximum value field in which we can set our minimum allowable value for linear movement. This can also be a negative value.

- **Mate alignment**: This adjusts how the two parts are aligned with each other. We can always try the different alignments to check which one matches our requirements.

4. Click on the *green check mark* to confirm the mate.

Once you confirm the mate, try moving the switch horizontally to double-check the functionality of the mate. The switch should only move within the slot built on the major joint part.

Now that we have finished setting up our distance range, we can set up the angle range for the arm next.

Applying the angle range mate

In this part, we will learn how to apply the angle range advanced mate. We will do that by limiting the range of angular movement of our level/arm to be between 0 and 180 degrees, as indicated in the following figure. In this exercise, we are continuing our work on the assembly that we started with the angle range mate:

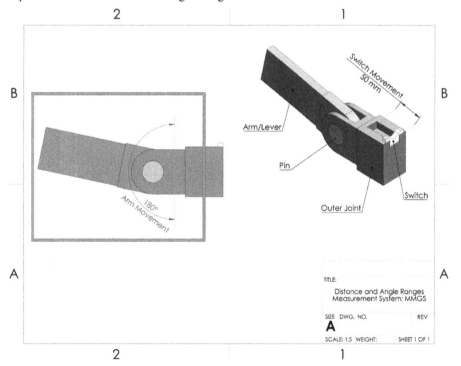

Figure 14.20 – The angle range we will apply in this exercise

Setting the angle range is similar to setting up a distance range. To set up an angle, we will require two surfaces or edges for two different parts. In this example, follow these steps to apply it:

1. Select the **Angle** mate by selecting the **Mate** command, the **Advanced** tab option, and then the **Angle** mate, as highlighted in the following screenshot:

Figure 14.21 – The location of the angle range mate

2. Set up the PropertyManager, as shown in the screenshot, similar to the distance range mate. Set up the angle between `0` and `180` with **Mate Selections**, as shown in the following screenshot:

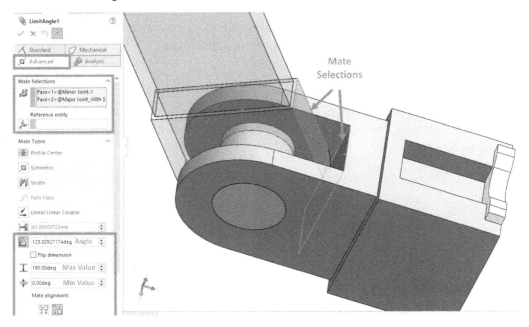

Figure 14.22 – The selection and PropertyManager for our angle range

> **Important Note**
>
> The additional field in this mate, **Reference entity**, is an optional entry in which we can select an edge, plane, or face as a base to measure the angle. In this exercise, we can leave it empty.

3. Click on the *green check mark* to apply the mate.

Once you confirm the mate, try moving the arm to double-check the functionality of the mate. The arm should have a range of movement that equals 180 degrees.

Now that we are familiar with most of the advanced mates, we can move to the last set that includes the path mate and linear/linear coupler advanced mates.

Understanding and using the path mate and linear/linear coupler mates

In this section, we will explore the **path mate** and **linear/linear coupler** advanced mates. Both mates will help us to introduce dynamic movements into our assemblies. We will learn what these mates are, what they do, and how to apply them.

Defining the path mate

The advanced mate's path mate allows us to restrict the movement of a specific part to follow a designated path. For example, we can have tiles follow a path, as shown in the following screenshot:

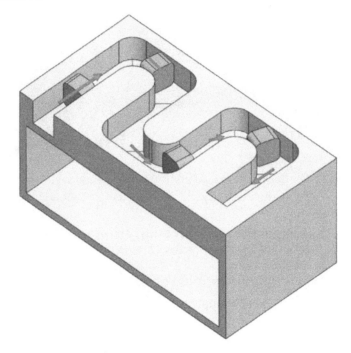

Figure 14.23 – The impact of the path mate

To apply the path mate relationship, we are required to have a path to follow and a point/vertex to follow that path. Next, we will apply the path mate command to an assembly.

Applying the path mate

In this section, we will learn how to apply the advanced mate's path mate. We will do that by limiting the movement of a tile to stick to a specifically designated path, as highlighted in the following figure. All of the parts and assembly files needed for this exercise are available for download with this chapter:

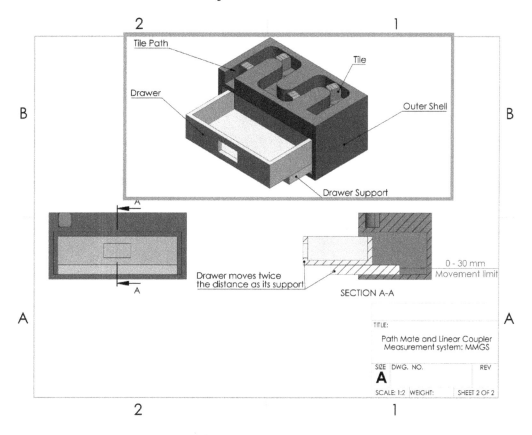

Figure 14.24 – The path mate we will apply in this exercise

To apply the path mate, we will be required to set up a path to follow. For this, we have the option of selecting an existing path that includes ones from existing sketches or existing edges. Another option is to create a new sketch to represent our path. However, when creating or selecting a path, note that a point within the moving part will have to coincide with the path at all times. For this exercise, follow these steps:

1. We can create the following additional sketch in the assembly, as shown in the following screenshot. Note that the line for the path goes through the middle of the carved path in the outer shell. After creating the sketch, exit the sketch mode:

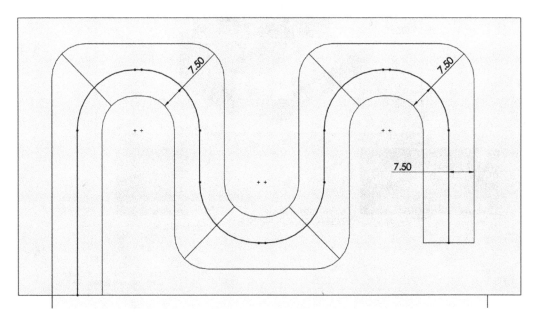

Figure 14.25 – To use the path mate, we must have a path

2. Select the **Path Mate** option by selecting the **Mate** command, the **Advanced** tab option, and then the **Path Mate** option, as highlighted in the following screenshot:

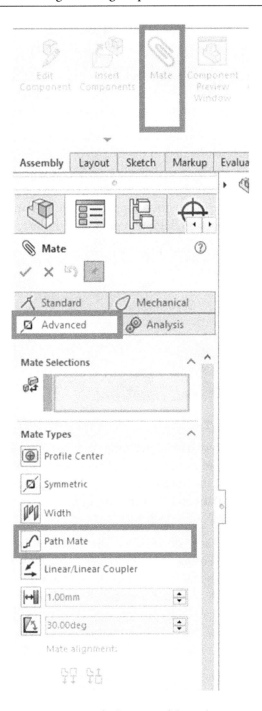

Figure 14.26 – The location of the path mate

3. Once we select the mate, we will be prompted to set it up. Here is a brief description of the different settings as well as what to select in this exercise:

-   **Component Vertex**: In this, we will select the vertex/point that would be following the path. Note that this selection will coincide with the path throughout the part's movement. We can select the midpoint shown in the following screenshot.

-   **Path Selection**: Here, we will select the path to follow. In the case of this exercise, we can select the sketch we created in the first step. To do that, we need to select the Selection Manager, and then we can click on **Select Open Loop** and select the path directly from the canvas afterward. Finally, click on the *green check mark* in the SelectionManager to confirm the open-loop selection. The open-loop selection will then show at the **Path Selection** field, as highlighted in the following screenshot:

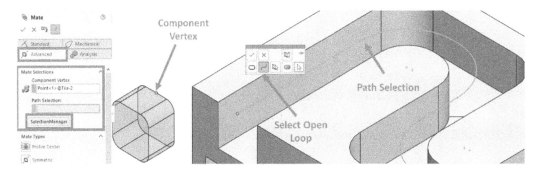

Figure 14.27 – The selection of the path and vertex

-   **Path Constraint**: This allows us to determine the movement of the part (tile) along the path. In this exercise, we can choose the **Free** option, which will allow us to simulate the movement of the tile along the path by dragging it. Other options include **Distance Along Path** and **Percent Along Path**. These will fix the tile in a certain position along the path, which we can determine by distance or percentage.

-   **Pitch/Yaw Control**: This has to do with the orientation of the object – the tile, in this case. We can pick the **Follow Path** option; this will allow us to constrain one axis of the moving part to be tangent to the selected path throughout the movement. Note that the selection of **X**, **Y**, and **Z** axes are all in relation to the selected path, not in relation to the part or assembly. Another option we can select for this control is **Free**, which will not impose any type of limitation. A good practice is to experiment with different options before confirming what fits your needs.

- **Roll Control**: This also has to do with the orientation of the moving object. We can pick the **Up Vector** option; this will constrain one axis of the moving part to align with a vector of our choosing. With this option, we can set one side of the tile to be facing one direction throughout its movement. In our example, we can pick the edge highlighted in the screenshot. Another option we can select for this control is **Free**, which will not impose any limitation related to rolling along the path. Similarly, when setting up **Pitch/Yaw Control**, a good practice is to experiment with the different options to see their effect. You might need to mix and match between the different options in **Roll Control** and **Pitch/Yaw Control** to see which combination of options fits your requirements. Keep experimenting until you get the orientation shown in the following screenshot:

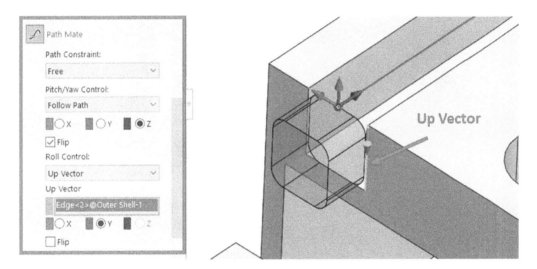

Figure 14.28 – Vectors allow us to define the orientation of the part while following the path

4. Click on the *green check mark* to confirm the mate. Once you confirm the mate, try moving the tile; you will notice that it will follow the path without any rotation around the path.

This concludes our application of the path mate. Next, we will cover the linear/linear coupler mate.

Defining the linear/linear coupler

The linear/linear coupler advanced mate allows us to create a linear transitional relationship between two different parts. Here, if one part moves, that other part will move as well according to a specific ratio. The following screenshot highlights the effect of the linear/linear coupler mate. Note that from the screenshot, the *Drawer* and *Drawer Support* are coupled together in such a way that the *Drawer* and the *Drawer Support* will follow each other's linear movements at a specific ratio. The screenshot highlights a 1:2 ratio where the *Drawer* will move twice as much as the *Drawer Support*. This ratio can be adjusted to our needs:

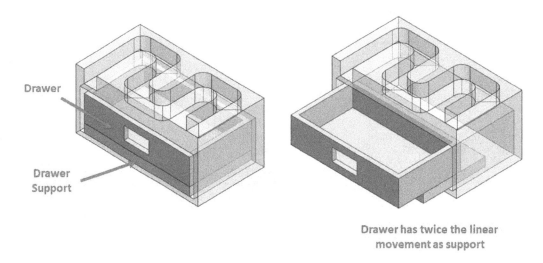

Figure 14.29 – The impact of the linear/linear coupler mate

Now that we know what the linear/linear coupler mate does, we can start applying it in our assembly.

Applying the linear/linear coupler

In this section, we will learn how to use the linear/linear coupler mate. We will do that by setting a transitional relationship between the *Drawer* and the *Drawer Support*, as highlighted in the following figure. In this exercise, we are continuing our work on the assembly we started with the path mate:

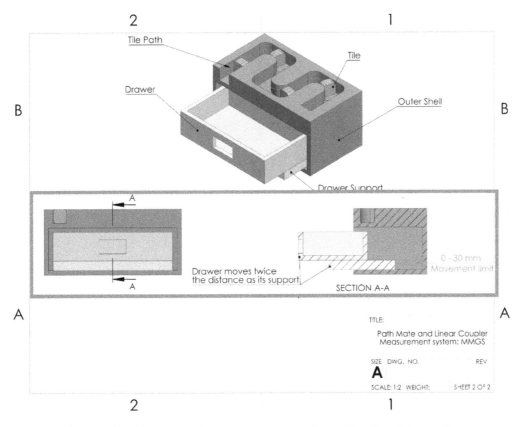

Figure 14.30 – The linear/linear coupler command we will apply in this exercise

To apply this mate, we are only required to have two different components that will move linearly in relation to each other. We can follow these steps to set it up:

1. Select the **Linear/Linear Coupler** mate by selecting the **Mate** command, the **Advanced** tab option, and then the **Linear/Linear Coupler** mate, as highlighted in the following screenshot:

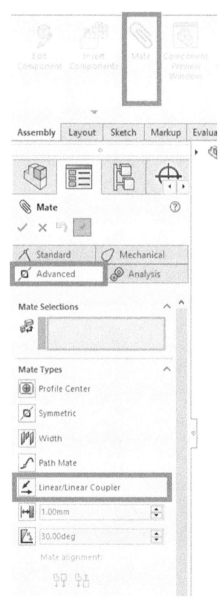

Figure 14.31 – The location of the linear/linear coupler mate

2. Once we select the mate, we will be prompted to set it up, as shown in the following screenshot. Here is a brief description of the different settings as well as what to select in this exercise:

- **Entity to mate**: In these two fields, we can select faces or edges upon which the relationship will be built. Note that this selection will also indicate the direction in which the movement will happen. For example, note the face selections we have in the following screenshot for this exercise; this indicates that the movement we want to couple/pair is in the normal direction to these surfaces.

- **Reference component to entity**: Here, we can select a part to act as a reference for the moving component. We can leave this field empty, in which case the movement will use the assembly origin as a reference. Since both the drawer and the drawer support movement are not referenced to any moving part, we can leave it blank.

- **Ratio**: In this, we can set the distance movement ratio. For example, in a ratio of 1:2, one part will move twice as much compared to others in the same linear direction of movement. As indicated in the screenshot, we can set the ratio to 1 : 2, where 1 refers to the drawer support movement and 2 refers to the drawer movement. Note that, in the SOLIDWORKS interface, the selection of the **Ratio** and **Entity to mate** are color-coded with blue and purple for easier reference. By default, both movements will be in the same direction. However, we also have them moving in different directions by selecting the **Reverse** option:

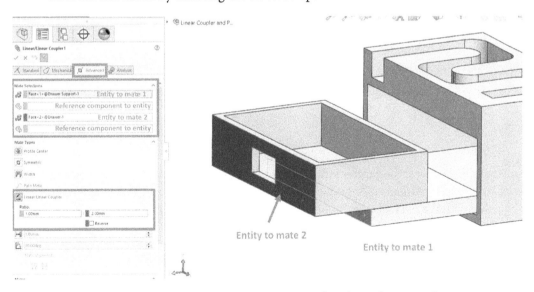

Figure 14.32 – The selection and PropertyManager of our linear/linear coupler mate

3. Click on the *green check mark* to confirm the mate. After that, try moving any of the mated parts; you will notice that the drawer will always move twice as much as the drawer support.

This covers the basic application of the linear/linear coupler mate. However, in practice, this mate is often coupled with other mates to help set up the positions that the moving parts start from and their movement limits. Let's do that next.

Fine-tuning the linear/linear coupler mate

If we look back at the resulting assembly we have after applying the linear/linear coupler mate, we will notice that the part does not have a starting point from the backend of the outer shell. Also, the linear movements are not restrained for the two parts. To set a starting point and limit the linear movement, we will utilize the coincident and distance range mates respectively.

Before creating those adjustments, we need to suppress the linear/linear coupler mate we applied earlier. To do this, we can click on the mate in the design tree and select the suppress icon, as highlighted in the following screenshot. We will un-suppress the mate after adjusting the other settings:

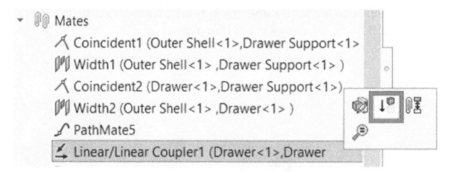

Figure 14.33 – Suppressing the linear/linear coupler allows us to adjust the starting position

Now, we will start by setting our starting point using the coincident standard mate. Here, we will create a coincident mate between the back of the drawer and the drawer support to coincide with the back of the outer shell, as indicated in the cross section shown in the following figure:

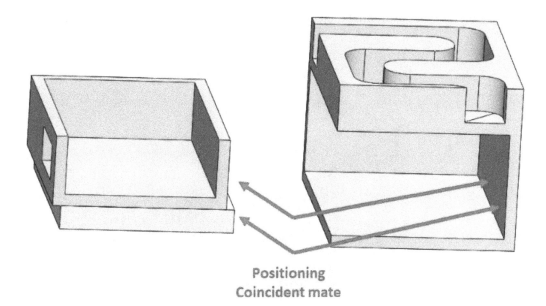

Positioning
Coincident mate

Figure 14.34 – We set up the initial positions of the parts according to the design of the mate

Note that we are applying this mate for positioning purposes only. To limit the mate to the positioning effect only, we can select the **Use for positioning only** option, as highlighted in the screenshot. We can find this option at the bottom of the PropertyManager when setting up the mate. Checking this option will not list the mate in the design tree as an active mate; rather, it will only bring the parts to the position as if the mate was applied. After applying this to both the drawer and the drawer support, you will notice that the coincident positioning mate will not be listed in the design tree with other mates:

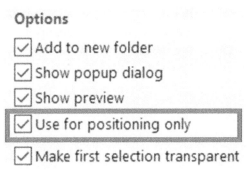

Figure 14.35 – We can set up mates for positioning purposes only

After applying the two positioning mates, our assembly should look like the following cross section:

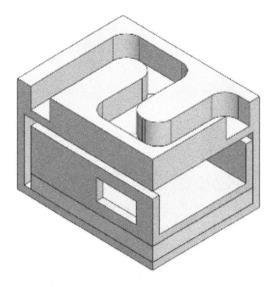

Figure 14.36 – The starting positioning of the drawing assembly

Now, we can use the distance range advanced mate to limit the movement of the drawer support from 0 to 30 mm. Make sure that you set the initial distance to 0 mm to keep the positioning. After that, we can un-suppress the linear/linear coupler mate we suppressed earlier. After this, you can start testing the functionality of the mate by moving the drawer or the drawer support. An alternative approach to this exercise is to set the positioning and the movement limitation first, and then apply the linear/linear coupler mate.

This concludes our coverage of the linear/linear coupler advanced mate. We covered what it is, how to set it up, and how to use different mates to support it. This mate can help to simulate different common sliding mechanisms, such as the ones in common drawers, making it an important tool in our mechanical design toolkit.

Summary

In this chapter, we learned about all of the advanced mates available in SOLIDWORKS assemblies. These include the profile center, symmetric, width, distance range and angle range, path mate, and linear/linear coupler mates. For each of the mates, we learned what the mate does and how to use it in the SOLIDWORKS assemblies environment. These advanced mates allow us to create more flexible assemblies that are closer to a variety of dynamic design intents, which is the objective of this chapter. At this point, we are well acquainted with both standard and advanced mates.

In the next chapter, we will continue working with assemblies to cover configurations and design tables, which will enable us to generate different assembly versions in a single SOLIDWORKS assembly file. This is in addition to different assembly tools such as collision and interference detection and assembly features.

Questions

Answer the following questions to test your knowledge of this chapter:

1. What does the profile center mate do?

2. What does the symmetric mate do, and what are the elements needed to apply it?

3. What does the path mate do, and what are the different constraints we can use with the mate?

4. What does the linear/linear coupler mate do, and what is a common application for this mate?

5. Using the part files provided for this question, generate the assembly highlighted in the following figure. Note that both sticks move symmetrically on the path in the base:

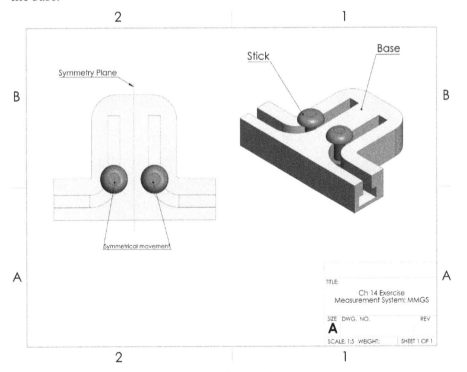

Figure 14.37 – The drawing for question 5

6. Using the part files provided for this question, generate the assembly highlighted in the following figure:

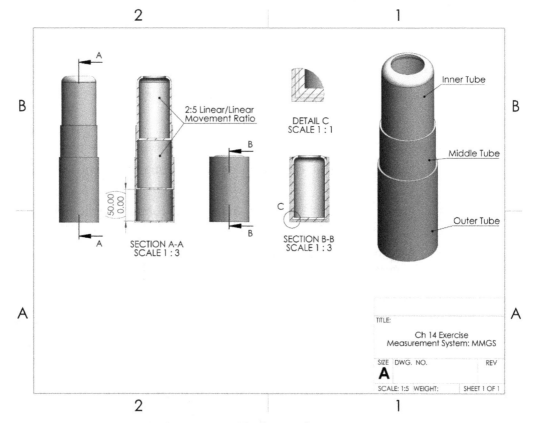

Figure 14.38 – The drawing for question 6

Important Note

The answers to the preceding questions can be found at the end of this book.

15

Advanced SOLIDWORKS Assemblies Competencies

Working with assemblies is more than just putting parts together and linking them with mates. In this chapter, you will learn about additional assemblies' competencies that can help us build more technically feasible products. First, we will learn about the **Interference Detection** and **Collision Detection** tools, through which we can identify when parts are interfering or colliding with each other. Then, we will cover assembly features that are applied in the assembly file rather than the part file. Finally, we will learn about configurations and design tables within the assembly's context. These will enable us to generate and manage multiple variations of an assembly within one file.

The following topics will be covered in this chapter:

- Understanding and utilizing the **Interference Detection** and **Collision Detection** tools

- Understanding and applying assembly features

- Understanding and utilizing configurations and design tables for assemblies

By the end of this chapter, we will be able to put together more advanced assemblies that are verified for interferences and collisions and include variations and features outside the scope of parts. The overall goal of this chapter is to acquire additional skills to make our assemblies more flexible and able to cater to more specific applications.

Technical requirements

This chapter requires access to SOLIDWORKS and Microsoft Excel on the same computer. The project files for this chapter can be found in the following GitHub repository: `https://github.com/PacktPublishing/Learn-SOLIDWORKS-Second-Edition/tree/main/Chapter15`.

Check out the following video to see the code in action: `https://bit.ly/3s6WgBI`

Understanding and utilizing the Interference Detection and Collision Detection tools

By now, we know that most of the products we deal with in our everyday life consist of more than one part. This is the case regardless of whether we are looking at a simple pen or a complex engine. As SOLIDWORKS professionals, we will require more tools than just mates to help us to evaluate our assembly. In this section, we will cover the **Interference Detection** and **Collision Detection** tools. We will cover what they are and how to use them in SOLIDWORKS assemblies. Throughout this section, we will be working with the assembly we linked with this chapter.

Interference Detection

Within an assembly, we might end up having different parts interfering with each other, meaning there might be one part inside another. This could happen because of a deficiency in our design or because it was intentional. Regardless, SOLIDWORKS assembly provides us with the **Interference Detection** tool to determine how much interference is taking place in our assembly. This tool lets us know whether or not any interference is taking place; it also allows us to determine the location of the interference, the parts involved, and how much volume is interfering.

To learn how to use the tool, we will apply it to the assembly shown in the following screenshot. You can download the parts as well as the assembly file with this chapter. Open up the assembly file after downloading it. Our task is to find out whether some parts are interfering with each other:

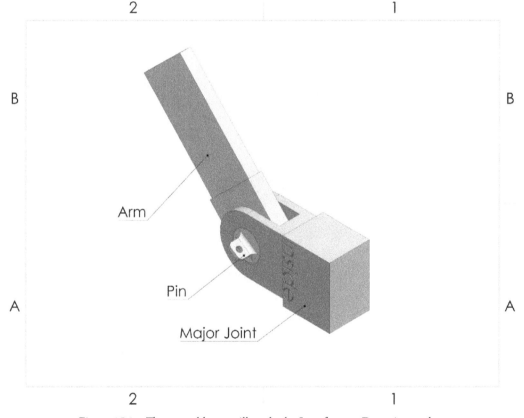

Figure 15.1 – The assembly we will apply the Interference Detection tool on

To apply the **Interference Detection** tool, we can follow these steps:

1.   Select the **Evaluate** tab, then select the **Interference Detection** command, as shown in the following screenshot:

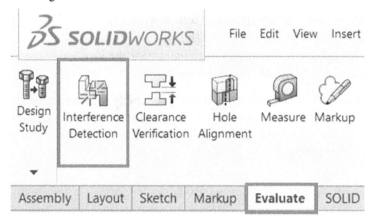

Figure 15.2 – The location of the Interference Detection command

2.   In the feature's **PropertyManager**, the assembly will be listed under **Selected Components,** as shown in the following screenshot. Click on **Calculate**. Then, the interference detection result will be shown in the space under **Results:**

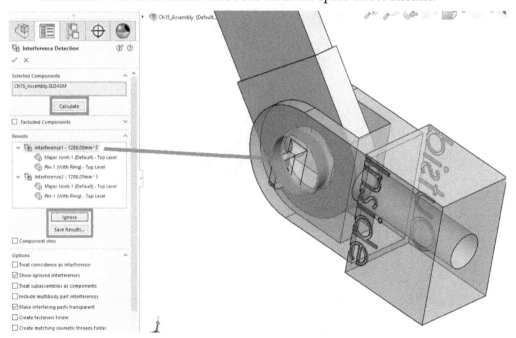

Figure 15.3 – The interference detection calculation and the result

Notice that the result will show how many interference incidents are taking place in the assembly alongside the interference volume. In this example, we have two interference incidents, each with an interfering volume of **1288.05 mm^3**. If you expand the interference incident menu, you will see a list of the parts involved in that particular interference. Also, if you select the interference incident, it will be highlighted in the canvas as a visual indication of where the interference is taking place.

Once we know where the interference is taking place, we can choose to address or ignore or save results to an external file, as indicated in *Figure 15.3*. Generally, we can address those interferences by adjusting the design of the parts or adjusting how the different parts are joined together. A common perception in practice is that interferences are always undesired when modeling products. However, that is not always the case. Let's address this practical point.

Interferences in practice

In practice, interferences are not negative in an absolute sense. Rather, they could also be the desired outcome, depending on the application. For example, in industrial machines such as pumps and compressors, ball bearings can have an interference fit with shafts to ensure both shaft and bearing will rotate together. Maintenance personnel will then use methods such as heating to install ball bearings with their interfering shafts. As SOLIDWORKS users, we need to be able to determine the interference instance in our assembly. It is also essential for us to understand that, in some instances, interference is desirable for fitting applications.

At the **Interference Detection PropertyManager**, we have different options that we can utilize depending on our needs. The options are highlighted in the following screenshot:

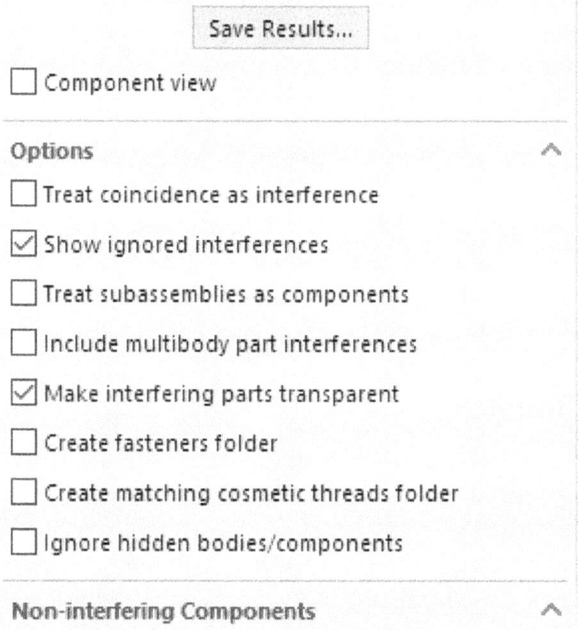

Figure 15.4 – More options for the interference detection calculation

Let's take a look at what each of those options does, as follows:

- **Treat coincidence as interference**: We can find this option if we expand the **Options** menu in the **PropertyManager**. This will list all instances with two parts touching each other in the interference result—that is, parts with a transitional fit.

- **Show ignored interferences**: When using the **Ignore** command with an interference, it will get hidden from the results in the **PropertyManager**; checking this option will have it shown.

- **Treat subassemblies as components**: This will have the software treating subassemblies as a single component, meaning that interfering parts within the subassembly will not be included in the interference detection study.

- **Include multibody part interferences**: Having this checked will ask the software to check if different bodies within one part are interfering and, if so, to report those interferences.

- **Make interfering parts transparent**: This will change the display of the interfering parts to be transparent, making it easier to visually study the interference.

- **Create fasteners folder**: This will create a folder under the results only for fasteners involved in the assembly.

- **Create matching cosmetic threads folder**: This will create a separate folder for components with matching cosmetic threads. Unmatching threads will not be listed in the same folder.

- **Ignore hidden bodies/components**: This will exclude hidden parts and bodies from the interference detection study.

This concludes our exploration of the **Interference Detection** tool. We learned what it does and how to use it. Next, we will learn about the **Collision Detection** tool.

Collision detection

The **Collision Detection** tool allows us to get notified when two parts collide with each other. This is commonly used with dynamic assemblies. As we are moving different parts within the assembly, the tool can notify us when the collision happens and highlight which parts are colliding.

To illustrate how to use the tool, we will apply it to the same assembly we used with the **Interference Detection** tool earlier. To start, make sure the assembly file attached to this chapter is open. To use the tool, we can follow these steps:

1. Select the **Move Component** command under the **Assembly** tab, as highlighted in the following screenshot:

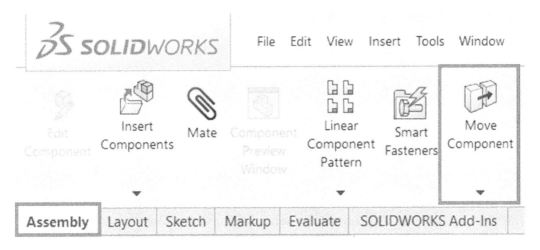

Figure 15.5 – The location of the Move Component command

2. On the **PropertyManager,** select the **Collision Detection** option. Also, make sure to check the **Stop at collision** and **Dragged part only** options and the **Highlight faces** and **Sound** advanced options, as shown in the following screenshot:

Figure 15.6 – Collision Detection is within the Move Component command

3. Start moving the arm; you will notice that the movement will stop when the *arm* hits the *major joint*. The software will also give you a sound indicator, notifying you that the collision took place. Also, notice that the colliding faces will be highlighted to help us to identify where the collision is happening.

Once we have the parts at the colliding position, we can use the **Smart Dimension** command to get the exact collision angle or distance measurements. For instance, in this exercise, we can measure the collision angle with the following steps:

1. Drag the arm clockwise until it collides with the major joint, as shown in *Figure 15.7.*

2. Use the **Smart Dimension** command in the **Layout** tab to measure the angle between the arm and the major joint, as in *Figure 15.7.*

This way, we will be able to identify the collision angle between the arm and the major joint as **90** degrees, as highlighted in the following screenshot. Those numerical measurements can then help us to make different design decisions:

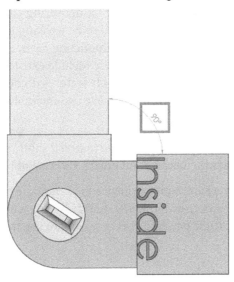

Figure 15.7 – Smart Dimension can be used to generate measurements relating to the collision

> **Note**
> Interfering parts will be flagged as being already colliding when using the **Collision Detection** tool. In the preceding example, if we keep the **Dragged part only** option checked, SOLIDWORKS will give us the message that the model is in a colliding position because the *pin* is in an interference relationship with the *major joint*, as explored earlier in this section.

This concludes our coverage of the **Collision Detection** tool. We learned what it is, what it does, and how to use it. We also learned about the **Interference Detection** tool in an earlier section titled *Interference Detection*. Next, we will learn about assembly features, which refers to applying features in the assembly's context.

Understanding and applying assembly features

Assembly features refer to applying features such as extruded cuts and fillets within the assembly file rather than the part file. In this, the features will be recorded and applied on the assembly file and listed in the assembly file design tree without impacting the design of the original part. In this section, we will learn about assembly features and how to apply them.

Understanding assembly features

We can look at assembly features as standard features that are stored in the assembly file rather than the part file. This makes them unique for delivering particular messages or design intent. To apply assembly features, we need to be working in the assembly file. Compared to the features we can apply as we are modeling parts, assembly features are limited and only include subtractive features. These include extruded cuts, swept cuts, revolved cuts, fillets, and holes. By default, assembly features do not propagate to the parts; however, we can make them propagate if needed. We use assembly features for two main reasons, outlined as follows:

- We use them to deliver the message that the material subtraction takes place after manufacturing the parts themselves and during the assembly process.

- They are used to apply a subtractive feature across multiple parts, which will make some design intents easier to accomplish, such as having a hole perfectly aligned and going through multiple parts. In this case, we can use the **Propagate feature to parts** option.

Next, we will apply the **Extruded Cut** assembly feature to our assembly.

Applying assembly features

Here, we will apply the **Extruded Cut** assembly feature to the assembly we generated earlier. We will use the feature to drill a hole in the major joint and the pin, as shown in the following diagram:

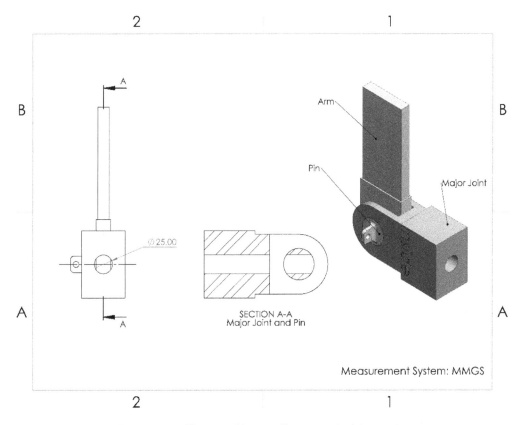

Figure 15.8 – The assembly we will generate in this exercise

To apply the **Extruded Cut** assembly feature, follow these steps:

1. Select the **Assembly** tab, then expand the **Assembly Features** command and select **Extruded Cut**, as shown in the following screenshot:

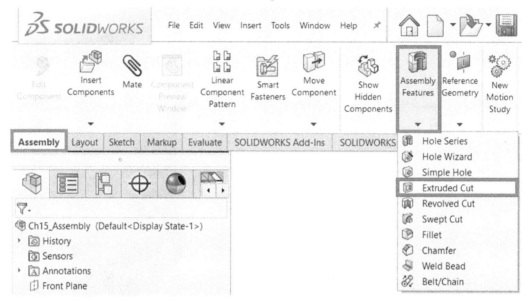

Figure 15.9 – The location of the different assembly features

2. Apply the features as usual by selecting the sketch face and drawing the circle sketch, then exit the sketch to set up the extruded cut.

3. Set the end condition to **Through All**; you will notice that the cut preview will go through all of the parts. However, we only require it to apply to the major joint and the pin. We can set that up on **Feature Scope**, as shown in the following screenshot:

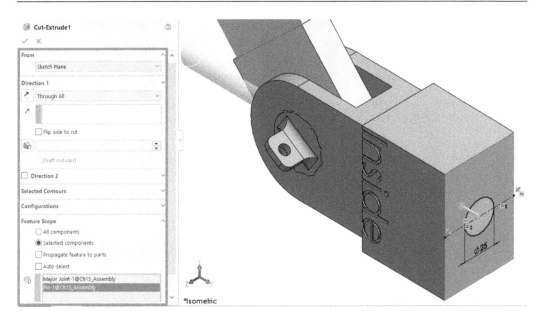

Figure 15.10 – Feature Scope allows us to apply the assembly feature to specific parts

4. Click on the *green checkmark* to apply the extruded cut.

We can view the cut on the two parts using the cross-section view, as shown in the following screenshot. Note that the hole only goes through the major joint and pin. One important aspect to note is that the holes on the two parts (major joint and pin) will always be aligned. If we rotate the pin then rebuild the assembly, the hole in the pin will be remade to be aligned with the one in the major joint:

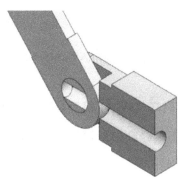

Figure 15.11 – The resulting cut did not apply to the arm

Note that, after applying the extruded cut, it will then be listed in the assembly's design tree and not under any particular part, as highlighted in the following design tree:

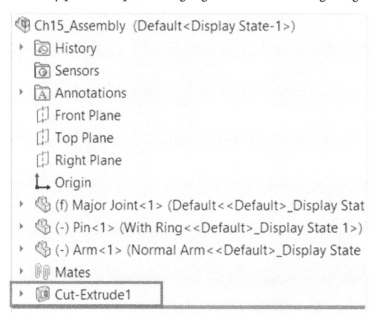

Figure 15.12 – The assembly feature is listed in the assembly design tree

Also, if we open the part file for the major joint or pin, we will notice that the new extruded cut is not there. If we want the extruded cut to reflect on the part, we can check the **Propagate feature to parts** option under **Feature Scope** when setting up the extruded cut.

This concludes this section on assembly features. We learned what they are, why we should use them, and how to apply them. Next, we will use the same model to learn about configurations and design tables in the assembly's context.

Understanding and utilizing configurations and design tables for assemblies

With configurations and design tables, we will be able to create different versions of an assembly in the same assembly file. This is very similar to what we learned previously when creating configurations for a part. Within an assembly, our configurations can vary based on the mates applied and the existing parts within the assembly. We can also link them to configurations within the individual parts themselves. In this section, we will generate different versions of the assembly file we worked on in the previous section. We will start by generating a configuration *touchpoint* by manually adding a configuration.

Using manual configurations

Here, we will create the shown configuration by manually adding a new configuration. Note that, in the configuration shown in the following screenshot, there is an additional mate compared to the default configuration, which fixes the arm at **90** degrees to the **Major Joint**:

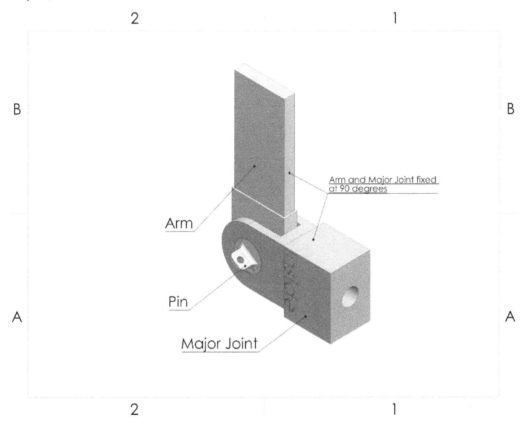

Figure 15.13 – The configuration we will generate in this exercise

The procedure for adding a configuration in assemblies is the same as in parts. The difference here is that we will be dealing with different elements, such as mates. To generate the indicated configuration, we can follow these steps:

1. Go to the **ConfigurationManager**, right-click, then select **Add Configuration...**, as shown in the following screenshot. Assign the name **TouchPoint** to the configuration. Then, make sure we are operating within this new configuration:

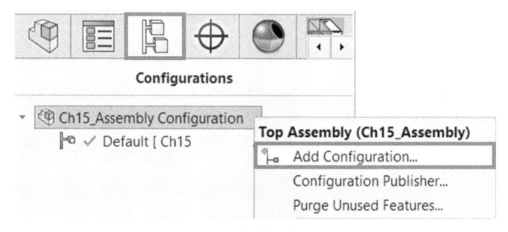

Figure 15.14 – The location of the Add Configuration… command

2. Apply the standard mate angle to set the *arm* at 90 degrees with the *major joint*. This will fix the *arm* at the position shown in the initial drawing.

This applies to the new **TouchPoint** configuration. As usual, go back to the **Default** configuration to double-check how the two configurations are different. You will notice that in the **Default** configuration, the *arm* can rotate freely, while at the **TouchPoint** configuration, the *arm* is fixed at 90 degrees. Also, you will notice that the additional angle mate we added will be shown as suppressed in the **Default** configuration.

> **Note**
>
> The **Default** configuration will have the angle mate suppressed if the **Suppress new features and mates** option is checked in its properties. You can find more information about this option in the next section, *Configurations' advanced options*.

Configurations' advanced options

When starting a new configuration, three notable advanced options are important for our workflow. These are highlighted in the following screenshot:

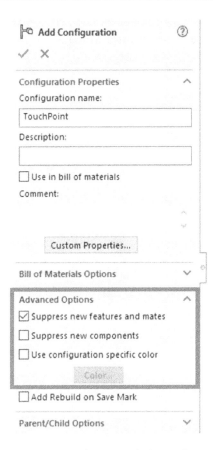

Figure 15.15 – Configurations' advanced options

Let's explore the impact of each option, as follows:

- **Suppress new features and mates**: Having this option checked means that any new features and mates applied to other configurations would be suppressed in this configuration.

- **Suppress new components**: This option means that any other components that get added to other configurations will be suppressed in this particular configuration.

- **User configuration specific color**: This allows us to apply a specific color appearance to the configuration for easier identification.

> **Note**
> The first two suppress advanced options relate to how the configuration at hand interacts with changes in other configurations, not to how other configurations interact with the configuration at hand.

After adding the configuration, we can access and adjust the advanced configuration by accessing **Properties…** at the **ConfigurationManager**, as highlighted in the following screenshot:

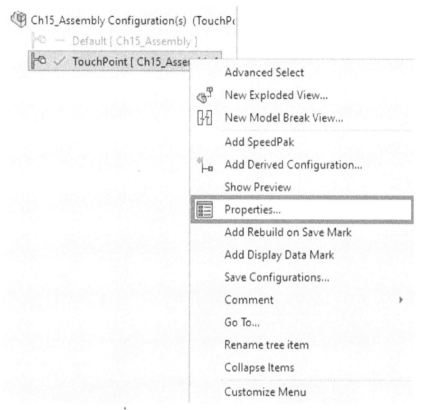

Figure 15.16 – We can access the advanced options through properties

This concludes how to add a new configuration within an assembly. Next, we will generate different configurations using design tables.

Design tables

Using design tables, we will generate the following configurations for our assembly. Note that all new configurations are based on the **Default** one, not the **TouchPoint** one. Also, the difference between each configuration is the shape of the *arm* and the *pin*. Both the *arm* and *pin* already have two-part configurations. For the *arm*, we have the **Normal Arm** and **Fork Arm** configurations, while for the *pin*, we have **With Ring** and **Plain**, as illustrated in the following screenshot:

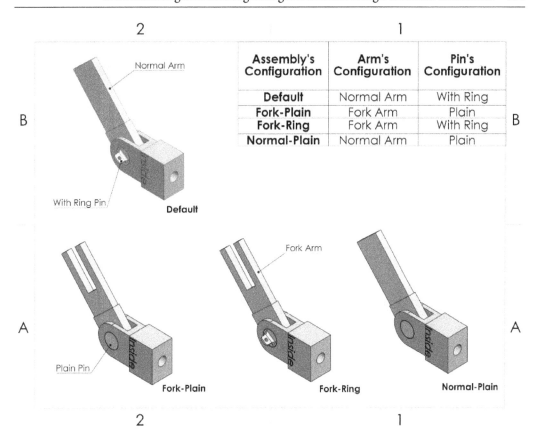

Assembly's Configuration	Arm's Configuration	Pin's Configuration
Default	Normal Arm	With Ring
Fork-Plain	Fork Arm	Plain
Fork-Ring	Fork Arm	With Ring
Normal-Plain	Normal Arm	Plain

Figure 15.17 – The design table we will generate in this exercise

We add a design table to an assembly in the same way we add it to a part. However, we can control multiple parts within the assembly. To generate the four configurations shown in the preceding diagram, follow these steps:

1. Insert a new design table by going to **Insert**, **Tables**, then **Design Table**. Set the table to **Auto-create**, then click on the green checkmark.

> **Note**
>
> Once at the design table, we will notice the two configurations we have are already listed. These are **Default** and **TouchPoint**. This is in addition to all of the differences between them, including the angle mate we generated earlier when working with configurations.

2. To recall the parts configurations, we can use the `$configuration@`
    `partName<instance>` title format. We want to call the configurations for two
    parts, the *arm* and the *pin*. We can find both the part name and the instance in the
    assembly's design tree. The design table will look like this:

1	Design Table for: Ch15_Assembly				
2		$DESCRIPTION	$STATE@Angle1	$configuration@Arm<1>	$configuration@Pin<1>
3	Default	Default	S	Normal Arm	With Ring
4	TouchPoint	TouchPoint	U	Normal Arm	With Ring
5	Fork-Plain		S	Fork Arm	Plain
6	Fork-Ring		S	Fork Arm	With Ring
7	Normal-Plain		S	Normal Arm	Plain
8					

Figure 15.18 – The format of the design table

3. Apply the design table by clicking anywhere on the canvas. Then, *double-check* the
    different new configurations generated. The configurations should match the ones
    shown in *Figure 15.17*.

This concludes how to generate a design table for part configurations within an assembly.
Note that the part configurations we used in this exercise were generated within part files.
We can also manually adjust the part configuration within the assembly. To do this, we
can simply click on the part listing in the design tree and change the configuration from
the drop-down menu, as shown in the following screenshot. We can also access the same
drop-down menu by clicking on the part in the canvas:

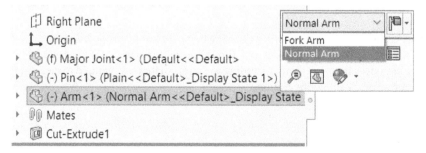

Figure 15.19 – We can adjust the active part configuration in the assembly from the design tree

In this section, we learned about configurations and design tables within an assembly context. This enabled us to generate multiple variations of an assembly within one assembly file.

Summary

In this chapter, we learned various skills relating to SOLIDWORKS assembly tools. We started with the **Interference Detection** and **Collision Detection** tools to determine whether two or more parts are interfering with each other in a static position or colliding with each other in a dynamic movement. Next, we learned about assembly features, which are subtractive features that are applied and stored with the assembly file rather than the part file. Finally, we learned about configurations and design tables, which enabled us to generate multiple variations of an assembly within one assembly file. Overall, the skills we have learned in this chapter will enable us to better meet different design intents by evaluating, adjusting, and varying our assemblies' outcomes.

Next, we will work on a 3D modeling project to create a remote-control helicopter model. The project will provide you with a comprehensive practice for many of the topics covered in this book.

Questions

Answer the following questions to test your knowledge of this chapter:

1. What do the **Interference Detection** and **Collision Detection** tools do?
2. What is the difference between assembly features and the features we add within parts?
3. What is the difference between dealing with configurations and design tables in assemblies versus parts?

4. Download the parts and assembly linked to this exercise. Find the interference volume between the **Helical Gear** and the **Fixture** and between the **Worm Gear** and the **Ball Bearing**. All parts are as indicated in the following diagram:

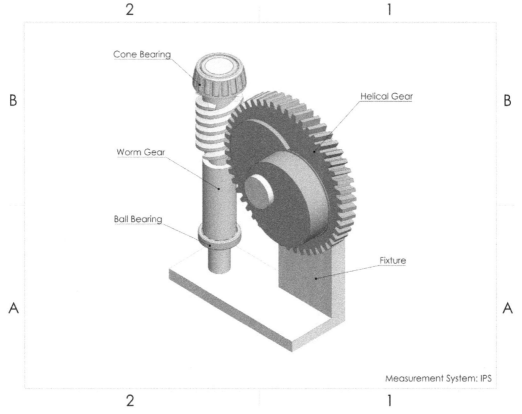

Figure 15.20 – Drawing for question 4

5. Using the assembly from the previous question, generate a design table introducing the following configurations:

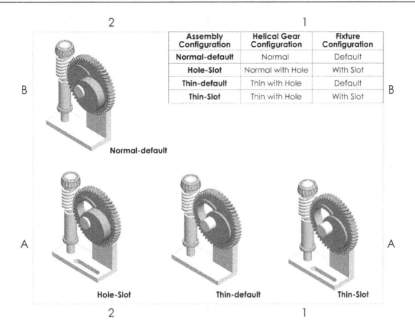

Figure 15.21 – Drawing for question 5

6. Download the parts and assembly files with this exercise. Use the assembly features to create the hole shown in the following diagram. Make the assembly feature propagate to the parts involved in their part files:

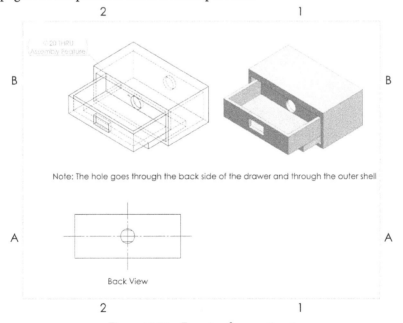

Figure 15.22 – Drawing for question 6

7. Using the assembly from the previous question, generate two additional configurations, one showing the drawer fully open and the other showing it fully closed, as shown in the following diagram. Make sure the default configuration remains flexible:

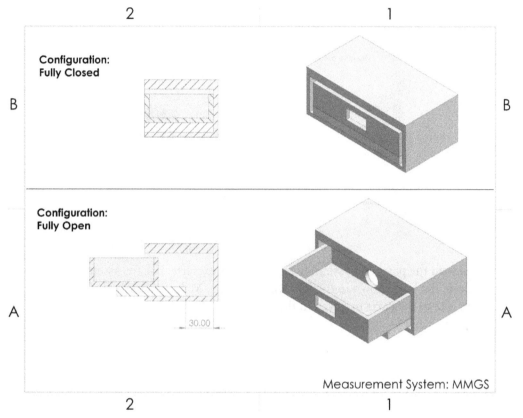

Figure 15.23 – Drawing for question 7

Important Note

The answers to the preceding questions can be found at the end of this book.

Project 2: 3D-Modeling an RC Helicopter Model

SOLIDWORKS is a 3D design tool. Just like all tools, the more you use it, the better you become at it. In this project chapter, you will be provided with project work that you can do to hone your skills. In this project, you will be 3D-modeling and assembling a helicopter toy model from a set of engineering drawings.

This project chapter will cover the following topics:

- Understanding the project
- 3D-modeling the individual parts
- Creating the assembly

By the end of this chapter, you will have more confidence in using the different SOLIDWORKS tools for practical projects.

Technical requirements

You will need to have access to SOLIDWORKS to complete the project.

Understanding the project

Understanding what the project entails is essential before starting the work. This will allow you to draw a plan and manage your work expectations towards completing the project. For this exercise project, you will be 3D-modeling a **Remote Control** (**RC**) helicopter toy, as shown in the following figure:

Figure P2.1 – The RC helicopter you will 3D-model in this project

The model consists of 16 parts, 12 of which are unique. The following figure highlights the bill of materials, showing the names of the parts, their quantity, and position in the assembly:

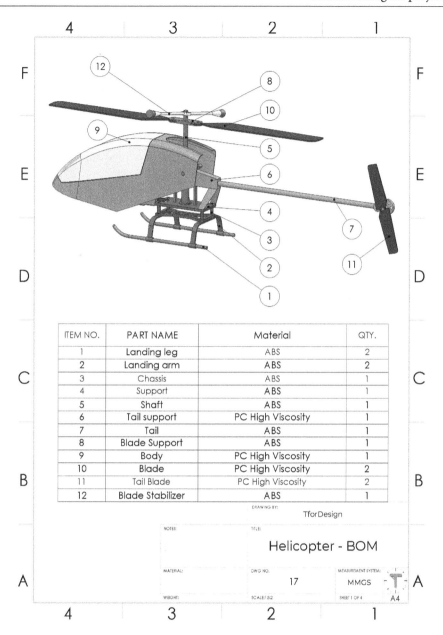

ITEM NO.	PART NAME	Material	QTY.
1	Landing leg	ABS	2
2	Landing arm	ABS	2
3	Chassis	ABS	1
4	Support	ABS	1
5	Shaft	ABS	1
6	Tail support	PC High Viscosity	1
7	Tail	ABS	1
8	Blade Support	ABS	1
9	Body	PC High Viscosity	1
10	Blade	PC High Viscosity	2
11	Tail Blade	PC High Viscosity	2
12	Blade Stabilizer	ABS	1

DRAWING BY: TforDesign

NOTES: TITLE:

Helicopter - BOM

MATERIAL: DWG NO.   17   MEASUREMENT SYSTEM: MMGS

WEIGHT: SCALE1:2   SHEET 1 OF 4   A4

Figure P2.2 – The RC helicopter and its bill of materials

At this point, you already have an idea of the project's outcome and the complexity of the needed parts and assembly. In the following section, we will provide you with the engineering drawings needed to replicate all the parts and assembly. Now that we have an idea about the project's final output, we can discuss how you can tackle it in the context of this writing.

> **Important Note**
> The drawings and 3D models presented in this project are for practice purposes rather than for manufacturing purposes.

There are two ways in which you can tackle this project, depending on your 3D-modeling level. They are as follows:

- **Moderate level**: Take a look at the drawings and the provided sample procedure to complete the project.

- **Advanced level**: Only take a look at the drawings without using the sample procedure.

Other than the two suggested approaches, you can also follow your own way to 3D-model the RC helicopter without utilizing the provided drawings and sample procedures. Keep in mind that the sample 3D-modeling procedures provided are meant as a sample guide. They are not meant to present an optimal procedure, rather, you can look at them as one possible procedure to generate the models and assemblies. To grow your own 3D-modeling style, you can experiment with modeling the project using different modeling procedures.

> **Tip**
> You can treat the project as your own and customize the provided RC helicopter to end up with your unique design.

In this project, we will first explore the individual parts, then move into the assembly. So, let's get started with the parts. We will also provide you with hints that can assist you with your work.

3D-modeling the individual parts

In this section, we will explore the different part drawings that represent the RC helicopter. The provided drawings have enough information for you to replicate all the parts to end up with an identical result to the one shown in *Figure P2.1*. Thus, one option for handling the project is to create an exact replica of the given drawings. However, you can also choose to customize and adjust different elements of the design to make it your own. Keep in mind that this is your project, so feel free to treat it as such.

Exploring the individual parts

The provided RC helicopter consists of 16 parts. However, 12 of those are unique, which you will need to 3D-model, as highlighted in *Figure P2.2*. The parts you will need to 3D-model are as follows:

1. Landing leg
2. Landing arm
3. Chassis
4. Support
5. Shaft
6. Tail support
7. Tail
8. Blade Support
9. Body
10. Blade
11. Tail blade
12. Blade stabilizer

> **Important Note**
> The names of the parts presented in the bill of materials might be different than practiced names in the industry.

Your task is to use the presented drawings to 3D-model the individual parts. As you are 3D-modeling the different parts, keep in mind that there is no one correct way of 3D-modeling any of the parts. However, we will provide you with some hints for one approach that can push you forward if you find yourself getting stuck. You can also feel free to customize your design using the given drawings as a base of inspiration. The order in which the following drawings are presented is arbitrary.

> **Important Note**
> All engineering drawings are presented using the third angle projection.

Let's start exploring the drawings one after the other. The first drawing is for the *landing leg*:

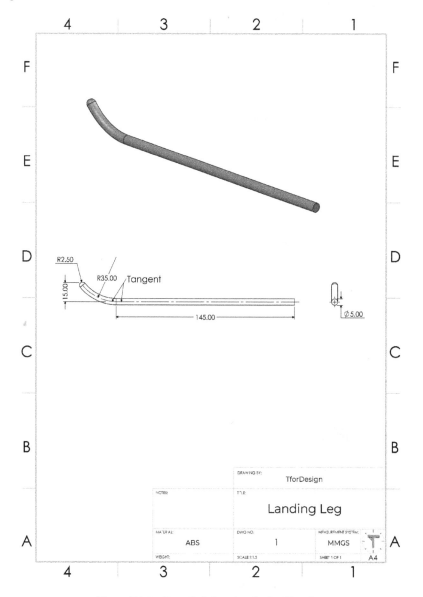

Figure P2.3 – Detailed drawing for landing leg

Here is a sample procedure for 3D-modeling the *landing leg*:

1. You can start with a swept boss to create the main shape, as shown in the following figure:

Figure P2.4 – Swept boss be used to start the landing leg

2. Apply fillets to get the rounded edge.

Next, we can look at the *landing arm*:

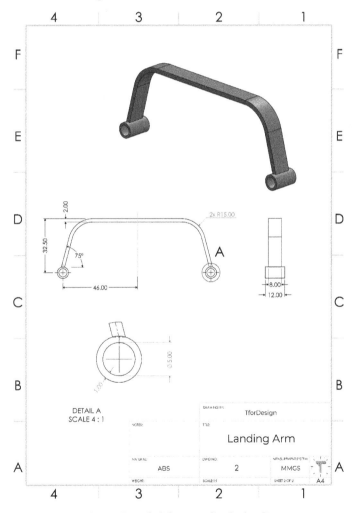

Figure P2.5 – Detailed drawing for the landing arm

Here is a sample procedure for 3D-modeling the landing arm:

1. Create a sketch highlighting the entire shape of the arm. Then, use an extruded boss with selected contours to extrude the long-connected shape:

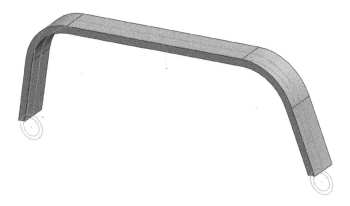

Figure P2.6 – A possible first step in creating the landing arm is using an extruded boss

2. Reuse the sketch from *step 1* to extrude-boss the rings to get the final shape shown as follows:

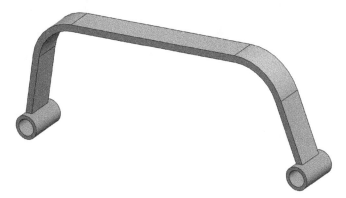

Figure P2.7 – One sketch can be used for more than one feature

After the *landing arm*, we can explore the *chassis*:

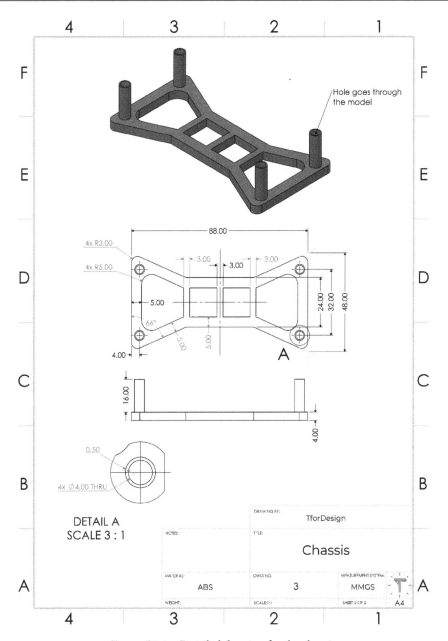

Figure P2.8 – Detailed drawing for the chassis

Here is a sample procedure for 3D-modeling the *chassis*:

1. Use an extruded boss to create the mainframe of the chassis, as shown in the following figure:

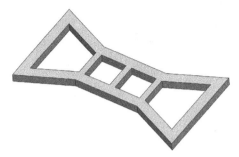

Figure P2.9 – An extruded boss can be used to create the mainframe of the chassis

2. Use the fillet feature to round corners. Then, use an extruded boss and extruded cut to create the extensions and holes on the corners. You will get the following final shape:

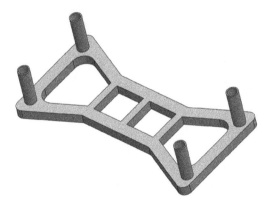

P2.10 – The chassis can be finalized with fillets, an extruded boss, and an extruded cut

> **Tip**
> In general, it is a good practice to add fillets and chamfers at the end of the modeling process.

After the *Chassis*, we can take a look at the *Blades*. There are two blades in the RC helicopter, top blades, and tail blades. We can utilize configurations or design tables to create both blades in one part file. This is because both blades have similar design features. We will first create the top blade and then generate the tail blade out of it. The drawing shows the default (top) blade:

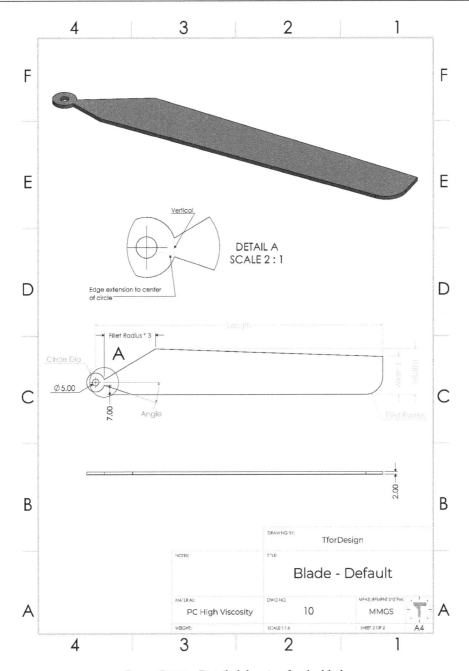

Figure P2.11 – Detailed drawing for the blade

Note the drawing has many variables without numerical values. The numerical values for both the default and tail configurations are presented in the following drawing:

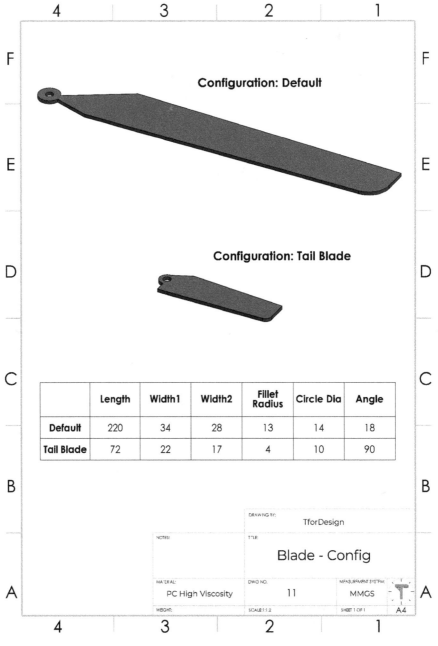

Figure P2.12 – The two configurations for the blade

Here is a sample procedure for 3D-modeling the two blades:

1. Use an extruded boss with the sketch shown in the following figure to create the main blade:

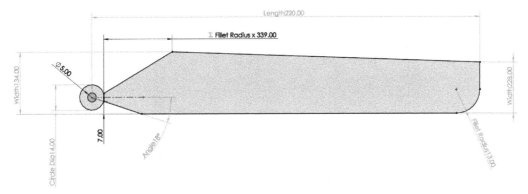

Figure P2.13 – step 2 An extruded boss can create the bulk of the blade

3. Introduce a design table, as shown in *Figure P2.14*, to create the tail blade. Use the variables named in *Figure P2.13*:

	A	B	C	D	E	F	G	H
1	Design Table for: Blade							
2		Length@Sketch1	Width1@Sketch1	Width2@Sketch1	Fillet Radius@Sketch1	Circle@Sketch1	Angle@Sketch1	
3	Default	220	34	28	13	14	18	
4	Tail Blade	72	22	17	4	10	90	
5								
6								

Figure P2.14 – step 3 A design table can be used to create the tail blade

4. After introducing the design table, double-check on the generated configurations for the default top blade and the tail blade.

Next, we can look into the *Shaft*, which will connect most of the helicopter parts:

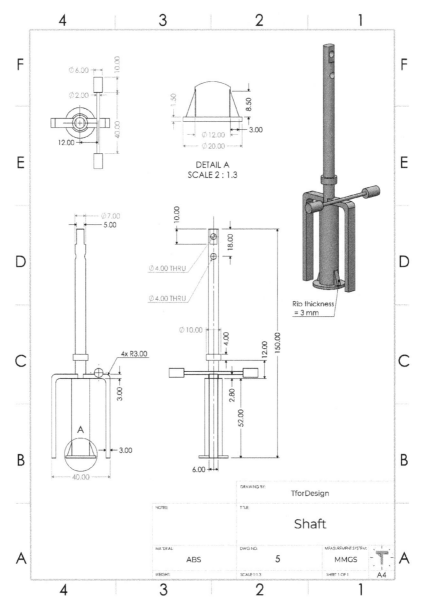

Figure P2.15 – Detailed drawing for the shaft

Here is a sample procedure for 3D-modeling the *shaft*:

1. Use revolved boss to create the long part of the shaft, as shown in the following figure:

Figure P2.16 – Revolved boss can generate the long rod of the shaft

2. Use the extruded boss and fillet commands to create the two legs on the sides of the shaft. You can create one side and then mirror the other side:

Figure P2.17 – Extruded boss, fillet, and mirror features can create the two leg-shaped figures

3. You can create the sideway-looking antenna using a revolved boss. You might need to generate a new plane for that, as shown in the following figure below. You can get the exact shape using multiple applications of the extruded boss feature as well:

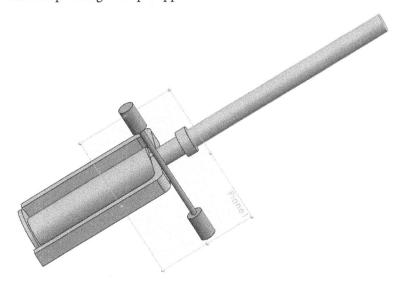

Figure P2.18 – A new plane was used as a base for a revolved boss

4. Using extruded cuts, you can create a straight surface on top of the shaft and the two holes, as shown in the following figure:

Figure P2.19 – The holes and straight surface can be made with an extruded cut

5. Using the rib feature, you can create the two ribs on the lower part of the shaft. You can apply the feature twice or apply it once and mirror the result to the other side:

Figure P2.20 – The rib feature can be used to create the ribs

After the *shaft*, we can start working on the *blade support*, which will connect the blades with the shaft. The details of the blade support are highlighted in the following figure:

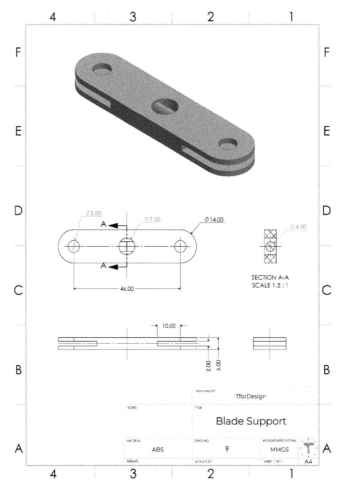

Figure P2.21 – Detailed drawing for the blade support

Here is a sample procedure for 3D-modeling the *blade support*:

1. Use the extruded boss feature to create the following shape. When doing so, the Mid Plane end condition might make it easier to do the cuts in *step 2*:

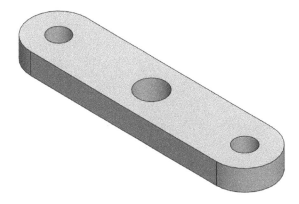

Figure P2.22 – One extruded boss application can generate most of the shape

2. Use an extruded cut to make the slot in the middle of the shape:

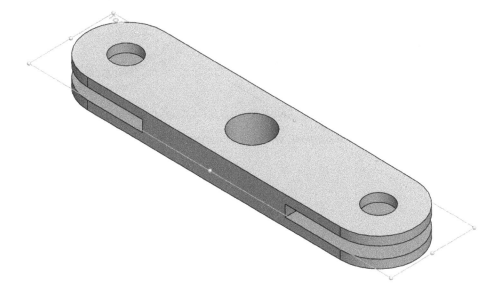

Figure P2.23 – The middle cut can be made with an extruded cut

3. Apply another extruded boss to create the cylindrical rod located in the middle hole. You can use the Up To Next end condition to have the circular extrusion bounded by the surface:

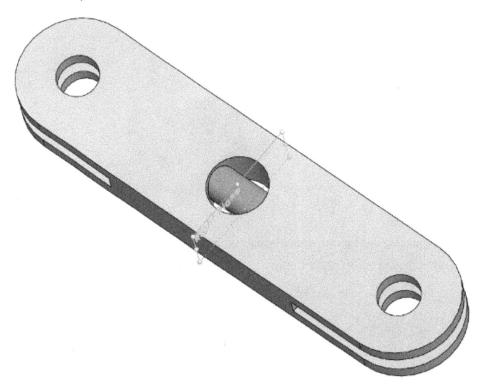

Figure P2:24 – The 90 mm-long cylindrical part can be created with an extruded boss

Next, we can start looking at *tail support*, which will connect the *tail* to the *support*:

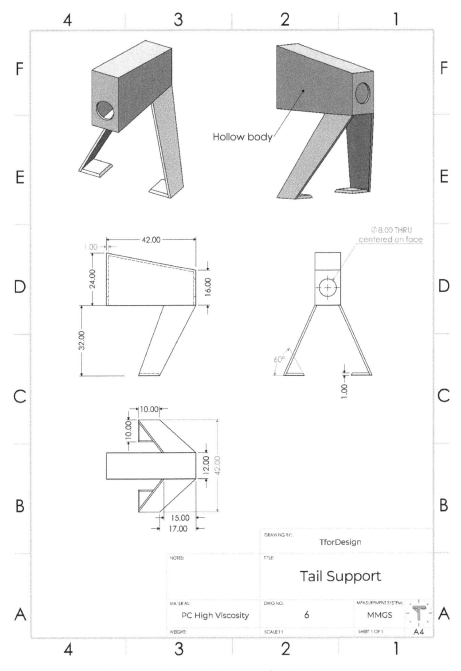

Figure P2.25 – Detailed drawing for the tail support

Here is a sample procedure for 3D-modeling the *tail support*:

1. We can start by creating the bulky hollow part shown in the following figure. This can be made with the extruded boss, shell, and extruded cut features, in that order. Note that the bottom surface is removed using the shell feature. Also, note that the circular cut goes through both sides of the *tail support*:

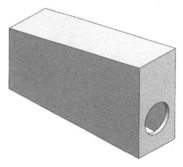

Figure P2.26 – The hollow part of the tail support can be made with the shell feature

2. The leg part of the support can be made with a lofted boss, as shown in the figure. Note that even though both ends are rectangular, they have different sizes, with the lower end not aligned with the side of the boxy area:

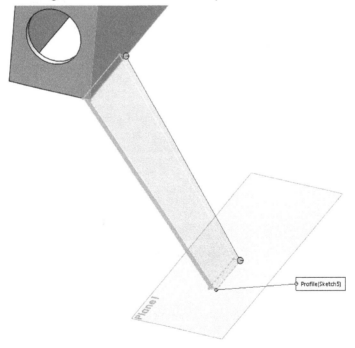

Figure P2.27 – Lofted boss can be used to create the long leg

3.   Use an extruded boss to create the lower straight foot indicated in the figure. Note that the foot is drafted. You can toggle the draft within the extruded boss feature or apply the draft as a separate feature:

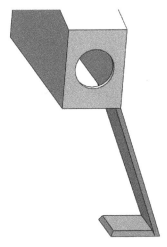

Figure P2.28 – A drafted extruded boss can be used to create the foot

4.   Use the feature mirror to mirror the lofted and extruded boss to the other side to get the shape shown in the following figure:

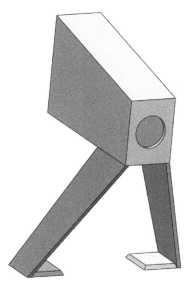

Figure P2.29 – The feature mirror command can quickly replicate features

Next, we can have a closer look at the *tail,* which will connect to the *tail support* at one end and the *tail blades* at the other end:

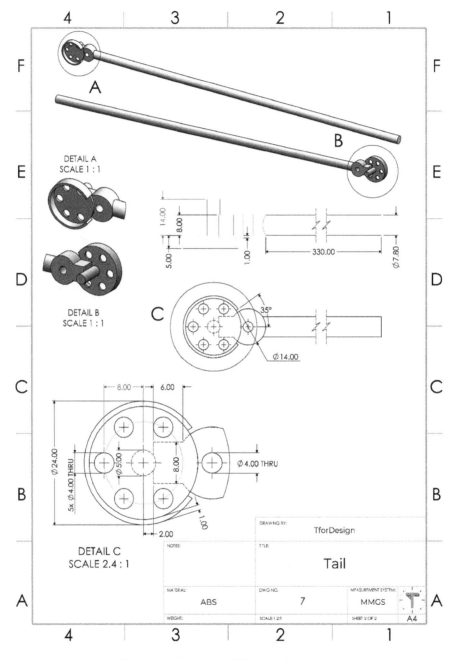

Figure P2.30 – Detailed drawing for the tail

Here is a sample procedure for 3D-modeling the *tail*:

1. Use two extruded bosses to create the base of the tail, getting the shown result. To make sketching simpler, you can create one sketch and use it for the extruded bosses:

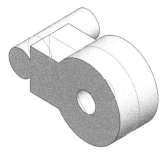

Figure P2.31 – The same sketch can be used twice for different features

2. Use two other extruded boss features to create the disk-shaped part and its boundary. We can also use one sketch two times for two extruded boss applications:

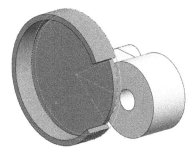

Figure P2.32 – Two extruded bosses with the same sketch can be used to create the disk

3. Use the extruded cut feature to create the five cuts shown in the following figure. Note the cuts follow a circular pattern of six instances, with the one on the far right being skipped:

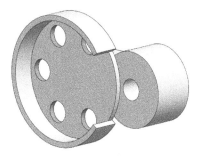

Figure P2.33 – The cuts on the disk follow a circular pattern

4. Use an extruded boss to create the long part of the tail. As shown in the following figure, you might need to create a new plane as a base for the extruded boss:

Figure P2.34 – An extruded boss with a new reference plane used to create the long rod

Next, we will take a look at the *support*, highlighted in the following figure. Note that this can be a multi-body part:

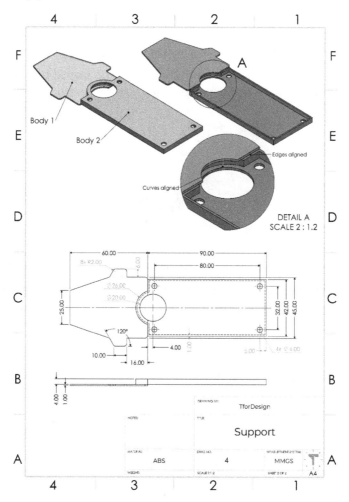

Figure P2.35 – Detailed drawing of the support

Here is a sample procedure for 3D-modeling the support:

1. Use an extruded boss to create the flat part of the support, as shown in the following figure. We can include the fillets in the sketch or apply them separately as a feature.

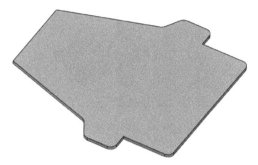

Figure P2.36 – Extruded boss can be used to create the shown body

2. Use another extruded boss to create the second body, as shown in the figure. Make sure that the bodies are not merged by unchecking the **merge result** option. Note the first body was made transparent for clarity only:

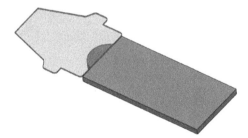

Figure P2.37 – Extruded boss can be used to create another body

3. Use the shell feature to shell the second body, as shown in the following figure:

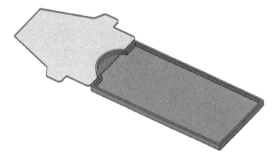

Figure P2.38 – Shell can be used to shell only one body

4. Use the extruded cut feature to create the final cuts. Note the larger cut applies to both bodies. Thus, we have to make sure both bodies are included in the scope of the cut. This should conclude the part, giving us the result shown in the following figure:

Figure P2.39 – An extruded cut applied to more than one body at once

Now, we can work on the *blade stabilizer*, as shown in the following figure:

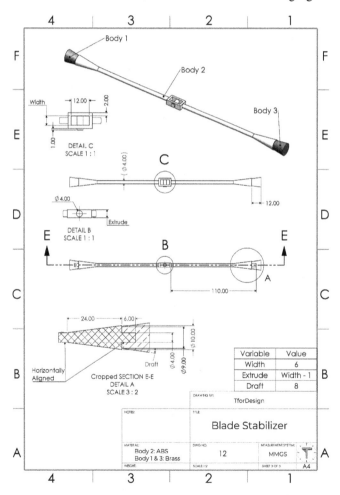

Figure P2.40 – Detailed drawing for the blade stabilizer

The *Blade Stabilizer* consists of three different bodies, two of which are identical. Also, some of the dimension of the indicated **Extrude** is related to the dimension indicated with **Width**. Here is a sample procedure for 3D-modeling the *Blade Stabilizer*:

1. Use two extruded boss features to create the middle part of the blade stabilizer, as shown in the following figure:

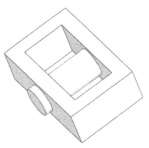

Figure P2.41 – The extruded boss can be used to create the middle part of the blade stabilizer

2. Use revolved boss to create the rod-like part, as shown in the following figure:

Figure P2.42 – Revolved boss can create the long rod with one application

3. Create the brass part toward the tip of the stabilizer. Note that the part is drafted by 8 degrees, as shown in the drawing in *Figure P2.40*. We can set the draft angle from the PropertyManager of the extrude boss feature. Since the tip is a different body, we can uncheck the **Merge result** option when applying the extruded boss:

Figure P.43 – A drafted extruded boss can be used to create the tip body

4.   Create the hole on the tip part, as indicated in the figure below:

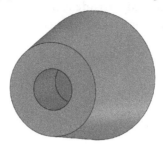

Figure P2.44 – The hole applied to a specific body

The location of the hole is covered by the main rod. To access the hidden location, you *right-click* on the body listed in the design tree and then select **Isolate**, as highlighted in the following figure. Alternatively, we can change the transparency of the other body:

Figure P2.45 – We can isolate bodies to make them easier to work with

5.   Mirror the revolved boss feature and the tip body to the other side to get the final result, as shown in the following figure. To do this, we will have to apply the mirror feature twice – once to mirror the revolved boss feature and once to mirror the body at the tip. This should conclude the part, giving the result shown in the following figure:

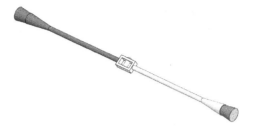

Figure P2.46 – Mirroring features and bodies can make modeling the part faster

After the *Blade Stabilizer*, we can move to 3D-modeling the last part of the RC helicopter, the *body*. The detailed drawing of the *body* is shown in the following drawing.

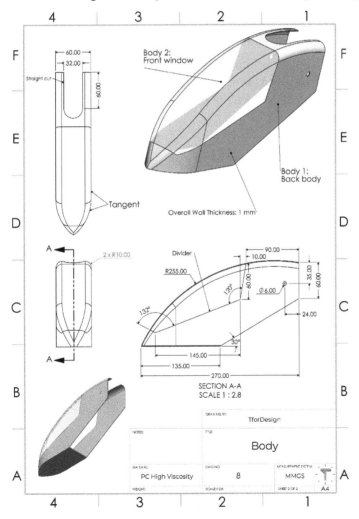

Figure P2.47 – Detailed drawing of the body

Note that the body part is a multi-body part with two bodies. One is transparent while the other is not. To 3D-model this part, we will 3D-model it as one part. Then, we are going to split the part into two. Here is a sample procedure for 3D-modeling this part:

1. Use an extruded boss to create the overall shape of the body:

Figure P2.48 – The bulk of the body can be made with an extruded boss

2. We can make the tip of the body using an extruded cut, as shown in the following figure. The sketch used for the extruded cut is shown in the following figure as well. We can then use the fillet feature to round the long top edges to a 10 mm radius, as indicated in *Figure P2.47*:

Figure P2.49 – The tip of the body can be made with an extruded cut

3. Use the shell feature to hollow the shape to 1 mm. Make sure to remove selected faces to end with the shape shown in the following figure:

Figure P2.50 – The shell feature can both hollow the body and remove unwanted faces at the same time

4. Use two extruded cut features to generate the cuts, ending up with the following figure:

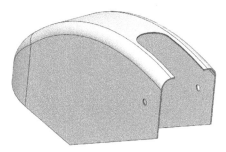

Figure P.51 – Extruded cuts can cut the holes and the top slot

At this point, we have created all the overall design features for the body. We are left with splitting the body into two to end up with the front window and the back solid body part. Next, we will explore how to split our body.

Splitting a body into two

To split a body, we will first create a sketch that highlights where the split is happening. Then, we will use the **Split** command to split the body. Let's do this by following these steps:

1. Create the cut sketch, as shown in the following figure:

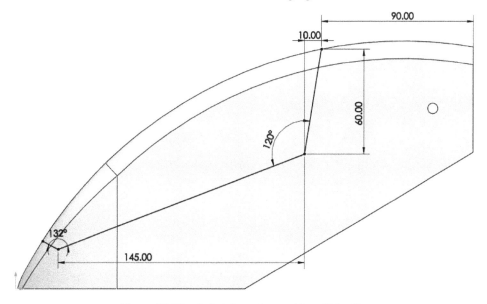

Figure P2.52 – A sketch can be used to split bodies

2. Select the **Split** command by going to **Insert | Molds | Split…**, as shown in the following figure:

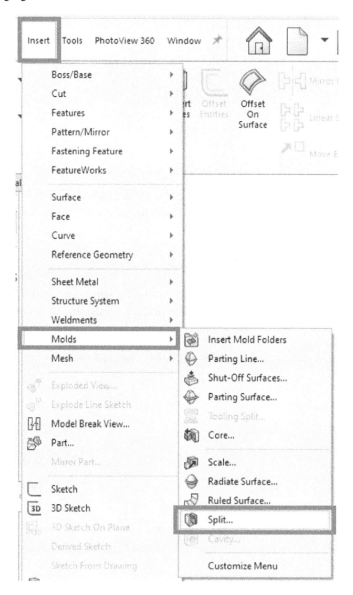

Figure P2.53 – The location of the split command

3. Under the **Trim Tools** file, select the cut sketch we created in *step 1*, and then click on **Cut Part**. This will create two **Resulting Bodies** under that title. Check the two bodies, and then click on the green check mark. The sequence is highlighted in the figure below:

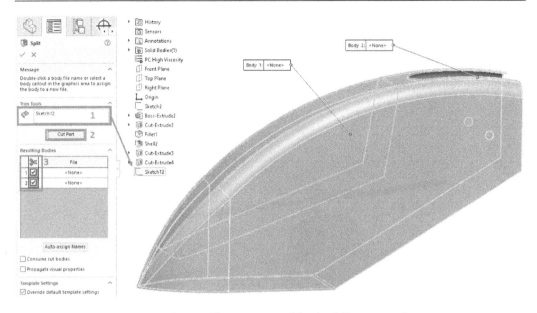

Figure P2.54 – The settings used for the Split command

> **Important Note**
>
> Giving names to the files in the PropertyManager, as shown in *Figure P2.54*,
> will save the named body in a separate part file.

At this point, we should have the body part complete like the one shown in the following
figure. We can assign different materials and appearances to each body as we see fit:

Figure P2.55 – The final body part

At this point, we are done 3D-modeling all the unique parts required for our RC
helicopter model. Next, we will work on assembling the parts.

Creating the assembly

Now that we have all the parts 3D modeled, we can start exploring the assembly and start joining all the parts together. We will do that in this section. The following drawings highlight the fully assembled RC helicopter model:

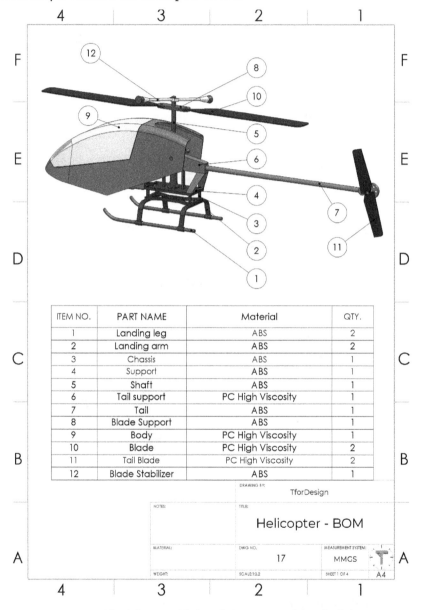

ITEM NO.	PART NAME	Material	QTY.
1	Landing leg	ABS	2
2	Landing arm	ABS	2
3	Chassis	ABS	1
4	Support	ABS	1
5	Shaft	ABS	1
6	Tail support	PC High Viscosity	1
7	Tail	ABS	1
8	Blade Support	ABS	1
9	Body	PC High Viscosity	1
10	Blade	PC High Viscosity	2
11	Tail Blade	PC High Viscosity	2
12	Blade Stabilizer	ABS	1

DRAWING BY: TforDesign

NOTES: TITLE:

Helicopter - BOM

MATERIAL: DWG NO.   MEASUREMENT SYSTEM:
17   MMGS

WEIGHT: SCALE:1:2   SHEET 1 OF 4   A4

Figure P2.56 – The fully assembled RC helicopter model with all the parts

Let's explore more drawings that showcase different mates within the assembly. The following drawing highlighted the connections between the shaft and the support. It also shows the relation between the tail support and the support parts. Note that it also highlights an additional hole created in the context of the assembly:

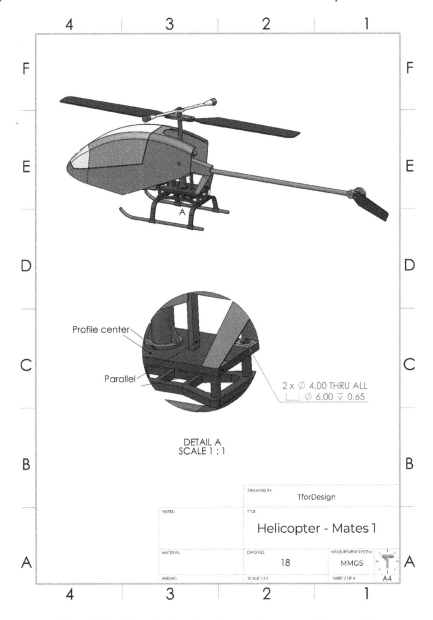

Figure P.57 – Drawing showing the specific mates in the assembly

The following drawing shows an exploded view of the RC helicopter as well as the selected mates. It also highlights an additional cut in the landing arm that was made in the assembly context:

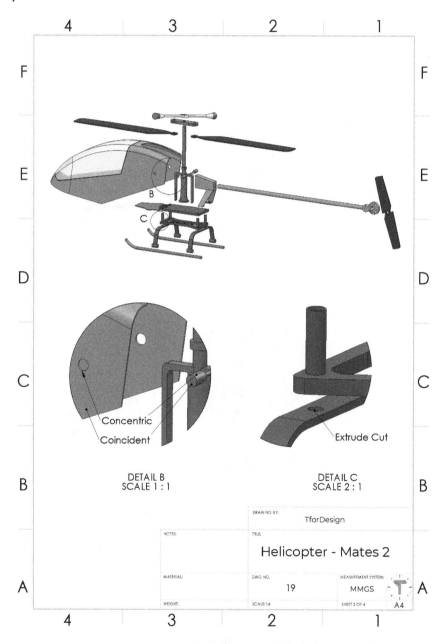

Figure P2.58 – An exploded view of the assembly

The following figure shows how the body part is angled with the chassis:

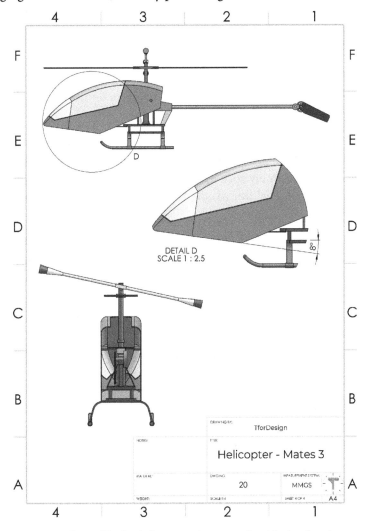

Figure P.59 – The body has an 8-degree angle with the chassis

The assembly figures explored previously contain all the major information needed to build the assembly. Note that not all the mates are specifically mentioned. Many of the mates can be concluded from the graphics of the overall assemblies. There are two different ways we can create the assembly, as follows:

1. Adding all the parts in one assembly file

2. Creating multiple subassemblies of fewer parts and then joining them together to form the larger assembly

In this text, we will follow the second approach. Let's explore some hints of one approach that can be adapted to build the assembly. Keep in mind that there is no one correct approach to generating an assembly such as this. Thus, treat the information presented as hints and food for thought. Feel free to adapt different approaches or experiment with them.

In the approach we are adapting, we will first build four smaller subassemblies, then join them to form the larger assembly. Each of the subassemblies we are building will consist of four or fewer unique parts. Building smaller subassemblies and then joining them to a larger one can help make them more manageable and easier to work with.

> **Important Note**
>
> A subassembly is an assembly file that is inserted into another larger assembly file.

The first subassembly consists of the *landing leg* and the *landing arm*, as indicated in the following figure. Overall, this subassembly consists of four parts, two of which are unique. The following figure also highlights the major mates used to build the subassembly:

> **Tip**
>
> You can experiment with using the linear pattern to build the shown subassembly.

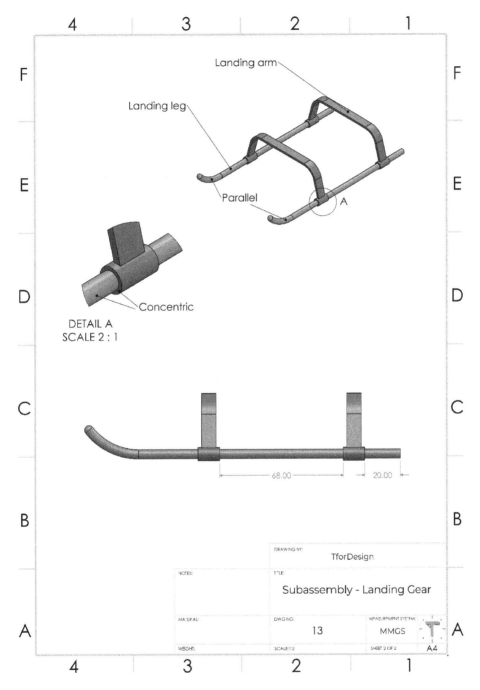

Figure P.60 – The landing gear subassembly

The next subassembly consists of the *support* and the *chassis*, as shown in the following figure:

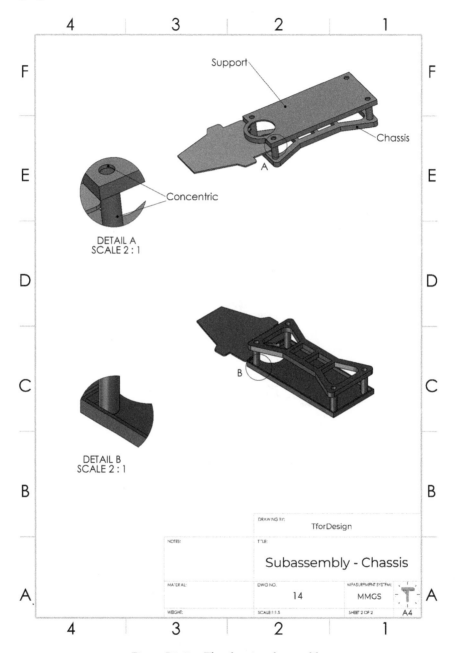

Figure P2.61 – The chassis subassembly

The next subassembly consists of the *shaft*, *blade stabilizer*, *blade support*, and *blades*, as shown in the following figure. We can set the shaft as the fixed part for this subassembly, as it is not a dynamic part in the final assembly. The following figure also highlights the selected mates that we can apply in the assembly:

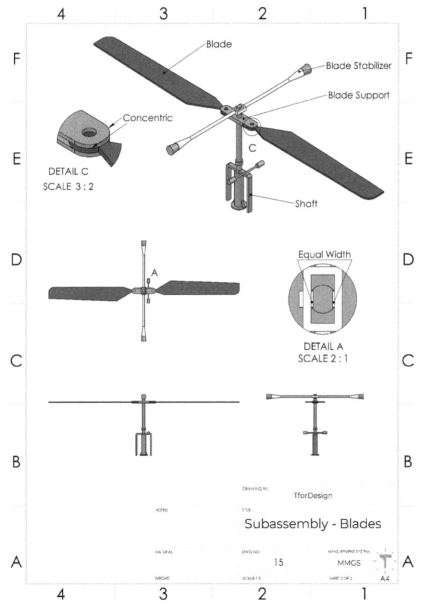

Figure P2.62 – The blades subassembly

The last subassembly consists of the *tail support*, *tail*, and *tail blade,* as indicated in the following figure. Overall, this subassembly consists of four parts, three of which are unique. The following figure also highlights the major mates used to build the subassembly. We can pick the tail support as the fixed part for this subassembly, as it is a non-moving part in the final assembly:

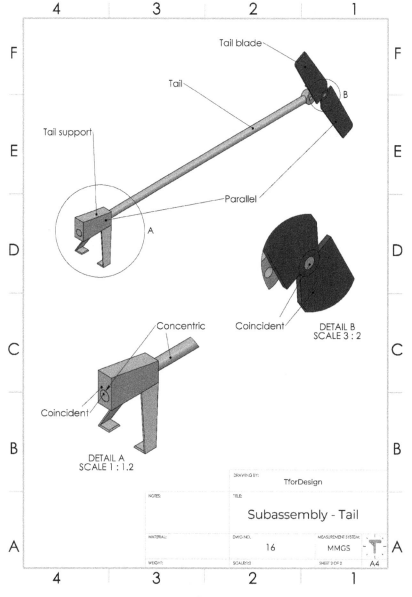

Figure P2.63 – The tail subassembly

At this point, we can create a new subassembly that will join the four subassemblies together with the body. The following figure highlights the major mates connecting the different subassemblies with the main RC helicopter body. You can revisit *Figure P2.56* to *Figure P2.59* explored earlier for more information about how the different parts in the assembly interact with each other:

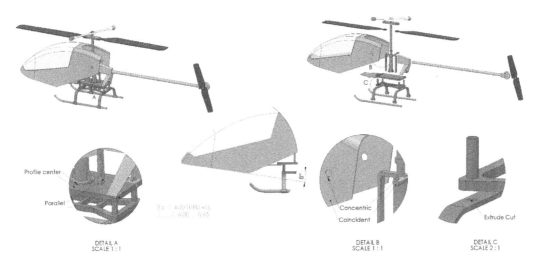

Figure P2.64 – The body connecting with the other subassemblies

By completing the assembly, you have completed the project work of 3D-modeling an RC helicopter. As additional activities, you can use different evaluation tools such as interference detection to find undesirable interferences and adjust the design as applicable. You can also further customize the RC helicopter model to make it your own.

Summary

In this project chapter, you worked to 3D-model an RC helicopter toy model. To achieve that, you had to interpret engineering drawings, 3D-model different parts, and then join them together in an assembly. The skills you used to construct this project include many advanced features often used by professional users; those include working with multibody parts, building configurations, using design tables, advanced mates, and assembly features. In the process, you have also constructed a project that can be included in your personal portfolio to highlight your 3D-modeling skills.

Congratulations on making it to the end of this book! This book was meant as a journey to build you a strong foundation in using SOLIDWORKS for 3D modeling. This foundation is the beginning of a new journey, whether you use the tools we explored in your next project or continue learning in more specialized areas such as surfacing, molding, weldments, drawings, sheet metals, or simulation. 3D modeling is a skill, just like any other; the more you use it, the better you will get at it, and the more you will develop your own style and a set of preferred tricks and shortcuts in the process. So, keep practicing, keep growing, and keep having fun in the process.

Index

A

active mates
  adjusting 281, 282
  viewing 281, 282
advanced mates 266
angle-distance chamfers
  applying 152, 153
angle mate
  applying 283-287
angle range mate
  applying 550-553
  defining 549
assemblies
  advantages 462
  configurations, utilizing 584
  design tables, utilizing 584
assemblies environment
  materials, assigning to parts in 288-290
assembly features
  about 580
  applying 581-584
  using, reasons 580
associate certifications, SOLIDWORKS
  about 12
  additive manufacturing (CSWA-AM) 12
  CSWA 12

electrical (CSWA-E) 12
  simulation (CSWA-Simulation) 12
  sustainability (CSWA-Sustainability) 12
Auto Balloon command 420
axonometric projections
  about 312
  dimetric 312
  isometric 312
  trimetric 312
  types 312

B

base cell 90
Bill of Materials (BOM)
  about 391-393
  assembly, inserting into
      drawing sheet 395, 396
  column category, modifying 403-405
  equations, utilizing 410
  information, adjusting 401
  information, sorting 406-408
  listed information, adjusting 401
  new columns, adding 408, 409
  of cap assembly 394

standard BOM, generating 395
title, modifying 402
bodies
    separating, into different parts 468-470

C

Canvas/Graphics Area 27, 28
centerlines
    about 54, 371
    adding 371, 372
center marks
    about 372
    adding 373, 374
Certified SOLIDWORKS
        Associate (CSWA) 6, 15
Certified SOLIDWORKS
        Professional (CSWP) 6, 15
chamfers
    about 66, 67
    sketching 69, 70
chamfers feature
    about 144
    applying 144, 149, 157
    creating, procedure 151, 152
    modifying 144, 157
    options 154
chamfers feature, types
    angle-distance 150
    distance-distance 150
    face-face 151
    offset face 151
    vertex 150
child views 354
circular pattern
    about 471
    applying 479-482

circular sketch patterns
    about 90, 100
    base sketch 102
    elements 101
    final result 102
    options 103, 104
    resulting sketch 105, 106
coincident mate
    applying 268-271
Collision Detection tool
    about 577
    using, steps 577, 578
column category
    modifying, in BOM 403-405
columns
    adding, in BOM 408, 409
Command Bar 24, 25
complex models
    versus simple models 121
complex sketches
    versus simple sketches 38-40
concentric mate
    applying 277
configuration
    about 508
    advanced options 586, 587
    applying 508-514
coordinate system
    about 238
    creating 240-243
    defining 238
    need for 239
cube 181
customization options, sketch
    Add dimensions 89
    Bi-directional 90
    Caps ends 90

Construction geometry 90
Reverse 89
Select chain 90

D

Delete Linear Pattern command 100
design intent
    about 9
    with equations 506, 507
design tables
    about 515, 516, 588
    editing 522, 523
    editing, by modifying model 524
    generating, for part configurations
        within assembly 589-591
    setting up 517-522
dimensions
    about 367
    adding 368, 369
    deleting 370
    modifying, with equations 504
    used, for defining sketches 43
dimetric, axonometric projections 312
display
    adjusting 359, 366, 367
    types 364
distance mate
    applying 283-285
distance range mate
    applying 550-553
    defining 549
document, measurement system
    about 29
    adjusting 30-32
drafted rib 460, 461
drafting 430

drafts
    about 430
    applying 431
    creating 432-434
    example 430
drawing
    exporting, as image 385, 386
    exporting, as PDF file 384
drawing scale
    adjusting 359
    modifying, for front parent
        view 360, 361
    modifying, of isometric
        child view 362, 363
drawing sheet
    information block, utilizing 378

E

Edit Linear Pattern command 100
engineering drawings
    about 300, 301
    interpreting 302
    lines, interpreting 302, 303
    views, interpreting 304
engineering drawings, views
    auxiliary view 307
    broken-out section views 309, 310
    cropped views 310, 311
    detail views 308, 309
    orthogonal views 305, 306
    section views 307
equation function
    applying 413-416
equations
    about 498
    applying, in parts 500-504
    demonstrating 499

design intent 506, 507
dimensions, modifying 504
inputting, in table 411-413
in SOLIDWORKS drawings 410
utilizing, in BOMs 410
within Equations Manager 505, 506
existing mates
modifying 282
expert certificates, SOLIDWORKS
about 14
Certified SOLIDWORKS
Expert (CSWE) 14
Certified SOLIDWORKS Expert
in Simulation (CSWE-S) 14
extruded boss feature
3D modeling, planning 126
3D modeling, sketching 126
about 123, 124, 504
applying 124-132
editing 140, 141
modifying 139
options 131, 132
extruded cut feature
about 123, 124
applying 132, 139
deleting 139, 142-144
model, planning 133
model, sketching 133
options 138
sketch, creating 134-137

F

FeatureManager Design Tree
about 26
Commands/Features 26
Default Reference Geometries 26
Materials 27
Others 27
feature patterns
about 471
circular pattern 471, 479-482
fill pattern 471, 483-489
linear pattern 471-478
feature scope applications 466-468
features mirroring
applying 450
fillets
about 66, 67
creating 68
fillets feature
about 144
applying 144-146, 149
modifying 144, 157
options 147, 148
fill pattern
about 471
applying 483-489
applying, to more than one
boundary 489, 490
Finite Elements Method (FEM) 12
first-angle projections 352
front parent view
drawing scale, modifying for 360, 361
Fully Defined 279
fully defined line 43, 44
fully defined sketches 72
functions
about 410
average 410
count 410
if 410
max 410
min 410

sum 410
total 410

G

geometrical relations
  about 44
  exploring 44, 45
geometry
  planes, defining 182-186
guide curves
  using 226-231

H

highlighted shell
  creating 437-439
hole callout
  about 376
  adding 377
  types 378
holes
  about 442
  identifying, in SOLIDWORKS 442
Hole Wizard
  about 442
  utilizing 443-449

I

image
  drawing, exporting as 385, 386
imperial system 29
Inch, Pound, Second (IPS) 102
information
  adjusting, in BOM 401
  storing, in BOM 406-408

information block, drawing sheet
  editing 379, 380
  information, adding 382, 383
  utilizing 378
Instances to Skip option 100
Interference Detection tool
  about 573
  applying, steps 574
  options 576, 577
  practical point, addressing 575
International System of Units (SI) 29
isometric, axonometric projection 312
isometric child view
  drawing scale, modifying of 362, 363
isometric views
  about 345
  generating 348-351

L

limit mates 549
linear/linear coupler mate
  applying 566
  defining 562
  exploring 556
  fine-tuning 566-568
linear pattern
  about 471
  applying 472
  creating 473-477
  options 478
Linear Sketch Pattern command 94
linear sketch patterns
  about 90, 91
  defining 92-96
  elements 92, 93
  options 96
  resulting preview 98, 99

listed information
    adjusting, in BOM 401
lock mate
    applying 278
lofted boss
    about 210-212
    applying 213-219
    Guide Curves 221
    modifying 226
    options 220, 221
    Start/End Constraints 221
    Thin Feature 221
lofted cut
    about 210-212
    applying 222-225
    modifying 226

M

Manual Balloon command 420-422
manual configurations
    using 585, 586
mass properties
    center of mass concerning new
        coordinate system, finding 250
    center of mass (millimeters), finding 250
    evaluating 243
    finding, task 248
    mass of model in grams, finding 250
    mass of model in pounds,
        finding 252, 253
    overriding 254, 255
    viewing 247-249
materials
    assigning, to parts 244-247
mates 265, 266
mathematical operations 410
measurements 44

measurement systems 29
mechanical mates 266
metric system 29
Millimeter, Gram, Second (MMGS) 81
Mirror command
    utilizing, to mirror features 451-455
mirrored entities
    defining 83, 84
Mirror Entities command 82
model
    selecting, to plot 346-348
multi-body parts
    advantages 462
    defining 462
    generating 463-466
    utilizing 461
multi-thickness shell 440, 441

N

non-value-oriented standard mates
    about 266, 267
    applying 266
notes
    about 375
    adding 375

O

offset
    deleting 89
Offset Entities sketch command 84
orthographic views
    about 345
    generating 348-351
Over Defined 279
over defined sketches 73, 74

P

pair of glasses project, 3D-modeling
  about 316, 317
  advanced level 317
  assembly, creating 335-339
  individual parts, creating 318-330
  mirrored part, creating 331-334
  moderate level 317
  parts, exploring 318
parallel mate
  applying 275, 276
parameter 8
parametric modeling
  about 8-10
  advantages 11
parent views 354
partial chamfers
  applying 158, 159, 162
partial fillets
  applying 158-161
parting line draft 434
parts callouts
  utilizing 416-420
path mate
  applying 557-561
  defining 556
  exploring 556
patterned entity 90
PDF file
  drawing, exporting as 384
perpendicular mate
  applying 272, 273
planes
  about 180
  defining, in geometry 182-186
  defining, in SOLIDWORKS 186-194
  sketching, for features 122, 123

power trimming
  using 108-111
professional advanced certifications,
      SOLIDWORKS
  about 14
  advanced drawing tools
      (CSWPA-DT) 14
  mold making (CSWPA- MM) 14
  sheet metal (CSWPA-SM) 14
  surfacing (CSWPA-SU) 14
  weldments (CSWPA-WD) 14
professional certifications, SOLIDWORKS
  about 13
  Certified PDM Professional
      Administrator (CPPA) 13
  CSWP 13
  CSWP-API 13
  CSWP-CAM 13
  flow simulation (CSWP-Flow) 13
  Model-Based Definition
      (CSWP- MBD) 13
  simulation (CSWP-Simulation) 13
profile center advanced mate
  about 534
  applying, to generate assembly 535-565
  defining 534
  using 534
projection style
  adjusting 352, 353

R

RC helicopter model, 3D-modeling
  about 596-598
  advanced level 598
  assembly, creating 628-637
  body, splitting 625-627

individual parts, exploring 598-625
moderate level 598
reference geometries 180, 238
relations
  about 44
  used, for defining sketches 43
revolved boss feature
  3D modeling, procedure 166
  about 163, 164
  applying 163-169
  modifying 174
revolved cut feature
  3D modeling, procedure 170, 171
  about 163, 164
  applying 163, 170-173
  modifying 174
Rib command
  applying 456-460
ribs 455, 456
rotational patterns 91

S

scale ratios 364
shell feature
  about 436
  applying 436
simple models
  versus complex models 121
simple sketches
  versus complex sketches 38-40
sketches
  defining 43, 44
  mirroring 80-83
  offsetting 84-89
sketching commands
  arcs, sketching 56-61
  circles, sketching 56-61

construction lines, using 61-66
ellipses, sketching 61-66
lines, sketching 46-51
origin 46
rectangles, sketching 52-56
squares, sketching 52-56
sketch patterns
  creating 90
  defining 90
  linear 91
  rotational 91
sketch planes 40, 41
Smart Dimension command
  using 579
Smart Dimension tool
  using 368-370
SOLIDWORKS
  about 4
  applications 4, 5
  core mechanical design 5, 6
  mates, categories 266
  planes, defining 186-194
  sample 3D models 6, 7
SOLIDWORKS assembly
  about 260, 261
  active mates, adjusting 281, 282
  active mates, viewing 281, 282
  coincident mates, applying 267-271
  concentric mates, applying 274, 275, 277
  coordinate system, setting for 287
  definition statuses, searching
    of parts 280
  definition statuses, selecting 281
  existing mates, modifying 282
  Fully Defined 279
  lock mates, applying 274, 275, 278
  mass properties, evaluating 290, 292
  mass properties, utilizing 287

material edits 288
materials, utilizing 287
opening 260
Over Defined 279
parallel mates, applying 274-276
parts, adding 260
perpendicular mates, applying 267-273
tangent mates, applying 274, 276, 277
Under Defined 279
SOLIDWORKS assembly file
parts, adding 261-265
working with 261, 262
SOLIDWORKS certifications
associate certifications 12
expert certifications 14, 15
exploring 11
professional advanced certifications 14
professional certifications 13
SOLIDWORKS drawing file
opening 342-344
SOLIDWORKS features
about 120, 121
planes, sketching 122, 123
role, in 3D modeling 120, 121
simple models versus
complex models 121
SOLIDWORKS files
assemblies 18-20
drawings 18-21
opening 21, 22
parts 18, 19
SOLIDWORKS interface, components
about 23, 24
Canvas/Graphics Area 27, 28
Command Bar 24, 25
FeatureManager Design Tree 26, 27
Task Pane 28

SOLIDWORKS sketching
geometrical relations 44, 45
mode 41-42
position 36-38
working with 41
standard BOM
generating 395, 398, 400
standard mates
about 266
options 266
step draft application 435
swept boss
about 195, 196
applying 197-201
Circular Profile 202
Curvature Display 203
Guide Curves 202
modifying 208-210
options 202
profile orientation 203
Profile Twist 203
Start and End Tangency 203
Thin Feature 203
swept cut
about 195-197
applying 204-207
modifying 208-210
symmetric advanced mate
about 539
applying 545-548
defining 543, 544

T

table
equations, inputting 411-413
tangent mate
applying 276, 277

Task Pane 28
title
    modifying, in BOM 402
trimetric, axonometric projections 312
trimming 107

U

Under Defined 279
under defined line 43, 44
under defined sketches 70-72

V

value-driven standard mates
    about 282, 283
    angle mates 283
    applying 282
    distance mates 283
vertex chamfer
    applying, special features 156
    applying, steps 155
Vertical Extrusion 9
views
    adding, via View Palette 354-358
    deleting 358, 359

W

width advanced mate
    about 539
    applying 540-543
    defining 539, 540
width dimension 539

Packt.com

Subscribe to our online digital library for full access to over 7,000 books and videos, as well as industry leading tools to help you plan your personal development and advance your career. For more information, please visit our website.

Why subscribe?

- Spend less time learning and more time coding with practical eBooks and Videos from over 4,000 industry professionals

- Improve your learning with Skill Plans built especially for you

- Get a free eBook or video every month

- Fully searchable for easy access to vital information

- Copy and paste, print, and bookmark content

Did you know that Packt offers eBook versions of every book published, with PDF and ePub files available? You can upgrade to the eBook version at packt.com and as a print book customer, you are entitled to a discount on the eBook copy. Get in touch with us at customercare@packtpub.com for more details.

At www.packt.com, you can also read a collection of free technical articles, sign up for a range of free newsletters, and receive exclusive discounts and offers on Packt books and eBooks.

Other Books You May Enjoy

If you enjoyed this book, you may be interested in these other books by Packt:

Increasing Autodesk Revit Productivity for BIM Projects

Fabio Roberti | Decio Ferreira

ISBN: 978-1-80056-680-4

- Explore the primary BIM documentation to start a BIM project.

- Set up a Revit project and apply the correct coordinate system to ensure long-term productivity.

- Improve the efficiency of Revit core functionalities that apply to daily activities.

- Use visual programming with Dynamo to boost productivity and manage data in BIM projects.

- Import data from Revit to Power BI and create project dashboards to analyze data.

Managing and Visualizing Your BIM Data

Ernesto Pellegrino | Jisell Howe | Manuel André Bottiglieri | Dounia Touil | Luisa Cypriano Pieper

ISBN: 978-1-80107-398-1

- Understand why businesses across the world are moving toward data-driven models.
- Build a data bridge between BIM models and web-based dashboards.
- Get to grips with Autodesk Dynamo with the help of multiple step-by-step exercises.
- Focus on data gathering workflows with Dynamo.
- Connect Power BI to different datasets.

Packt is searching for authors like you

If you're interested in becoming an author for Packt, please visit `authors.packtpub.com` and apply today. We have worked with thousands of developers and tech professionals, just like you, to help them share their insight with the global tech community. You can make a general application, apply for a specific hot topic that we are recruiting an author for, or submit your own idea.

Hi!

I am Tayseer Almattar, author of *Learn SOLIDWORKS Second Editon*. I really hope you enjoyed reading this book and found it helpful to kickstart your journey in SOLIDWORKS 3D Modeling.

It would really help me (and other potential readers!) if you could leave a review on Amazon sharing your thoughts on *Learn SOLIDWORKS Second Edition*.

Your review will help me to understand what's worked well in this book, and what could be improved upon for future editions, so it really is appreciated.

Best Wishes,

Tayseer Almattar